Springer Theses

Recognizing Outstanding Ph.D. Research

Aims and Scope

The series "Springer Theses" brings together a selection of the very best Ph.D. theses from around the world and across the physical sciences. Nominated and endorsed by two recognized specialists, each published volume has been selected for its scientific excellence and the high impact of its contents for the pertinent field of research. For greater accessibility to non-specialists, the published versions include an extended introduction, as well as a foreword by the student's supervisor explaining the special relevance of the work for the field. As a whole, the series will provide a valuable resource both for newcomers to the research fields described, and for other scientists seeking detailed background information on special questions. Finally, it provides an accredited documentation of the valuable contributions made by today's younger generation of scientists.

Theses may be nominated for publication in this series by heads of department at internationally leading universities or institutes and should fulfill all of the following criteria

- They must be written in good English.
- The topic should fall within the confines of Chemistry, Physics, Earth Sciences, Engineering and related interdisciplinary fields such as Materials, Nanoscience, Chemical Engineering, Complex Systems and Biophysics.
- The work reported in the thesis must represent a significant scientific advance.
- If the thesis includes previously published material, permission to reproduce this must be gained from the respective copyright holder (a maximum 30% of the thesis should be a verbatim reproduction from the author's previous publications).
- They must have been examined and passed during the 12 months prior to nomination.
- Each thesis should include a foreword by the supervisor outlining the significance of its content.
- The theses should have a clearly defined structure including an introduction accessible to new PhD students and scientists not expert in the relevant field.

Indexed by zbMATH.

More information about this series at http://www.springer.com/series/8790

Monika Snowdon

Methods of the Alignment-Relay Technique for Nanosystems

Optimization and Innovation

Doctoral Thesis accepted by the University of Waterloo, Waterloo, Canada

Springer

Monika Snowdon
Department of Chemistry
University of Waterloo
Waterloo, ON, Canada

ISSN 2190-5053 ISSN 2190-5061 (electronic)
Springer Theses
ISBN 978-3-030-84412-7 ISBN 978-3-030-84413-4 (eBook)
https://doi.org/10.1007/978-3-030-84413-4

This Springer imprint is published by the registered company Springer Nature Switzerland AG
The registered company address is: Gewerbestrasse 11, 6330 Cham, Switzerland

To those who came before me, to those with whom I shared the struggles and the curiosity, and to those who would like to continue this work.

Supervisor's Foreword (by Dai-Ming Tang)

Current transistors rely upon silicon metal-oxide semiconductor field-effect transistors (MOSFETs), which follow the trend of doubling in number on a chip every 2 years (Moore's law). By mid-2021, IBM has developed a 2-nm transistor, which can significantly increase the calculating speed and processing power. However, the trend is heading toward a plateau regarding the number of transistors and power of performance. Therefore, we must explore other approaches for either replacements or enhancement of Si MOSFETs.

One alternative is to use single-walled carbon nanotubes (SWNTs) as a channel material for transistors. Due to their atomic thickness, they require lower operating voltage, meaning less power consumption. Nevertheless, it is currently challenging to purify the SWNTs between their semiconducting or metallic types in a cost-effective way.

Prof. Schipper's group developed the *alignment relay technique* (ART) to solve the carbon nanotube's alignment and sorting problem. The method is based on iptycene small molecules, which have been shown to have π-π stacking interactions with SWNTs. In addition, iptycenes have a concave structure to coordinate with the nanotubes' cylindrical nature, acting like a tweezer that can pick them out in a solution. The molecules can bind and stay on a wide range of surfaces, and it cannot be easily washed away. ART presents a unique molecular tweezer approach to aligning the SWNTs and selecting specific diameters simultaneously.

Monika's work aimed to identify the different types of selectivity of the SWNTs with ART using Raman spectroscopy and diversify the functioning surfaces such as Al_2O_3 through synthetic changes to the iptycene tweezer backbone. She also explored different SWNTs functionalizations and other nanosystems such as nanowires and graphene nanoribbons, surfactants, and sonochemical treatments.

At the National Institute for Materials Science (NIMS), under my supervision, Monika identified a way to reproducibly obtain below 1 μm gaps in source-drain patterns using LED photolithography. This patterning approach was subsequently used for the fabrication of carbon nanotube field-effect transistors (CNTFETs). The photolithography prepared devices were subsequently compared to the performance of devices with nanometer scale gaps, fabricated with electron beam lithography

(EBL). The results presented in this thesis demonstrate clearly the effects of contact resistance and the molecular layer to the transport properties of the transistors, by comparing the performance of the top and back gated, random and aligned CNTs. Monika also worked on ionic gel gates, layer-by-layer and Langmuir-Blodgett depositions, gold nanoparticle inks for printing electronics, and so on.

Monika's work is unique in bridging organic synthesis, nanofabrication, assembly, and electronic devices. Springer has now recognized her work through this publication, which will serve as an excellent resource and inspiration to upcoming researchers.

Functional Nanomaterials Group
International Center for Materials Nanoarchitectonics (MANA),
National Institute for Materials Science (NIMS)
Tsukuba, Ibaraki, Japan

Dai-Ming Tang

Supervisor's Foreword (by Derek Schipper)

I am very pleased to introduce this work by my former doctoral student, Monika Snowdon. She optimized a technique for nanoscale alignment recently coined by the Schipper group as the *alignment relay technique* (ART). Her thesis reported a unique approach to the alignment of carbon nanotubes (CNTs) using iptycenes as molecular tweezers and liquid crystals. Dr. Snowdon described carefully and clearly a multi-step methodology and its effectiveness in performing simultaneous alignment and selectivity of single-walled carbon nanotubes for transistor applications. This work contributes evidence for the selectivity and stability of the method, as well as the first proof-of-concept that ART can be used to create CNT field-effect transistors (FETs). The thesis encloses the synthesis of the tweezers, the concept and actualization of the technique, the various factors that can influence deposition and selectivity, alongside the material fabrication with both photolithography and electron beam lithography. The research is significant for progress in transistors and the applications of small organic molecules in the materials science field.

Overall, the key contributions of this work are:

1. An extensive overview of ART, its components, and limitations
2. New materials chemistry methods for the optimization of semiconducting CNT selectivity.
3. A comprehensive understanding of CNTFETs fabrication and characterization for future applications.

Department of Chemistry
University of Waterloo
Waterloo, ON, Canada

Derek J. Schipper

Research Funding

Several institutions financially contribute to the research study:

- Queen Elizabeth II Graduate Scholarship in Science and Technology (2020)
- Natural Sciences and Engineering Research Council of Canada (NSERC) Postgraduate Scholarship—Doctoral (2017–2020)
- National Institute for Materials Science International Cooperative Graduate Program (ICGP) (2020)
- Paul Bridger Graduate Award in Advanced Applied Science (2017)
- Waterloo Institute of Nanotechnology—Nanotechnology Fellowship (2017, 2019)
- University of Waterloo—Provost Doctoral Entrance Award (2016)

The author gratefully acknowledges the mentioned institutions for their financial support, essential for the fulfillment of the Thesis.

Parts of This Thesis Have Been Published in the Following Journal Articles

1. Snowdon, M. R., Wang, S., Mashmoushi, N., Hopkins, S. W., & Schipper, D. J. (2021). Carboxylic acids as anchoring components on aluminum oxide for the alignment relay technique of single-walled carbon nanotubes. *New Journal of Chemistry*, *45*(12), 5340–5349.
2. Snowdon, M. R., Selmani, S., & Schipper, D. J. (2019). Sonication-Enhanced Alignment Relay Technique for the Orientation of Single-Walled Carbon Nanotubes. *ACS Applied Nano Materials*, *2*(10), 6637–6645.
3. Kulak, M. R., & Schipper, D. J. (2019). Carbon nanotube alignment and sorting: Attempting a sulfur moiety as anchoring component. *Phosphorus, Sulfur, and Silicon and the Related Elements*, *194*(7), 760–763.

Tribute to Nanotubes

Tiny carbon tubes
All straight in a line
Takes some time
To perfectly align

To get rid of the small and metallic
Sonication worked quite fantastic
To eliminate the surfactant
I had to use an oxidation tactic
To diversify the surface
It required some extra synthesis with purpose
Tried to also increase the density
Had to test some fillers quite intensely

Other systems were tried
But the π-π stacking wasn't satisfied
"Let's stick with the SWNTs"
Derek said, not at all mystified

Nearing the end of the time
An opportunity was sublime
So the tubes went to Japan
To be tested as transistors with expert Dr. Tang

Before I can exult
Here is the result
読んでくれてありがとう

Acknowledgments

Writing my doctoral thesis would not have been possible without help from my "work team" and my "home team."

On the work side, I thank my supervisor, Dr. Derek Schipper, for his patience throughout my Ph.D. His innovation for the project made it a pleasure to conduct research in the Schipper lab. I wish him good luck with the future directions of his research program.

Additionally, I am forever grateful to Dr. Dai-Ming Tang for allowing me into the Functional Nanomaterials Group at the National Institute for Materials Science. I have learned so much, and a significant portion of my work would not have been possible without him. His patience, enthusiasm, and innovation are without contest. I wish him, too, good luck with the future directions of his research program.

I would like to thank the members of my doctoral advisory committee: Drs. Michael Chong, Aiping Yu, and Dmitriy Soldatov, as well as examining members Drs. Benoît Lessard and Anna Klinkova.

My acknowledgements would not be complete without addressing my colleagues at the University of Waterloo and the Functional Nanomaterials Group (MANA—NIMS). I have worked with many outstanding people.

Thank you to NIMS, NSERC, WIN, the Royal Society of Chemistry, Bruce Power, and the UW Faculty of Science for providing financial assistance.

Special thanks to Professors Adrian Schwan (Guelph), Graham Murphy (Waterloo), David Wunsch (UMass Lowell), Tong Leung (Waterloo), Na Young Kim (Waterloo), and Praveen Saxena (Guelph).

Additional thanks to Drs. Aan-san, Li-san, Sun-san, Bai-san, Lulu-san, Namiki Foundry Cleanroom, and Sawabe-san, my coaches Jordynn, Teri, Jon, Ally, and Hossein; Drs. Rob, Irfani, Inna, Parisa, Joanne, Leanne, Tristan, HeeBong, the WINGSS team of Drs. Jon, Anand, Shahid, and Mohammed; the WIN team of Drs. Sushanta, Lisa, and Oleg; WATLab, Drs. Lei and Nina, the NanoFab Clean Room team, Jan, Cathy, Kim, Sarah and Vlad, Slavica, Steve, Julie, Kristina and Laura, the UW writing team: Mary, Jane and Stephanie; and the Centre for Teaching Excellence, Svitlana and Monika—I look forward to collaborating in the coming years.

Thank you to my home team. To Verena, Michael, Percy, Nancy, and John, for their continuous support.

Contents

Abbreviations

[O]	Oxidation
1D	One-dimensional
5CB	4-cyano-4′-pentylbiphenyl
8CB	4′-octyl-4-biphenylcarbonitrile
AC	Alternating current
AFM	Atomic force microscope
AIST	[National Institute of] Advanced Industrial Science and Technology
ART	Alignment relay technique
ATP	Aqueous two-phase
ATR	Attenuated total reflectance
BG	Bottom-gate
BSE	Backscattered electrons
C	Capacitance
c-AFM	Conductive atomic force microscope
CNTs	Carbon nanotubes
CoMoCAT	Cobalt and molybdenum catalysts
CP	Catalytic pyrolysis
CVD	Chemical vapor deposition
D	Drain
D-band	Disorder band
DC	Direct current
DDQ	2,3-Dichloro-5,6-dicyano-1,4-benzoquinone
DEP	Dielectrophoresis
DGU	Density ultracentrifugation technique
DLSA	Dimension-limited self-assembly
DOC	Sodium deoxycholate
DOS	Density of states
DS	Directional sticking
DWNT	Double-walled carbon nanotube
EBL	Electron beam lithography
EDS	Energy dispersive X-ray

EDX	Energy dispersive X-ray
E_f	Fermi level
E_g	Bandgap
EISA	Evaporation induced self-assembly
FESA	Floating evaporative self-assembly
FETs	Field-effect transistors
FTIR	Fourier transform infrared
G_0	Quantum conductance
G-band	Graphene band
G_m	Transconductance
GNRs	Graphene nanoribbons
GQDs	Graphene quantum dots
HipCO	High pressure carbon monoxide
HMDS	Hexamethyldisilazane
HR	High-resolution
I_D	Drain current
I_G	Gate current
I_{OFF}	Off-state current
Ion	On-state current
IPA	Isopropanol
Ipt	Iptycene
ITO	Indium tin oxide
I-V	Current-voltage
LB	Langmuir-blodgett
LCs	Liquid crystals
LED	Light emitting diode
LS	Langmuir-schaefer
m-	Metallic
MOS	Metal oxide semiconductor
MWNTs	Multi-walled carbon nanotubes
NIMS	National Institute for Materials Science
NIR	Near infrared
NMP	*N*-methyl pyrrolidine
NMR	Nuclear magnetic resonance
NTs	Nanotubes
NWs	Nanowires
PEG	Polyethylene glycol
PL	Photoluminescence
PMGI	Polydimethylglutarimide
PMMA	Polymethylmethacrylate
PR	Photoresist
S	Source
SAMs	Self-assembled monolayers
SB	Schottky barrier
sc-	Semiconducting

SC	Sodium cholate
SDBS	Sodium dodecyl benzene sulfonate
SDS	Sodium dodecyl sulfate/sodium lauryl sulfate
SE2	Secondary electrons
SEM	Scanning electron microscope
ShA	Shrinking alignment
SMU	Source measurement unit
SPMs	Scanning probe microscopy
SS	Subthreshold slope
SW	Stone-wales
SWNTs	Single-walled carbon nanotubes
TEM	Transmission electron microscope
TG	Top-gate
TMDs	Transition metal dichalcogenides
UHDA	Ultra-high-density assembly
UV	Ultraviolet
VdW	Van der Waals
V_G	Gate voltage
VHS	Van Hove singularities
VLS	Vapor-liquid-solid
V_{SD}	Source-drain voltage
VSS	Vapor-solid-solid

Chapter 1
Introduction

Abstract This work was motivated by the need to overcome some of the problems that SWNTs present before they can be integrated into devices. The main difficulty for satisfactory integration has been the tendency of SWNTs to configure in a variety of patterns that renders them unreliable. The main goal of the project was to optimize their alignment by using the alignment relay technique (ART) and to select tube conformations with the desirable semiconducting properties. The introduction provides the key concepts in understanding the theory and components of ART, as well as detailing its experimental techniques. This work was motivated by the need to overcome some of the problems that SWNTs present before they can be integrated into devices. The main difficulty for satisfactory integration has been the tendency of SWNTs to configure in a variety of patterns that renders them unreliable. The main goal of the project was to optimize their alignment by using the alignment relay technique (ART) and to select tube conformations with the desirable semiconducting properties. The introduction provides the key concepts in understanding the theory and components of ART, as well as detailing its experimental techniques.

Keywords Alignment · Carbon nanotubes · Single-walled · Field-effect transistor · Characterization · Theory · Nanotechnology · Introduction · CNTFET

Innovative approaches to increasing the power of in-silico devices can significantly impact the fields of aeronautics, drug discovery, and general daily life. Specifically, transistors are critical electronic components that demand constant improvement for device acceleration. Silicon-based transistors follow the general pattern of doubling in quantity in an electronic system approximately every 2 years [1, 2]. This trend is depicted in Fig. 1.1 from microprocessor data, with future projections showcasing a plateau in performance. This stipulation is because only a limited number of Si transistors can fit on an electronic chip, and decreasing their size can lead to problems with electron transport and current dissipation [1]. One alternative to repeatedly increasing the number of transistors per chip that also improves transport and

M. Snowdon, *Methods of the Alignment-Relay Technique for Nanosystems*,
Springer Theses, https://doi.org/10.1007/978-3-030-84413-4_1

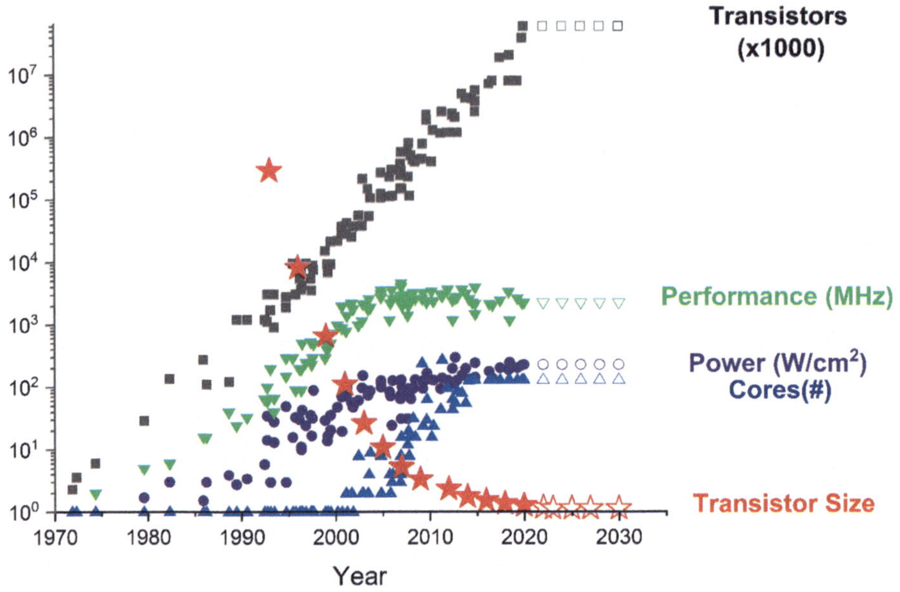

Fig. 1.1 Graphical microprocessor trend data showing Moore's law progressiong from the 1970s toward 2020, with a prediction of performance stagnation due to transistor limitations thereafter. From [3]

energy storage is to utilize single-walled carbon nanotubes (SWNTs) as a silicon substitute [4].

SWNTs are stable nanostructured carbon allotropes that can have either conducting or semiconducting metallicities [5]. Because transistors are semiconducting systems, the SWNTs must be semiconducting as well. Therefore, it is essential to obtain only semiconducting (sc-) SWNTs and eliminate any metallic (m-) SWNTs. This goal is a challenge, as SWNTs are typically created in a mixture of both sc- and m-SWNTs; and this mixture is difficult to separate. Additionally, SWNTs should not be superimposed on one another in a device, and achieving single layers of neatly aligned, appropriately spaced SWNTs is an additional hurdle in producing practical nanotube transistors, as SWNTs typically bundle [6–8]. Hence, it is desirable to sort out the different types of SWNTs as well as align them in a set orientation simultaneously. A schematic example of a desirable outcome in SWNT alignment is shown in Fig. 1.2. It is estimated that a target of 125 SWNTs/μm^2 or at least a density range between 100–200 carbon nanotubes/μm^2 in the device channel is necessary for field-effect transistor (FETs) applications [9, 10]. In order to accommodate for billions of transistors per chip, the channel length is targeted at below 100 nm.

The alignment relay technique (ART) developed by the Schipper group targets simultaneous surface organization and sorting of sc-SWNTs. This multidisciplinary

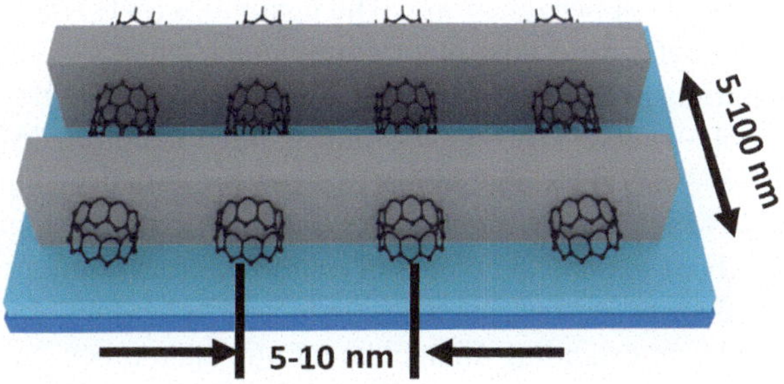

Fig. 1.2 Demonstration of aligned sc-SWNTs between two electrodes with target values for density and channel length

method combines organic chemistry and materials technology incorporating liquid crystals to promote the propagation and improvement of electronic nanodevices. The ART is the focus of this work; its background and the reasoning behind its development will be described in Chap. 1, as well as the characterization methods used for its optimization. A detailed introduction of CNTs will pre-face the ART primer. The various characterization techniques used are also outlined.

The primary contributions of this thesis are to investigate further and adapt ART, guided by the following question: what alterations can be done to improve or adapt the method? The theory was applied at various stages to address the following sub-questions: (1) how can the technique be modified to improve the alignment? (2) can the molecule be altered? (3) what other structures can ART work with? and (4) does this technique work in practice to create functioning devices?

Chapter 2 describes the effects of adding steps to ART to increase alignment and the number of tubes on the surface. These include simple modifications such as post-deposition sonication and the inclusion of filler molecules. Chapter 3 presents changes in the anchoring group and synthetic modifications to ART for different surfaces. This concept is followed in Chap. 4 with an assessment of the effects of oxidized carbon nanotubes for ART, and other nanosystems attempted with ART, such as multi-walled carbon nanotubes (MWNTs), graphene quantum dots (GQDs), graphene nanoribbons (GNRs), and nanowires (NWs). Chapter 5 describes the project's primary contribution, the successful creation of CNT transistors using ART. For comparison, it also describes the less-successful attempts. Chapter 6 summarizes the entire work and suggests possible directions for future research. The last Chap. 7 provides all the experimental details.

1.1 Background and Literature Review

1.1.1 Primer on Carbon Nanotubes

In 1991, Ijima observed and reported to the scientific communities a tube-like carbon structure, described now as a carbon nanotube (CNT) [11]. The assembly of a CNT can be represented as a layer of graphene being folded into a cylinder composed of benzene rings. The nanotubes (NTs) have a significantly larger length-to-diameter ratio than other compounds, categorizing them as 1-dimensional (1D) materials [12]. These NTs can be classified as single-walled nanotubes (SWNTs), double-walled (DWNTs) or multi-walled nanotube (MWNTs). The two main categories—SWNTs, and MWNTs—are presented in Fig. 1.3 [13]. From the image, MWNTs possess a much larger diameter than the single-walled version. The CNTs backbone can be comprised of many different configurations, which ultimately dictate their properties, such as conductive behavior.

The freestanding diameters of the nanotubes can vary from 0.4 to 100 nm depending on the structure. For SWNTs, the widths are typically in the range of 0.4–4 nm [14, 15], although 0.8–2.5 nm is what is commercially available. The range of MWNTs' diameter is quite broad, from 1.4 to 100 nm because of the concentric nature of the material. The smallest MWNT in terms of outer diameter is 1.4 nm (DWNT), and the largest is at the maximum defined range of nanomaterials at ~100 nm. However, some MWNTs can attain a diameter of up to 200 nm [16]. Interestingly, the smallest diameter tube that has been reported was a 0.3 nm SWNT grown inside the cavity of a MWNT and found to possess both metallic and semiconducting states [17]. The lengths of any of these carbon nanotubes can extend to many microns or even millimeters [18].

Additionally, the nanotubes can be separated by being either metallic or semiconducting, even though commonly carbon itself is not typically associated as being a conducting element. This electrical behavior is based on the electron orbital formation and carbon structure. It is the reason why diamond, an sp^3 hybridized carbon material, is an insulator, whereas graphene and CNTs, sp^2 allotropes, can display metallic or semiconducting behaviour [19]. Unsurprisingly, these nanomaterials are

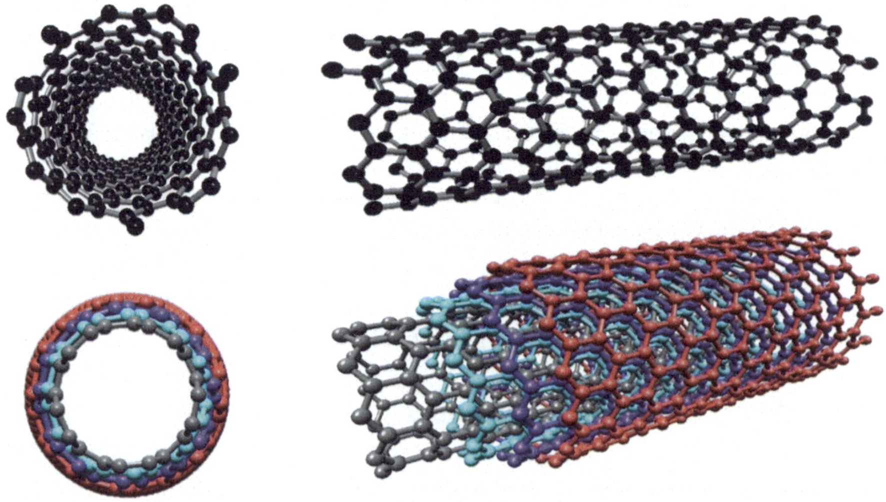

Fig. 1.3 Comparison of a single-walled (10,3) (top) and multi-walled (bottom) carbon nanotube from the front and side

the main topic of much research worldwide, as they can have such varied applications due to their structural and conductive differences. Nevertheless, for transistor-based applications, semiconducting-SWNTs are the predominant choice.

Figuratively, the development of CNTs can be described by the "rolling" of graphene. This process generates different kinds of SWNTs structures, all of which have their respective metallicity associated with them. CNTs are represented more specifically as (n, m), which are chiral indices that indicate the means by which the graphene sheet is wrapped, as Fig. 1.4a shows [13]. The CNT conductivities are subdivided into three forms based on *n* and *m*: armchair ((n, n), i.e., n = m), zig-zag

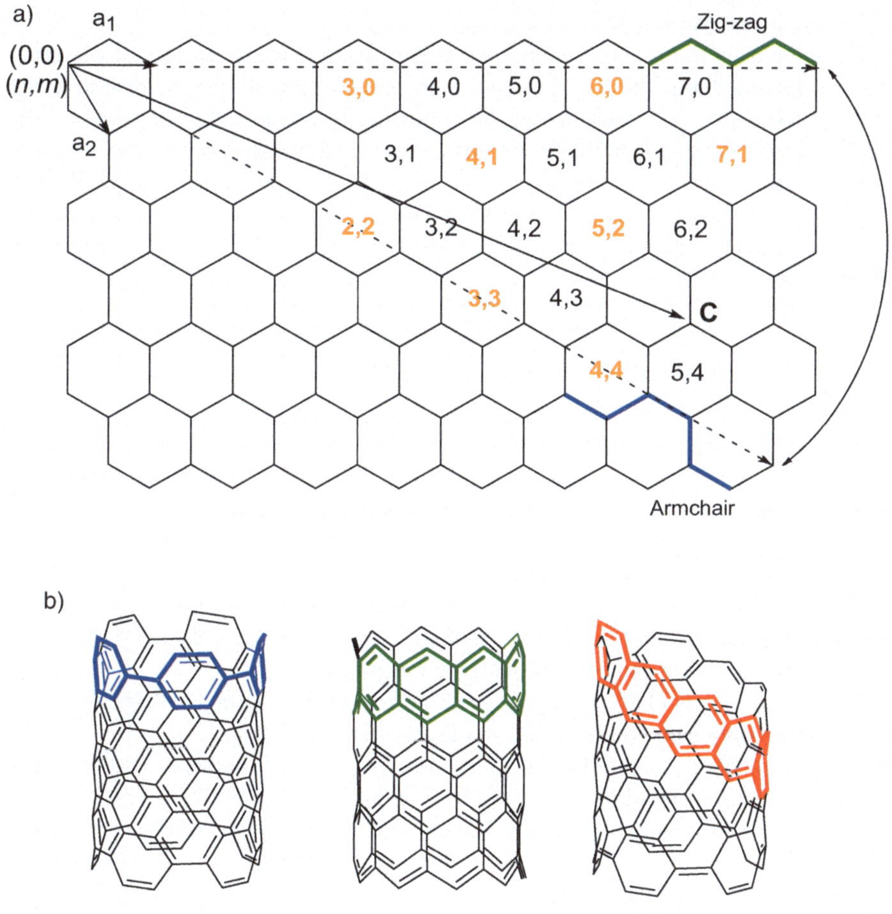

Fig. 1.4 (**a**) Illustration of the chirality map of a carbon nanotube on a sheet of graphene with sample n,m values, adapted from Lei et al. [20]. The basis vectors (a1, a2) are shown and drawn to show the formation of the chiral angle C. The (n,m) begins at (0,0). Orange colored (n,m) values are metallic CNTs, while the black integers are examples of semiconducting chiralities. (**b**) Examples of armchair (blue), zigzag (green), and chiral (red) nanotube final structures

(n, 0) or helical ((n, m), synonymously described as chiral in most literature sources) that possess a helical twist in their conformation (Fig. 1.4b) [20–22].

The two integers are representative of the graphene sheet rolling up and connecting at points n and m. The **C** (chiral) vector in Fig. 1.4a is defined by the two wrapping integers (*n*, *m*) and a duo of basis vectors (a_1, a_2) as demonstrated. Furthermore, the SWNTs with identical chiral vectors can take on diverse *handedness* in terms of orientation, e.g., right-handed R-(8,5) and left-handed L-(5,8) [23, 24]. Essentially, if the nanotube is neither zig-zag nor an armchair, then it is a chiral tube that will possess a directional twist. The point group of the helical CNTs is also distinguished differently than the armchair or zig-zag ones. The point group of chiral (*n*, *m*) CNTs is defined as D_Q—where the Q value is indicative of the quantity of hexagonal structures in the unit cell of the CNT, and it is always an even integer. For (*n*, *n*) and (*n*,0) CNTs, the point group is D_{2nh} [25].

Contingent on the *n* and *m* numerical values, even if the diameters of the tubes are closely matching, the SWNTs can display either metallic or semiconducting characteristics [26, 27]. Variance in the conductivity of the CNTs depends on the band structure and bandgap that is directly correlated to the differences in molecular bond structure, determined by the chiral properties of the starting material [27]. Hence, the electronic character of the SWNTs are reliant on the chiralities [28]. The diameter is also dependent on the (*n*, *m*) values, as per Eq. (1.1) designated in picometers:

$$d_t = \frac{a\sqrt{m^2 + mn + n^2}}{\pi} \approx 79.2\sqrt{((m+n)^2 - nm}\,[\mathrm{pm}] \tag{1.1}$$

where a is the lattice constant and is expressed as 2.49 Å. This expression originates from the lattice of graphite, as has been derived by Saito et al. [19]. Therefore, the conductive behavior and diameter are interrelated with (n, m). A more straightforward rendition of Eq. (1.1) is expressed as the circumference length of the nanotube (L) divided by π (i.e., $d_t = L/\pi$) [19]. The equation is used as an estimate, especially for CNTs where *n* and *m* are below 10 (i.e., small values). Additionally, Eq. (1.1) does not consider the thickness of the nanotube walls. The key takeaway is that the diameter of SWNTs is intertwined with their chirality.

Additionally, it is challenging to create pure semiconducting MWNTs, as they must contain only semiconducting SWNTs. If any of the NTs are metallic, then the entire system becomes metallic. This preference of conduction means that MWNTs are almost always metallic. It is possible to identify if the MWNTs are of mixed metallicity or purely metallic through Raman spectroscopy [29]. Some cases of pure double-walled and triple-walled semiconducting nanotubes exist [30, 31], but they are challenging to synthesize, making them unideal for commercial applications. These conductivities are further addressed in Sect. 1.1.4

Furthermore, compiled in Table 1.1 are some of the properties of CNTs. The data of the CNTs is paralleled with that of other materials and nanomaterials. The CNTs have a Young's modulus—which is the terminology used to express the stiffness of

Table 1.1 Examples of CNT strength and electrical properties as compared to other systems

Material	Properties							Ref.
	Mechanical		Electronic	Thermal	Electrical			
	Young's modulus	Tensile strength	Bandgap (eV)	Conductivity (W/m K)	Resistivity	Max. current density	Quantized conductance	
SWNTs	1 TPa	60–100 GPa	n-m/3 = integer, zero eV	1750–6600	10^{-6} Ω m	10^{7}–10^{9} A/cm^2	12.9 kΩ^{-1}	Lu and Han [32] Berber et al. [33], Wang et al. [34]
			n-m/3 ≠ integer, 0.4–2 eV					
MWNTs	1–1.2 TPa	0.15 TPa	0	>3000	10^{-6} Ωm	10^{7}–10^{9} A/cm^2	12.9 kΩ^{-1}	Wang et al. [34]
Diamond	1.2 GPa	1.2 GPa	5.45	2000	10^{11}–10^{14} Ω-m	1.3 A/mm	–	Spear and Dismukes [35], Hirama et al. [36]
Steel (structural)	200 GPa	400 MPa	0	45 (1% carbon)	10^{-7} Ωm	10^{-6} A/cm^2 (Avg)	–	www.engineeringtoolbox.com. Accessed May 14, 2020
Graphene	0.5 TPa	130 GPa	0	3000–5000	10^{-8} Ωm	10^{7}–10^{9} A/cm^2	12.9 kΩ^{-1}	Tombros et al. [37], Lee et al. [38]
GNRs	1 TPa	137 Gpa	Width dependant	3000–5000	0.02 Ωm (unzipped)	10^{8} A/cm^2	Edge defect dependant	Duryat [39], Li and Lu [40]
ZnO NWS	25–200 GPa	4–6 GPa	3	20	0.01 Ωm	10^{-3} A/cm^2	–	Hoffmann et al. [41], Yuldashev et al. [42], Lu [43], Lee et al. [44]

a material—that is nearly five times stronger than the one of steel [32]. The CNTs' tensile strength, meaning the strength required to break the nanotubes, is about 100 times stronger than structural steel. Additionally, the CNTs have demonstrated a high thermal conductive behavior in ambient temperature approximately three times higher than that of a diamond. Unlike carbon's famous form of diamond, the CNTs are also excellent electrical carriers. Indeed, CNTs are characterized by their high conductivity, specific surface, mechanical strength, and interfacial effects.

The extent of CNT applications depends on their quality, cost, and orientation. For instance, unidirectional arrays of CNTs apply to solar cells, optical antennas, and photonic crystals. Vertical CNTs can be used as flat panel displays [45], X-ray generators [46], and microwave devices [47]. Horizontal CNTs have shown potential as transistors [48], sensors [49], and photonic crystals [50]. However, these abilities are rendered mediocre, and all desirable applications disappear if their positioning, length, diameter (i.e., aspect ratio) and type are not controlled. Hence, the key to carbon nanotube architectonics and technologies is the development and control of CNTs with specific properties at scalable proportions. This control can be during CNT synthesis or post-processing, through various purification methods.

1.1.2 Fabricating Carbon Nanotubes

The synthesis of the CNTs can be through an assortment of approaches. The popular methods are laser ablation, chemical vapor deposition (CVD), and [electric] arc-discharge (synonymous with arc-evaporation). All three of these can fabricate MWNTs and SWNTs at an over 75% yield [51]. Electrolysis and sonochemical or hydrothermal formation of CNTs are also prevalent. As with any experimental preparation, each method possesses its respective benefits and drawbacks, producing different growth results. Based on the target properties of the CNTs, the synthetic methodology used to fabricate them can be chosen.

In particular, CVD is one of the most popular synthetic methods as it can create large batches of nanotubes with properties depending on the catalyst used, as the dimensions of the catalyst dictate the diameter of the tubes. This diameter dependence is because the particles nucleate the tube growth. The feedstock gas and substrate surface used can also affect the quality of the CNTs. An illustration of the procedure is represented in Fig. 1.5. The carbon source is input as gas, and plasma or a heating coil are implemented as an energy source for the gaseous carbon along with a carrier gas, like $N_{2(g)}$ or $Ar_{(g)}$. Hydrocarbons sources like methane are used as the source for ^{12}C. The sample surface is enveloped with metal catalysts such as Ni, Co, or Fe to above 700 °C (usually). Mainly, CVD is used to dissociate the carbon atoms from molecules, and the appropriate temperature, atmosphere, and catalyst dictate the yield and types of the CNTs obtained [52]. The SWNTs are formed at higher temperatures than MWNTs [53].

The mechanism for CVD growth is separated into two different versions: either a vapor-liquid-solid (VLS) model or a vapor-solid-solid (VSS) model. The VLS

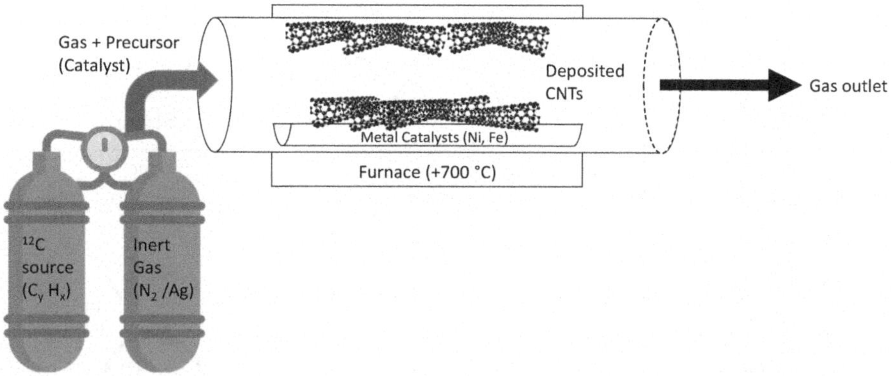

Fig. 1.5 Diagram for CNTs synthesis using common hot-wall CVD

method is defined by carbon atoms diffusing from a decomposed area and then precipitating in solution. The chosen surface and the catalyst dictate the different areas of exposed metal that absorb heat and precipitation areas that give away heat. Therefore, the interactions between the surface and the metal catalyst are significant in forming the CNTs. Weak catalyst-substrate interactions result in a tip-growth (Fig. 1.6a) mechanism, whereas strong interactions result in base-growth (Fig. 1.6b). The tip-growth model contains a metal that has an acute contact angle with the catalyst. The hydrocarbon will decompose on the exterior of the metal, and the carbon atoms diffuse across the metal particle. The CNTs then precipitate from the bottom of the metal surface, and the entire metal catalyst is pushed up and off the substrate. The CNT will exhibit continuous growth so long as the catalyst has areas that remain uncovered for the hydrocarbon atoms decomposition [54].

The end growth of the CNT occurs once an excessive amount of carbon encloses the catalyst. A favorable interaction among the substrate and the metal is also possible, resulting in the base growth of the CNTs [55]. Carbon diffusion and the hydrocarbon decomposition still occurs, but instead of CNT precipitation pushing the metal particle up, the CNT pushes itself up and away from the substrate [56]. The CNTs grow outward, with the catalyst particles remaining at the base. These two models are part of the VLS mechanism as they begin with hydrocarbon vapor, have a metal-with dissolved carbon atoms (liquid carbide) intermediate, and then output a crystalline carbon solid [57].

On the other hand, the alternative mechanism considers the carbon atoms to only diffuse on the very surface of the catalyst particle (Fig. 1.6c). This process describes the VSS mechanism, which indicates that the carbon precursor dissociates from the hydrocarbon vapor, then diffuses exclusively on the catalyst particle solid surface, followed by CNT growth (carbon solid). When growing SWNTs from transition metals like iron or nickel, the VLS is the accepted approach [56, 58]. When growing SWNTs on other nanoparticles, such as SiO_2, the VSS mechanism has been observed to be accurate [59].

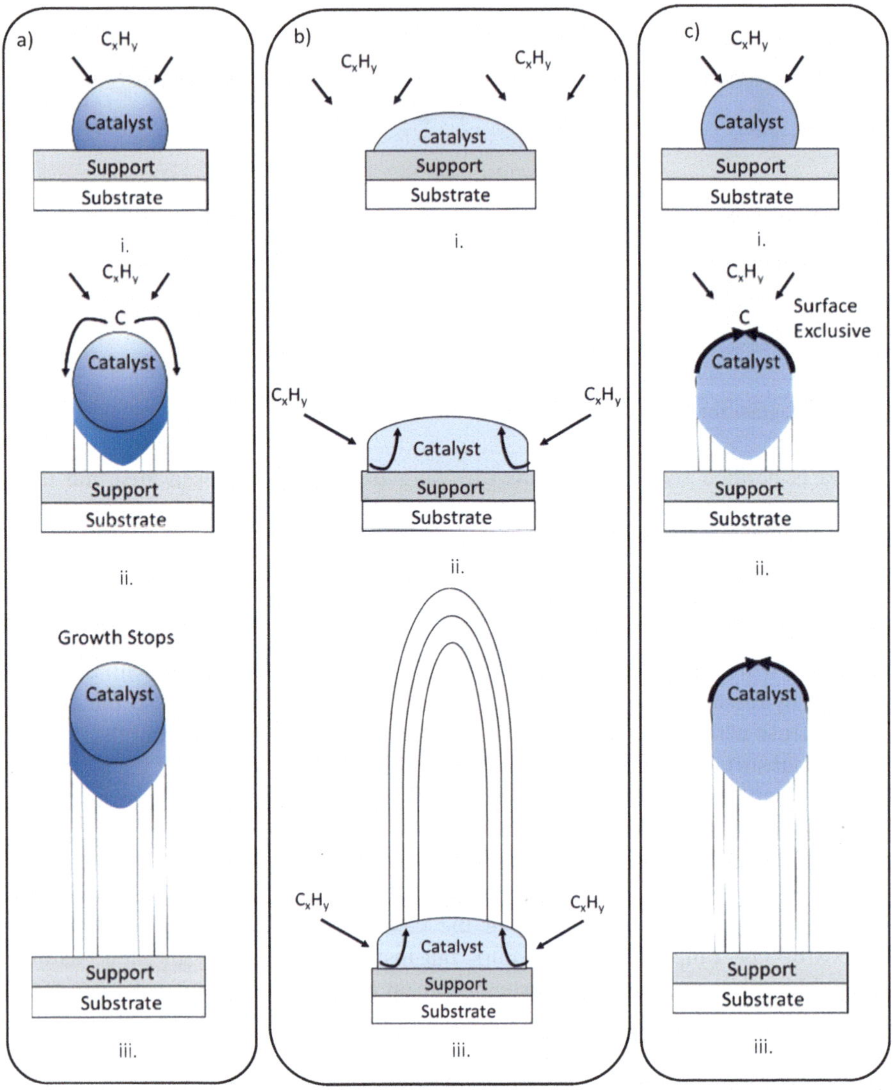

Fig. 1.6 Illustrations representing the growth mechanisms of CNTs using CVD (**a**) VSL tip-growth model, (**b**) VSL base-growth model, (**c**) VSS tip-growth. Adapted from Kumar and Ando [60], and Tessonnier [61]

There are many different versions of CVD that can all be differentiated based on pressure (low, high, or merely atmospheric), reactor (hot or cold-walled), precursors being solid, liquid or gas sources, and heating sources that can be thermal or plasma-enhanced; yet they all share the same fundamental growth aspects. That is, either the VLS or VSS models, and their dependence on catalyst type and concentration, and supported substrate for growth control. The negative aspect of the CVD technique is the poisoning of the catalysts's surface from other forms of carbon that prevent further CNT growth. The cost of pre- and post-fabrication preparation and cleanup are also considered drawbacks. One solution is to incorporate water vapor with the hydrocarbon feedstock gas to clean up the amorphous carbon buildup [54]. This addition keeps the surface clean and allows for the growth of nanotubes on the millimeter-long scale [60]. The two common subcategories of CVD are through the cobalt and molybdenum catalysts and CO gases (CoMoCAT) and the high-pressure carbon monoxide (HipCO) methods for extensive and scalable production. Synthesis of NTs with the CVD technique is commonly applied by Sigma-Aldrich,[1] frequently using CoMoCAT for their many SWNTs variations.

Conversely, another sub-category of CVD is catalytic pyrolysis (CP). This method is chemically the same as CVD by involving the irreversible decomposition of organics at high temperatures. However, CP possesses more straightforward processing steps. Catalyst molecules or organometallic precursors are injected into the hydrocarbon gas [62]. The CNT growth then occurs on a specific substrate, such as quartz, or on the reactor wall. So, the growth is generated indirectly from the metal catalysts eliminating the need for intensive metal purification. It is estimated that OCSiAl's Graphetron 1.0 that generates high purity Tuball brand SWNTs produces them in this manner [15]. Additionally, CP is capable of forming CNTs on a large scale from recycled plastics, such as polypropylene [63].

The other dominant techniques for CNT fabrication are arc-discharge and laser ablation. They function on similar principles by both incorporating the condensation of heated carbon atoms in the gas phases that are produced from solid carbon evaporation. The reason CVD is significantly more popular than these latter two methods stems from the concept that controlled synthesis, such as horizontal arrays of SWNTs on various substrates, is regularly achievable. In contrast, arc-discharge and laser methods can currently only create nanotube bundles in powdered form [64]. However, Nanointegris Inc., a sub-company of Raymor Industries, creates their SWNTs through arc-discharge.[2] The company is reputable for its high purity SWNTs, which are purified through a density gradient ultracentrifugation (DGU) technique after their fabrication.[3]

For arc discharge, the gas is electrically torn apart to generate plasma and it was the original method for CNT production [11]. Two carbon-containing rods, identified as cathode and anode, are held in a chamber that is either filled with a gaseous or liquid environment. Formation of nanotubes can occur with a direct-current (DC) or alternating-current (AC) as the power source, as well as through a pulsed arc-discharge method. The arc is created via close contact of the electrodes that can generate plasma at very high temperatures (~5700 °C) that can sublime the starting material carbon source that resides inside the anode [65]. The carbon atom vapors

---

[1] https://www.sigmaaldrich.com/materials-science/material-science-products.htmL?TablePage=16376687.

[2] http://nanointegris.com/wp-content/uploads/2020/03/NanoIntegrisCatalog_2020.pdf.

[3] http://nanointegris.com/wp-content/uploads/2017/10/NanoIntegris-Brochure-ENG.pdf.

cluster together and are attracted toward the cathode, where they cool down and remain. This process occurs in a few minutes, and the CNTs can be collected from the cathode as well as the walls of the chamber.

MWNTs are formed on the cathode when pure graphite rods are used, with fullerene soot produced inside the chamber. In contrast, SWNTs formation uses a catalyst, typically Fe or Co, on the graphitic anode in addition to a pure graphite cathode. The mechanisms for growth are unclear [65]. The SWNTs are then formed on the chamber walls as soot, as shown in Fig. 1.7. Similarly to CVD, the dimensions of the catalyst metal particles dictates the size of the SWNTs.

For laser ablation, sublimation of graphite on quartz occurs by using high powered laser vaporization under an inert atmosphere, typically Ar or He [51, 54]. The energy required is produced by the laser, yet the technique functions very similarly to arc-discharge. The CNTs are collected on the surface of a water-cooled collector rod [66]. A metal catalyst is also required for the production of high purity SWNTs, but not on as large a scale as its two other counterparts. Usually, Co or Fe catalysts in some form of a bimetallic combination are used and mounted on a pure graphite pellet. Additionally, the quality of the CNTs is also influenced by the power of the laser and its wavelength. For instance, as the power of the laser pulse is augmented, the thickness of the NTs decreases, and this treatment results in thinner tubes [67, 68]. Hence, the key benefit of this procedure is the control of the parameters to generate many different high purity CNTs with low amounts of metallic impurities, as the metal impurities evaporate from the end of the tube.

However, branching of the SWNTs is a significant drawback, along with the scalability regarding the temperature (>1200 °C), pressure (250–550 Torr, though MWNTs are favorable at 750 Torr), and the laser source. Temperature is particularly challenging to tailor for SWNTs formation as the growth zone is over 1200 °C, but the number of defects increases at over 900 °C. Yet, a low temperature (>400 °C) produces a minimal number of tubes, and heat applied below 200 °C forms no target product [66]. Some cases report the usage of two lasers or double pulsing of a high powered laser [69, 70]. Furthermore, frequent pulse repetition is required, but the larger the diameter, the fewer CNTs can be made, and the target surfaces must be thoroughly smoothed as roughed targets inhibit the evaporation of the catalysts. The laser-to-target interaction must also be fine-tuned to allow for efficient evaporation of particles into the gap phase for nanotube formation.

Overall, there are many different categories and sub-categories for SWNTs fabrication. Nevertheless, optimization of the main techniques is still desirable as CNTs are sought for many academic and industrial projects. As well, minimizing the cost for CNT production is another main challenge for all the methods described, and the post-synthesis sorting and purifying of the different SWNTs types is time consuming and costly. Some chiral control has been shown by having a planar metal surface rather than catalyst particles in some cases [71], and attempts at synthesizing CNTs from nanobelts or small molecule armchair orzig-zag segments is also an area of investigation for precise chiral control [72].

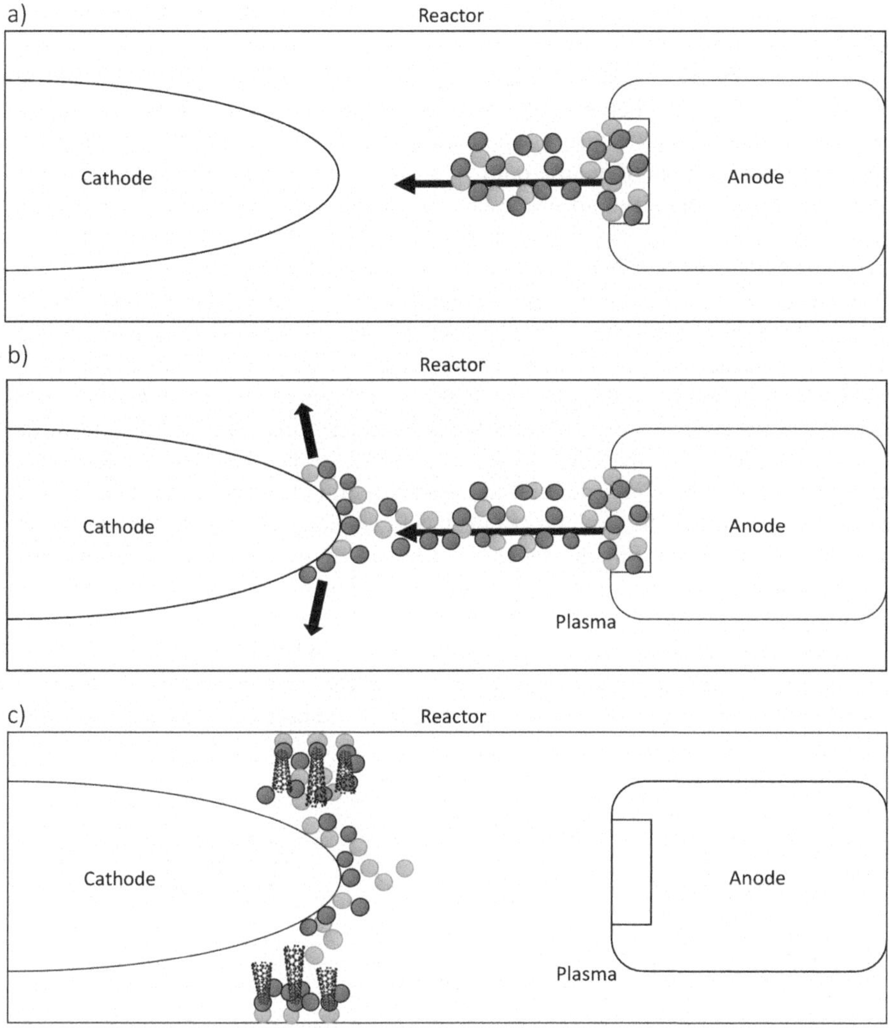

Fig. 1.7 Illustrations are representing the growth of SWNTs using arc-discharge. Carbon gaseous particles are the black particles and metal catalysts in the grey particles, (**a**) the catalysts and carbon vapors are released from the anode and move toward the cathode, (**b**) because of their fast momentum, the catalyst particles do not stick to the cathode and move away from the generated plasma, (**c**) the carbon particles nucleate at the surface of the metal particles on the walls of the chamber, producing SWNTs. Adapted from Arora and Sharma [65]

1.1.3 Carbon Nanotube Defects

Defects in CNTs can cause deviations in their strength, optical and electronic assets [73]. Depending on the synthetic method, various defects can be present in the tube structure. The CoMoCAT and high melting point alloy catalysts in CVD are methods for achieving high purity SWNTs. However, minor defects are still present, usually originating from the nucleation or capping step during CNT growth. Larger tubes such as DWNTs or MWNTs have a higher likelihood of having defects from the synthesis process, whereas SWNTs are nearly independent of structural defects during fabrication [74, 75]. Most deformities will arise from post-fabrication treatments [76]. The surrounding environment during synthesis and processing can primarily affect tube quality and properties. If SWNTs are exposed to high e-beam irradiation, such as with electron microscopy techniques, then the loss of an atom can occur that leads the CNTs to attempt to reconstruct their structure [77].

However, these defects can be advantageous for functionalization, wherein CNTs with various side-chains can be used for biological [78] or sensor applications [79]. Alternatively, they can also have a negative impact on conductivity, causing SWNTs breakage. Although not a prominent parameter of this work, it is essential to be aware of the different kinds of defects that can be inherent in the SWNTs structure. Mainly, simulations are used to study CNT defect effects, and their presence can be confirmed through transmission electron or scanning tunneling microscopy [80]. The curvature of the CNTs is induced due to process manufacturing in conjunction with the large aspect ratio of the materials and their low bending stiffness [81, 82]. The appearance of prominent CNT defects are depicted in Fig. 1.8.

For devices, reduced variations in performance and SWNT properties can be overcome by having multiple aligned nanotubes in the same channel [83]. However, it is also important to indicate that even if the CNTs are largely defect-free, substrate or atmospheric interactions can also interact and affect SWNT properties [75].

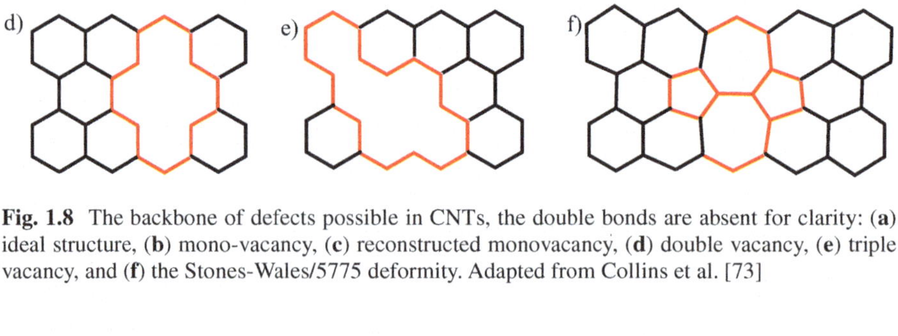

Fig. 1.8 The backbone of defects possible in CNTs, the double bonds are absent for clarity: (**a**) ideal structure, (**b**) mono-vacancy, (**c**) reconstructed monovacancy, (**d**) double vacancy, (**e**) triple vacancy, and (**f**) the Stones-Wales/5775 deformity. Adapted from Collins et al. [73]

1.1.4 Carbon Nanotube Electronic Properties

The bandgap (E_g) is crucial as it dictates the possible electrical conductivity of the material. The bandgaps for sc vs. metallic and insulator materials is shown in Fig. 1.9. From the schematic, insulators have too large a gap for electrons to travel toward the valence band from the conduction band. In contrast, metals have an unlimited exchange of electrons between the two. In semiconductors, energy can be added to move electrons across the gap, and hence the sc-system can be switched ON and OFF to control the electron flow [84]. The electrons must be excited to move above the Fermi level (E_f) to conduct electrical signals. The E_g requirement must also be no more than 3.2 eV, that is, the energy from sunlight, otherwise the gap is too challenging to overcome, and the material is an insulator. For CNTs, the bandgap of the sc-tubes is inversely proportional to their diameter. This property makes targeting diameter separation of CNTs a very high priority.

Therefore, CNTs have unique electrical transport properties due to their band structures in conjunction with their tiny dimensions. The electronic band structure for SWNTs is formulated with respect to the one of graphene with the addition of imposing a boundary condition due to the cylindrical structure [85]. Two wave vectors are related to the SWNT: a wave vector that is parallel to the tube $\mathbf{k}_{\parallel}$; and the circumference wave vector $\mathbf{k}_{\perp}$. Due to the planar structure of graphene, it is described by k_y and k_x wavefunctions of an electron in reciprocal space, whereas CNTs have $\mathbf{k}_{\perp}$ and $\mathbf{k}_{\parallel}$ because of their tubular morphology. The parallel vector is assumed infinitely long, while the circumference vector must repeat itself as it rotates 2π times around the SWNT diameter(d_t). Therefore, the $\mathbf{k}_{\perp}$ is quantized as it rotates 2π around the SWNT. It is represented by the below Eq. (1.2) and Fig. 1.10.

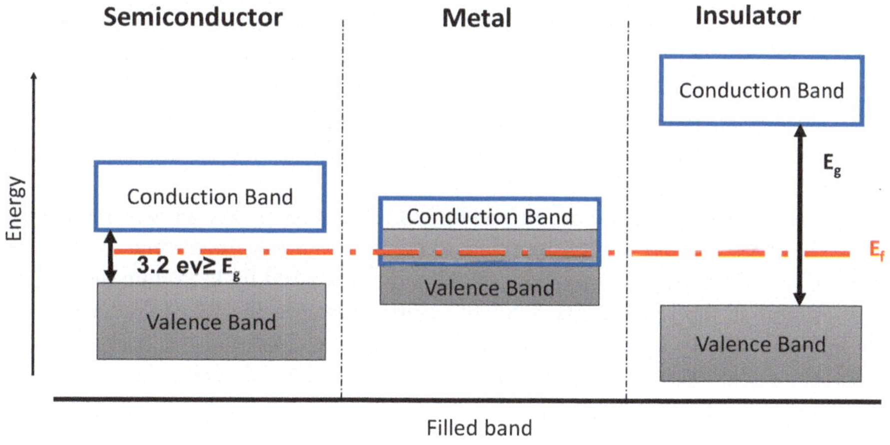

Fig. 1.9 Energy bands gaps for different materials. E_g is the material bandgap, E_f is the Fermi energy level. Semiconductors must have a band gap less than 3.2 eV in energy for electrons to move to the conduction band. A greater gap and the electrons are unable to be excited into the next band level. There are no insulating CNTs, but the insulator gap is shown for reference

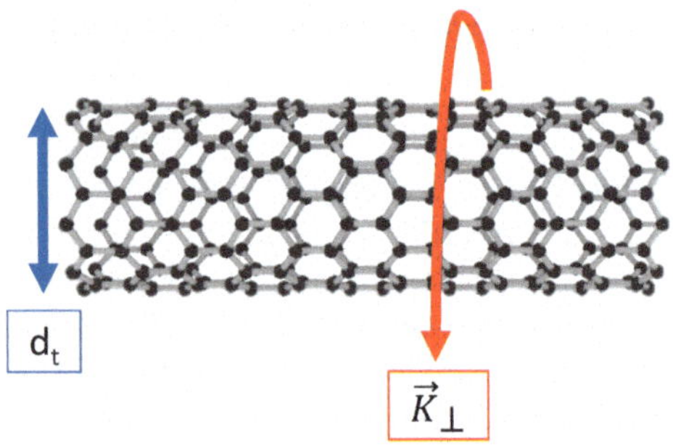

Fig. 1.10 Wave functions and diameter illustrated on a SWNT

$$\mathbf{k}_{\perp} \cdot \vec{C} = \pi d_t \mathrm{k}_{\perp} = 2\pi m \quad (1.2)$$

Subsequently, as shown in Fig. 1.11a, the 1D band construction of the CNT is attained from a cross-sectional analysis from slicing the graphene energy dispersion diagram model with the allowable $\mathbf{k}_{\perp}$. In Fig. 1.11b, c, the zone-folding schemes are shown for the m- and sc-SWNTs [17, 86]. A 1D sub-band is attained for each cross-sectional cut. The 1D bandgap constructions of SWNTs are resolved by the spacing among the allowed perpendicular K-values and their angles concerning the surface of the Brillouin zone of the graphene. Moreover, this is all determined by the (n, m) index of the SWNT, relating to their diameter and chirality. The essential transport properties are determined by the band structure close to the E_f. This property is specified by the permissible $\mathbf{k}_{\perp}$ environments that are nearest the K point. The K point is defined as being the node where the valence band encounters the conduction band. When the permitted $\mathbf{k}_{\perp}$ state passes through the K point (Fig. 1.11b), there is no bandgap. The two-energy dispersions are linear bands that intersect at the Fermi level, displaying metallic behavior. On the contrary, if the permissible $\mathbf{k}_{\perp}$ states do not transit across the K point, then a bandgap is present. This bandgap is shown by two parabolic 1D bands, and the distance between the bands can be in the range of 0.1–2 eV for SWNTs [87]. As the diameter of the CNTs increases, their bandgaps decrease.

Hence, two separate types of conductivities are possible within the CNTs. Moreover, these conductivities are relate to the diameter as $\mathrm{k}_{\perp}$ is ultimately affected by it [85]. No bandgap occurs when the K point is crossed by $\mathrm{k}_{\perp}$. When there is no cross-section at the K point, the CNT is a semiconductor. In order to determine an SWNT's metallicity from n and m values, a metallic condition has been extrapolated. The metallic condition can also be demonstrated numerically (Eq. (1.3)).

$$\frac{2m+n}{3} = integer(i) \text{ or } (\mathrm{n}-\mathrm{m}) = 3i \quad (1.3)$$

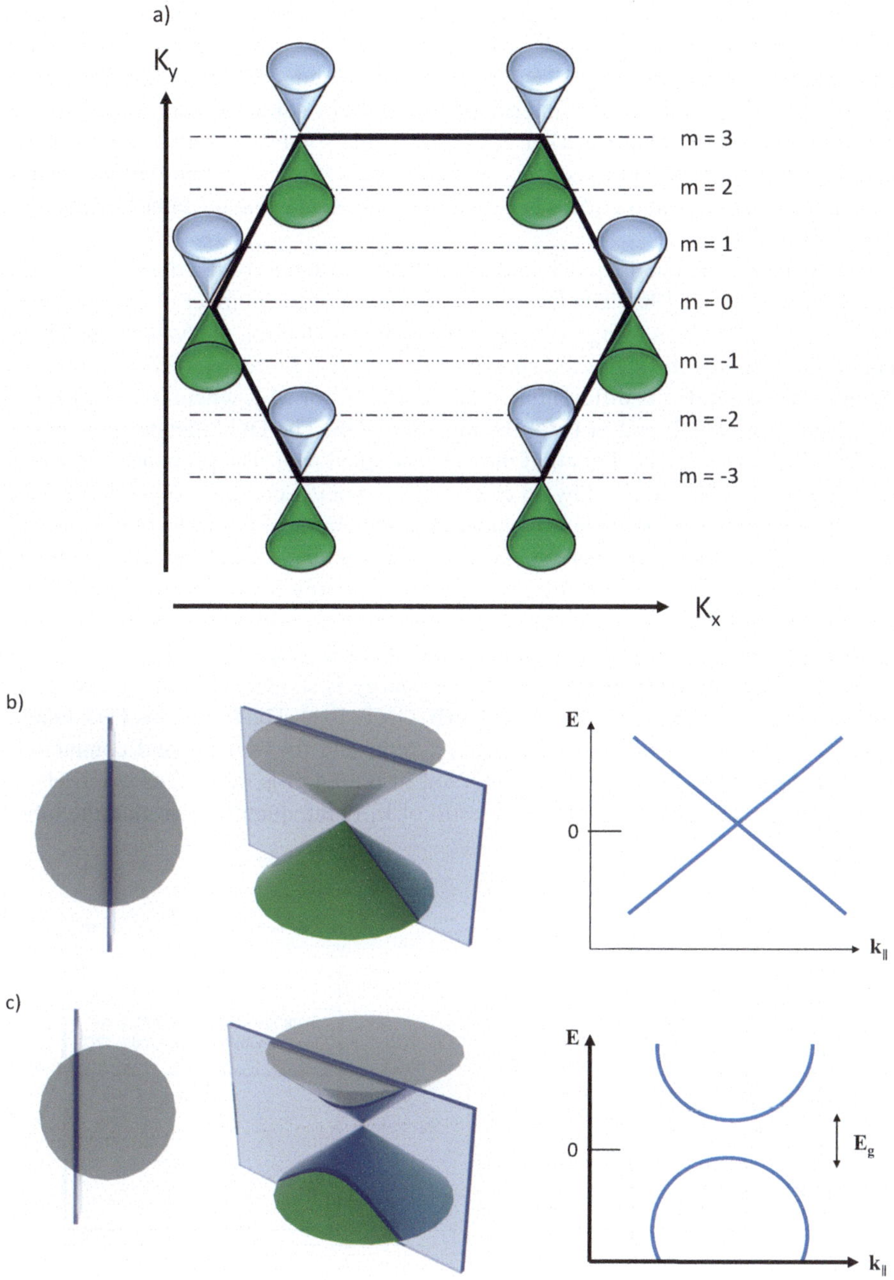

Fig. 1.11 (**a**) Representation of the energy dispersions at six K points for the initial Brillouin zone for graphene, where the SWNT $k_{\perp}$ allowed states are the dashed lines, (**b**) 1D energy dispersion for the metallic SWNT band structure and the cross section of the energy dispersion, (**c**) 1D energy dispersion for the semiconducting SWNT band structure and the cross section of the energy dispersion. Adapted from Park [85]

Equation (1.3) indicates that if there is an integer of 3 when subtracting the *n* and *m* values, then at least one of the $k_{\perp}$ values will coincide with the K point. Any other value not corresponding with "3" will predict a semiconducting nanotube. However, the concept if n = m or n-m = integer of 3 does have some exceptions due to the curvature of the CNTs with smaller thicknesses, and some CNTs also exhibit semi-metallic properties [88]. For instance, small diameter zigzag and chiral tubes exhibit finite gap behavior, and the (5,0) NT possesses metallic behavior rather than semi-conductive [89].

Furthermore, the bandgap is based on the band structure of the density of states (DOS) of the SWNT [90]. The DOS is an indication of the number of energy states that can exist at a given energy (E) concerning the conduction and valence bands of a material. Usually, the states are shown as E_{11} and E_{22}. For convenience in the following discussion, E_{11} is differentiated into either S_{11} or M_{11}, where the S is for the semiconducting DOS, and M is for the metallic DOS. The DOS dictates how many energy states there can be, but not if they are occupied [84]. The probability that one of the states will be occupied by an electron at a specific energy level is defined by the Fermi function if the system is under equilibrium. The Fermi level is when, at absolute zero temperature, there are no energy states above this level. It is the highest energy state at 0 K. No electrons have enough energy to surpass the Fermi level under cold conditions. Therefore, even a semiconductor will behave as an insulator, as no electrons have the energy to jump across the bandgap.

In Fig. 1.12, the DOS as it relates to the energy is shown for both the semiconducting (Fig. 1.11a) and metallic (Fig. 1.11b) CNTs [91]. The Fermi level is situated at zero, and the left and right of the graph represent the valence and conduction bands, respectively. A peak that is observed in the DOS is defined as a van Hove singularity (VHS) [90]. The VHS is a result of $k_{\perp}$ being quantized around the tube.

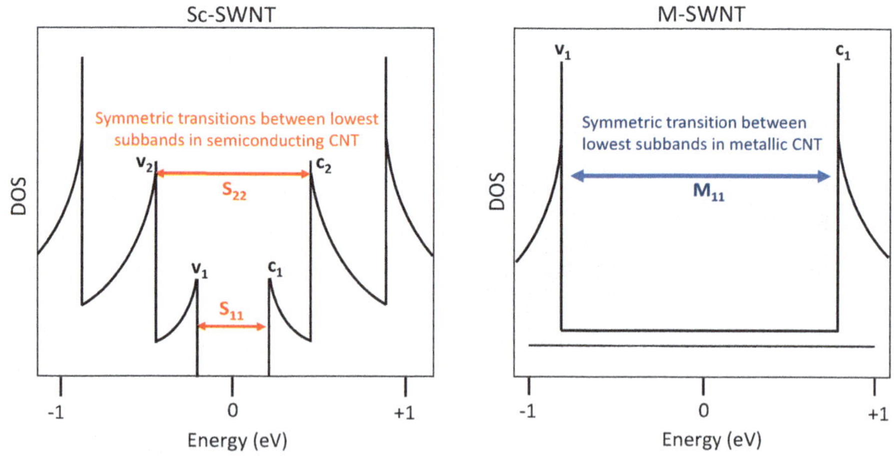

Fig. 1.12 (**a**) DOS energies for sc-SWNT, (**b**) DOS energies for m-SWNT. Adapted from Hagen and Hertel [91]

An optical transition is possible when an electron or hole obtains energy and can then jump or displace itself from one energy level to a different one [84]. A map of interband transition energies can be extrapolated from the information regarding the valence and conduction bands position within the Fermi energy and the diameter of the tube (d_t), which can allow for the identification of m- and sc-SWNTs [92].

In SWNTs, there are two transitions to recognize [84, 90]. Firstly, as per Fig. 1.12, the excitation of S_{11} describes the bandgap change from the valence to the conduction band symbolized as v_1 to c_1. There is also S_{22}, which indicates the changeover from the valence toward the conduction band of v_2 to c_2. Where S_{11} is standard in semiconducting materials and describes fluorescent emission, S_{22} describes a photon-induced excitation of the electrons in the CNT [93]. The presence of S_{22} allows for the identification of different SWNTs. The S_{11} and S_{22} values are unique for each sc-SWNT chirality [91]. In metallic tubes, there is only M_{11}, which is equivalent to S_{11} transition.

Primarily, the absorption and emission spectra work together for SWNT identification in solution. The m-SWNTs are not observed in the same manner, as they have no fluorescence due to the non-existent E_g. However, it is possible to pinpoint the S_{11} or M_{11} values of either m- or sc-SWNTs [91, 93]. The extrapolated trend for each CNT diameter is $M_{11} > S_{22} > S_{11}$, and this observation is further used in the Kataura plot that is incorporated into Raman spectroscopy for SWNT chiral identifications (see Sect. 1.3.3.1) [94–96]. It is because of these transitions that CNTs can be observed and characterized using Raman, optical absorption, and photoluminescent spectroscopy.

1.1.5 Carbon Nanotubes as Transistors

SWNTs possess all the properties for creating the perfect FET. They are promising for fabricating dense, high performance yet low power circuits [9, 97]. In particular, the SWNTs can provide 1D ballistic transport of either electrons and holes. This description means there is no energy wasted in scattering particles. They also possess high drive currents and large transconductance, along with high-temperature resilience and strong covalent bonding. In CNTs, the electrons travel across the tube axis [87]. The gate controls the current intensity in the transistor channel. If no gate voltage (V_G) is applied, then the transistor should be in the OFF state.

Unlike in MOSFETs, where channel width (W_{ch}) and length (L_{ch}) affect the drain current (I_D), the CNTFETs have spatial independence [26]. The CNT device current is also independent of carrier mobility. They do, however, depend on the charge in the channel, the Fermi, or carrier velocity. The carrier velocity describes the ability of the electron transmission between the two contacts [98]. The current is contingent on the source (S) and the drain (D) Fermi levels and how they can input the carriers into the system. This aspect is known as bias dependence [9, 97].

The current flow can trail two different types of pathways in the CNTFET. It can be either classified as "MOSFET-like" or work on the principle of a Schottky barrier (SB).

The SB indicates that there is a potential energy barrier residing in the center of the sc-material and the source/drain metal electrodes. The barrier occurs because there is a difference in Fermi energy between the bands of the sc-material and the metal that the current carriers must overcome. The SB based CNTFET describes direct tunneling through the SB at the source-channel junction [99]. The transistor functions by modulating the transmission coefficient in the device [100]. This property indicates that a significant barrier will limit the current in the channel at low gate bias. As the gate bias is amplified, the barrier width decreases, which then increases quantum tunneling along the barrier. The current flow in the transistor channel then increases. The barrier width and the transconductance—the amount of current formed when applying a voltage across the device gate- are ultimately dependent on V_G [99, 101]. The metal electrodes transfer charge to the CNT that leads to band bending and, therefore, SB behavior. Here, the metal electrode contacts can affect the I-V curves and results [102, 103].

Alternatively, heavy doping of the source and drain terminals creates CNTFETs that mimic MOSFET behaviour. The drain current (I_D) is controlled by the charge that is introduced in the channel through the gate terminal. Chemically modifying the surface of the CNTs is also possible, but it is difficult to control defects or defects scattering that can occur from the dopant species [104]. Therefore, the source or drain material results in an SB CNTFET or an electrostatically doped NT gives a MOSFET-like CNTFETs. In either system, the carrier transport region occurs within the CNT. The devices fabricated in this work were SB CNTFETs.

Nonetheless, a few problems must be solved before SWNTs can be integrated into working devices. Reliability and production cost must be consistent and accessible. The main difficulty is that the SWNTs can orient themselves in several arrangements on a surface and are a mixture of different chiralities, which is ineffective for FET applications. Therefore, having a highly uniform orientation of SWNTs with selected semiconducting properties is a big obstacle in this field [5]. Purification, placement, and performance need all be consistently and easily reproducible for the full acceptance of CNTFETs [9].

The placement and controlled position for CNTs is particularly problematic as the nanotubes are flexible structures and have a high tendency to bundle. They have strong π-π and Van der Waals (VdW) interactions within themselves and need some form of surfactant to disperse them in aqueous solutions, which ends up affecting their properties. Techniques need to target way of creating straight, equally dispersed tubes that will not be removed by solvent cleaning and retain a rigid positioning. A high density of CNTs is also desirable as FETs are inefficient if designed with only a single nanotube. In addition to this already difficult task, sc-SWNTs or better even, single chirality sc-SWNTs should be aligned for consistent device performance. And then, the technique needs to be scalable for mass fabrication in electronics.

For CNTFETs fabrication, there are a few variations in the overall structure of the system. Different geometries for the CNTs and the electrodes can be applied. The structure can be either top-gated, back-gated, or coaxial-gated, which means that the gate surrounds the CNT. Back gated CNTFETs have the nanotube placed on the surface of the source and drain. They were the first CNT devices attempted through depositing laser-ablated nanotubes on an already pre-patterned surface. The device was functional, but only with a very high gate voltage, with no current saturation, and low transconductance [105]. Improved data was obtained through the use of a top-gated CNTFET structure as the overall geometry of the material was scaled depending on the nanotube size, and the current flowed with less resistance through the tube itself [48]. Adding an extra feedback gate connecting to the D-electrode in a top-gated CNTFET has also been shown to improve the I_{ON}/I_{OFF} ratio and reduce current leakage in the OFF state, as well as decrease ambipolar conduction [106–108]. Coaxial-gated devices provide even more electrostatic control over the source-drain channel but are increasingly difficult to fabricate [97].

Table 1.2 indicates some of the CNTFETs fabricated thus far. Many examples are top gated SB operating devices but doping like the MOSFETs is also observed. The I_{ON}/I_{OFF} is affected by changes in the type of dielectric layer and their thickness, as well as the electrode metals and their thickness. Ideally, high current with a low V_{DS} below 1 V should be obtainable for CNTFETs. The results are also affected by the external treatment of the CNTs, as in, annealing and contact with external surfaces or atmosphere can impact the nanotubes and the device. Metal contacts can be formed usually using Ti/Au, Ag, Pd, or Al for SB formation at the CNT-Metal interface [109]. Pd has been shown to have a very similar work function to the CNTs and adhere to their surface [4], but other metals are also still investigated, as shown in Table 1.2. Both electron and hole channels can be formed [110]. Usually, the CNTs can be treated prior to metal electrode deposition to improve contact with the surface, as was done by Liu et al. in 2019 and 2020 [10, 107]. Annealing can also occur after metal deposition but at lower temperatures so as not to affect the thin electrode surface. It has been observed that when the gate voltage is equal to 50% of the drain voltage, the minimum current in the system is dissipated as an equal number of electrons and holes [109]. Therefore, all details when building and fabricating CNTFETs should be noted for the devices as any slight deviation can impact the results. This is also why CNTFETs are still continuously being investigated due to the continuous changes in not only the device fabrication but also alignment and sorting. All these efforts are for the possibility of creating wafer scaled CNTFETs to enable high-performance carbon electronics.

1.1.6 Carbon Nanotube Sorting

Post-processing of CNTs is often necessary as the nanotubes are created with catalytic and other carbon allotrope impurities, as discussed in Sect. 1.1.2. Separation by conductivities (electronic enrichment) or chiralities (monochiral enrichment) or directional handedness (enantiomeric enrichment) is desirable depending on the application. Over the three previous decades leading to 2020 since the popularization of CNTs, their sorting can be classified into either centrifugation like DGU,

Table 1.2 Summary of some CNTFET devices over the years

Transistor material	Type of gate	SWNTs diameter (nm)	SiO_2 layer	Electrodes (thickness)	Channel L	Channel W	Post process?	I_{ON}/I_{OFF}	Vos (V)	Ref.
Single sc-SWNT	Back-gated	1.4	300 nm	Pt (15 nm)	300 nm	–	No	–	–	Tans et al. [105]
Mixed sc-/m-SWNTs	Top gated	1.4	500 nm	Ti/Au (20/40 nm)	1–4 μm	3.6 μm	Yes—electrical breakdown of m-SWNTs; 300 °C anneal	10^5	1	Li et al. [111]
Single sc-SWNT	Top-gated	1.5	15 nm (HfO_2)	Sc (60 nm)	120 nm	5 μm	No	10^3	1	Zhang et al. [112]
Single sc-SWNT	Bottom-gated	1.65	300 nm	Cr/Pd/Au (1/10/50)	3 μm	100 μm	No	10^3	−4	Zhang et al. [113]
N-doped multi-sc/m--SWNTs	Bottom-gated	1	600 nm	Ti/Pd (2/40 nm)	7 μm	20	Yes—O_2 plasma etch (100 W, 100 Pa 1 min)	10^6	2	Kim et al. [114]
Single sc-SWNT	Top-gated	1	20 nm	Co-Mo alloy (20 nm)	11.2 nm	50 nm	No	10^3	0.5	Cao et al. [115]
Single sc-SWNT	Top-gated	1.3	300 nm	Graphene/Pd (1 layer/30nm)	20 nm	200 nm	No	10^3	0.4	Qiu et al. [48]
Sc-SWNTs film (aligned array)	Feedback gated	1–1.5	Un-specified thickness	Pd/Au (40/40 nm)	375 nm	3.6 μm	Yes—Anneal 3 h 600 °C, Ar	10^4	−2	Liu et al. [10]

aqueous two-phase (ATP) extraction, polymer, or small molecule separations, dielectrophoresis and gel chromatography.

Because the different chiralities of nanotubes have all slightly differing densities, DGU is a highly scalable method for sorting. The larger the diameter, the heavier the weight of the CNTs, and therefore, SWNTs can be separated based on diameter size [116]. These differences in diameter buoyancy allow for highly accurate SWNT purification that can range from conductive to enantiomeric separations [117]. The main challenge is identifying the appropriate conditions as this varies depending on the CNT mixture. Parameters vary in terms of density-gradient of the dispersants and medium, nanotube concentration, temperature, and centrifuge power.

Conventional surfactants including sodium dodecyl sulfate (SDS), sodium dodecyl benzyl sulfate (SDBS), sodium cholate (SC), or triblock copolymers can be used for preparing aqueous dispersions of SWNTs. Unbundling and uniform coverage of the SWNT sidewalls with the surfactant is necessary for successful DGU. If the SWNTs are not solo-suspended (individualized), then uniform sidewall surfactant coverage cannot occur, and the sorting process is hindered [116]. However, the CNT buoyancy is dependent on its surface coating's extrinsic and intrinsic electrostatic interactions, which can make DGU results challenging to interpret. Additionally, the centrifugal force that DGU is dependent on is weaker than the molecular interactions in the CNTs [88].

Gel chromatography is very similar to DGU, wherein a mixture of CNTs is placed in a gel media, and the sc-CNTs bind strongly to polysaccharide gels, and m-CNTs do not [118]. This technique has continued to expand, with much recent progress from the Kataura group [119–121]. A combination of hydrogels and different surfactant wrapped CNTs can extract sc-SWNTs for their helicity and handedness [122, 123]. The SWNTs can also be frozen or centrifuged in the gel for separation [88]. By tuning the concentration of SDS in the mixture, the specific chirality of (5,4) and (6,4) SWNTs can be purified from a mixture [121]. Additionally, investigation on a dextran gel column has shown that room temperature separations result in small diameters of SWNTs being purified. As the temperature increases, so does the SWNTs' diameter that leaves the column. Further work has also demonstrated that gel chromatography with HCl treatment can quickly remove m-SWNTs entirely from a mixed metallicity solution [119]. Then, by changing the concentration of the acid, specific chiralities of sc-SWNTs can then be eluted from the mixture [119].

A competing scalable technique is ATP extraction that has been shown to obtain large batches of high purity sorted metallic and semiconducting SWNTs through molecular force [124]. The method is based on polymer-polymer phases separations. A duo of immiscible aqueous phases are formed because there is a slight difference in the polymers physical properties. Through homogeneous molecular interactions and differences in the solvation energy of a mixture, the extraction of different components in the solution is possible. Different structures will have different affinities in either solution phases, enabling separation. The SWNTs must be either DNA- or surfactant enfolded. Notably, DNA-wrapped SWNTs input into the ATP extraction processes has led to R- and L-handedness and chirality separations

[23]. The method has high reproducibility and resolution as both phases are homogeneous, so many heterogeneous molecular interactions are eliminated, and the separation can be tailored based on the specific surfactants [88, 125].

Dielectrophoresis (DEP) has also been used for separating and depositing specific metallicities of SWNTs. The first report was by Krupke et al. on SDS wrapped SWNTs [126], and the work was subsequently upscaled [127]. Using DEP, researchers have fabricated CNTFETs based on a semiconducting aligned network [128]. Although this process requires high-temperature oxidation (1100 °C) and chemical post-processing with N-methyl pyrrolidine SWNTs, the protocol was highly scalable, using only a single layer of photolithographic treatment. Additionally, the authors indicate the system could be optimized if the SWNTs were individually aligned and separated between the electrodes, rather than overlapping [128].

Diameters of SWNTs can also be controlled through the use of flavin terminated silane SAMs [129]. The isoalloxazine derivative was chosen for interacting with the CNTs as it has favorable π-π stacking, H-bonding, and solvent interactions with the SDS wrapped SWNTs. The surface was functionalized for 12 h and then left in a solution of SDS-SWNTs for another 12 h. It was observed through photoluminescence and Raman spectroscopy that SWNTs were below 1 nm in diameter and of majority metallic chirality. It was speculated that the smaller diameter SWNTs are fully wrapped by the flavin, whereas the larger diameter tubes only have partial interaction with the molecular surface. Therefore, once the sample is washed with 1% SDS, the larger tubes are removed, while those below 1 nm in diameter remained on the substrate. This study displayed a preferential binding for diameters to the treated surface, but not the alignment of the SWNTs. Electrical studies were conducted to demonstrate low resistance in the network. The authors indicated future work in sensor applications [129].

1.1.7 Carbon Nanotube Alignment

The effective alignment of individual CNTs is desirable in order to harness their outstanding electronic properties fully [130]. Additionally, singular CNT properties were able to be analyzed due to efforts of separation and alignment [131]. Alignment methods can be divided into manipulating the SWNTs after the synthesis or controlling their orientation directly during growth [132].

1.1.7.1 After-CNT Synthesis Alignment

Subsequent work on the horizontal alignment of SWNTs showed that imprinting patterns of amines and carboxylic acids on a SiO_2 wafer can align SWNTs exclusively in the functionalized areas [133]. The SWNTs are in solution, and the functionalized wafer is dipped for 10 s to achieve functionalization. Additionally, high I_{ON}/I_{OFF} ratios at 10^5 were achieved without the removal of the organic layer underneath the electrodes. The work is reproducible on a large scale; however, the functionalization occurs through dip-pen lithography, which can be both time and labor-consuming when placing the individual molecules [133].

Acid-functionalized SWNTs have also been shown to adsorb to slightly basic hafnia (HfO_2) surfaces selectively [134]. Therefore, patterned channels of HfO_2 were placed onto a SiO_2 wafer and left in an aqueous hydroxamic acid-functionalized SWNTs solution for three hours. These CNTs became water-soluble due to acid modification. Then, the sample was annealed under argon at 600 °C to eliminate the acid functionality and obtain unfunctionalized SWNTs. This process depicted selectivity for the functionalized SWNTs to deposit and aligne only within the pre-patterned trenches on the sample. However, based on the electronic measurements, the SWNTs were mostly metallic. This observation was concluded to be because the m-SWNTs react 70% faster with the acid, and therefore were readily functionalized and dissolved in the water. There were only a few sc-SWNTs deposited. Nevertheless, of the CNTFETs tested with the aligned sc-SWNTs, an I_{ON}/I_{OFF} ratio of 10^7 was ultimately observed [134].

Alternatively, work by Sharma and Strano showed that an alternating pattern of polar and non-polar SAMs on an Au surface can result in a highly aligned SWNTs monolayer [135]. The process was possible due to droplets formed from a SWNTs thin film placed on the functionalized surface. The droplets of the film selectively migrated to only the polar parallel patterned areas exclusively [135]. Once the first layer of thin SWNT film was deposited, additional layers could be added to increase the density. Although the deposition of the patterned SAMs on only Au can be time-consuming, it is possible to align single SWNTs in this way.

The Langmuir-Blodgett (LB) technique can also be applied toward SWNTs' alignment. The LB method defines the transfer of a monolayer adsorbed at the air-water boundary by the vertical plunging of a solid substrate. Films can be compressed or expanded on the surface to control density and thickness, and the process can be repeated to create multiple thin layers. Through this process, thin films of quantum dots, Ag NWs, and CNTs are possible [136]. For SWNTs alignment, a highly concentrated thin film has been observed [137, 138].

Work by Li et al. showcased densely packed networks of polymer wrapped SWNTs [138]. And further progress has been made using high purity mixtures and adapting the LB technique. For instance, a high purity solution of 99.9% polymer wrapped sc-SWNTs was dissolved in chloroform and underwent the LB transfer to a hexamethyldisilazane (HMDS) patterned SiO_2 wafer [137]. A FET was created from the sample, depicting an I_{ON}/I_{OFF} ratio of 10^6 for the highly aligned network [137]. However, the LB method can be challenging to scale up, as the water-air interface is susceptible to impurities, which can affect the final film deposition. Additionally, with the network, sc-SWNTs do overlap, which also has an impact on their electronic efficiency. This overall process was named *floating evaporative self-assembly* (FESA). Subsequent work has shown that concentration and rate of dipping can be tailored and optimized for consistent densities on larger wafers [139]. It was also demonstrated that evaporation of the chloroform in the SWNTs is also not necessary for the FESA process and that the diameters and gaps between the tubes can also be controlled. The method is also dependent on the CNTs' size and width of the functionalized patterned areas. If the substrate is pulled too quickly from the solution, the tubes are bent and not straight on the surface [139].

A further adaptation to FESA for quicker SWNT alignment was conducted by Chai et al. [140]. Long-range SWNT alignment was achieved with the application of direct-current and the LB dipping method with additional gold patterns on a substrate [140]. The SWNTs become polarized between the substrate and Au electrodes and are drawn into alignment on hydrophilic channels. The SWNTs used were a mixture of sc- and one-third m-SWNTs but did depict a low resistance when tested as a device [140]. It would be interesting to observe the effects of the current on the sc-SWNTs vs. m-SWNTs.

The Langmuir-Schaeffer method has also been investigated, with a higher degree of SWNTs alignment than with the LB films and the FESA method [141]. Aligned 99% purity SWNTs in DCE were transferred to a substrate surface, achieving a network of 500 SWNTs μm^{-2}. The fabricated device displayed an I_{ON}/I_{OFF} ratio on the order of 10^3. Although the distribution and alignment in the thin SWNTs film is high, it does not portray any selectivity of diameter or conductivity, nor is there any selection for constant SWNTs lengths [141].

Single SWNTs have also been aligned using dielectrophoresis onto doped SiO_2 [142]. This use of external electric forces has proved useful in creating individual CNTFETs on a large scale with high accuracy. Moreover, the method eliminates CNT bundling. But both m- and sc-SWNTs are distributed on the surface, and there is no selection between the two due to the electrical parameters used [142].

Self-assembly is guided by the SWNTs, usually wrapped in SDS, being deposited to hydrophilic functionalized surfaces. Both experimental studies [137] and recent computational investigations [143] corroborate the observation that surfactant wrapped SWNTs bind to hydrophilic surfaces. Work by Kataura has shown that through functionalization and nitrogen blowing, SWNTs can also be moderately aligned [144]. Electrostatic attractions are responsible for the SWNTs' self-assembly, whereas VdW forces are responsible for the lack of SWNTs binding on hydrophobic surfaces [143]. A further analysis of all the underlying forces in SWNTs alignment has been reviewed by Goh et al. [145].

The majority of the developed alignment methods rely on a solution of SWNTs. These methods are categorized into using external forces [142], thin-film technology [141], self-assembly [143], or a combination as shown by FESA [137] and the current induced FESA [140]. In all the solution-based methods, the SWNTs alignment generally begins with creating a homogeneous suspension with the NTs. A surface is functionalized, and then the dispersed SWNTs solution is transferred onto a functionalized surface. These directions show the high alignment of either individual or connecting SWNTs. However, the selectivity of different types can be minimal, and there can be defects incurred to the SWNTs, in particular, if the SWNTs are functionalized covalently and not just with surfactant.

1.1.7.2 In Situ Alignment

The surface of the substrate itself can also be used to align SWNTs. In situ alignment can also avoid forming defects on the SWNTs as opposed to the post-synthetic processes. However, the chirality must be entirely controlled for CNTFETs, as the mixture of metallicities needs to be avoided for these systems for optimal application. The controlled alignment of SWNTs during the growth process has been identified on sapphire [146] and quartz [147]. Using a catalyst under CVD conditions, SWNTs alignment is observable along with specific directions of the surface cut axis. The quartz surface utilizes a Fe catalyst (600 °C), whereas the sapphire surface uses Co-Mo to allow for a more extensive temperature range for the CNT growth (750–900 °C).

The α-Al_2O_3 surface is also selective for different metallicities based on the growth axis. A majority of zigzag SWNTs were exclusively aligned along the A-plane, whereas armchair SWNTs were along the R-plane of the sapphire substrate [146]. The different atomic configurations of the sapphire surface seem to affect the SWNTs' chiralities [146]. Although this selectivity was not observed on the quartz surface, very densely packed arrays of 5 SWNTs/μm^2 were achieved efficiently [147].

Both in-situ and post-fabrication alignment is very valuable, and the scope of aligned SWNTs was summarized by Li and Zhang [148]. However, for CNTFETs, individual uniform SWNTs need to be aligned and not connecting on a surface for optimal performance. For this application, methods need to be both universally consistent in the production of aligned SWNTs and scalable. Current alignment methods require either high-temperature reaction conditions [149]; others are expensive [150], time-consuming [151], or fail to demonstrate high scalability [152]. Additionally, they do not simultaneously filter different properties and align the SWNTs in a mass-marketing feasible manner. It would be necessary then, to have a method that can orient and select for the different SWNTs types simultaneously. The following section details these efforts. Simultaneous synthesis and alignment of CNTs is a desirable research target as it produces the SWNTs without any post-processing. The in-situ type of methodology can also be used for creating vertically aligned SWNTs [153]. However, it is problematic to regulate the specific metallicity of the as-grown SWNTs.

1.1.8 Carbon Nanotube Alignment and Sorting

Work by Lemieux et al. developed a method for nanotube sorting between sc- and m-SWNTs by selectively functionalizing surfaces with either phenyl or amine-terminated SAMs. The functionalization occurred in an inert atmosphere at room temperature within a few hours [154]. A mixed metallicity blend of SWNTs was dropcast onto the functionalized surface while in motion on a spin coater. An extensive degree of CNT alignment was accomplished, and the phenyl groups favored m-SWNTs deposition, while the $-NH_2$'s had dominated sc-SWNTs, as was evidenced from the I_{ON}/I_{OFF} ratios and Raman spectra. Unfortunately, the center where the CNTs were dropcast did not have alignment. Additionally, the alignment radiated outward from the center of the wafer and was not uniform in one direction across the wafer [154].

The density amplification technique has also shown a significant degree of alignment of sc-SWNTs [155]. Specifically, m-SWNTs were selectively removed from a polymer film of polymethylmethacrylate (PMMA) through a thermocapillary flow method [156] in a single step. The sc-SWNTs-PMMA film was then stretched and transferred onto desirable substrates. Functioning CNTFETs were fabricated with a concentration of 10 SWNTs μm^{-2} and an I_{ON}/I_{OFF} ratio of 10^4, indicating that the transfer and procedure did not decrease performance [155]. However, specialized equipment was necessary, and the sorting did not co-occur with alignment; yet the method was demonstrated to have a promising high scalability and reproducibility [155]. The first report for the thermocapillary flow by Du et al. eliminated the m-SWNTs very successfully but showed a decrease in aligned tube density on the surface [156]. Further work by Si et al. then demonstrated increased density and device performance [155].

In-situ alignment and selection for different conductivities and diameters have also been proved further on a quartz surface [157]. During CVD synthesis using a copper catalyst, the quartz morphology was used for SWNTs alignment and growth, as discussed in Sect. 1.1.2. During this process, some random arrays are formed, and catalyst residues, as well as m-SWNTs, are present. To circumvent these issues, a water vapor purification treatment can be implemented. It was observed that the water vapors etch away the majority of m-SWNTs and remove any un-aligned CNTs from the surface [158].

Additionally, by optimizing five growth cycles, the density of the aligned SWNTs reached 10 μm^{-2}. Results indicate that sc-SWNTs in the diameter range of 1.5–2.0 nm remained on the surface, while m-SWNTs of this diameter were removed. All SWNTs with diameters below 1.5 nm were also removed regardless of conductivity. The reasoning for this removal is twofold: the presence of oxidant water vapors removes un-aligned SWNTs as they lack the rigidity and structural integrity. Then the concentration of exposed vapors can be targeted toward the m-SWNTs that were observed to be more sensitive to the oxidizing environment and were removed, along with the smaller diameter CNTs that have increased water sensitivity [157]. A CNTFET was fabricated but with a low I_{ON}/I_{OFF} ratio (i.e. below 100) indicating that the method still requires optimization for electronic systems, as the oxidizing treatment could be affecting the sc-SWNTs performance.

These are methods of direct growth, transfer, and directed assembly that attempt to achieved simultaneous sorting and alignments of sc-SWNTs for large-scale FET applications. High uniformity sc-SWNTs in parallel arrays on substrates free of bundles are required for efficient device performance. Drawing inspiration from Lemieux's work in using functionalized SAMs, the ART allows for a controlled assembled monolayer of aromatic tweezer molecules by using the aligning properties of liquid crystals. The selectivity of sc-SWNTs has shown preliminary evidence based on the size of the tweezer molecule, which demonstrates favorable binding to specific diameters. The ART attempts to be the reproducible, scalable, non-destructive process for CNTFETs fabrication.

1.2 Alignment Relay Technique

The Alignment Relay Technique (ART) methodology utilized herein can contribute practical solutions to the alignment and selection of SWNTs while avoiding some of the disadvantages of the aforementioned deposition processes. It integrates the inherent aligning nature of liquid crystals, intermolecular interactions of SWNTs with an iptycene structure, and surface chemistry between the substrate and the iptycenes. Because of the extensive interdisciplinary nature of the technique, its many components are separated and discussed individually prior to introducing the full process. The components include: the interactions of the ART molecule with the CNTs, how the molecule binds to the surface, and how the molecule aligns to the surface

1.2.1 Intermolecular Interactions with SWNTs

As discussed in Sect. 1.1.1, the structure of CNTs is that of many benzene rings joined into a cylindrical shape. The sp^2 bonds and aromaticity in the nanotubes can allow for interesting noncovalent interactions. The aromatic stacking refers to the electrostatic potentials of sp^2 ring systems, where the positive potential of one ring interacts with the negative potential of another ring. Variations of π-π stacking are illustrated in Fig. 1.13, representing electron-rich aromatic interactions [159, 160]. The strength of π-stacking is contingent on the geometry of the structures forming the interaction, and contributions can also come from other VdW and particle electric charges present in the material. The parallel slipped stacking (face-face) or direct stacking interactions are stronger than those of the T-shaped.

However, they are typically not any stronger than H-bonds (~4 to 40 kJ/mol) when modeling single molecules [159]. The extended π-system present in CNTs can have multiple areas for stabilizing π-π interactions that accumulate along the length or diameter of the tube. These interactions have been shown to work efficiently for separation [161, 162], and solubilization [163] of SWNTs. For example, the separation of SWNTs with aromatic polymers proved to remove m-SWNTs from sc-SWNTs selectively [164].

In particular, studies using porphyrin tweezers were able to access high purity (67% enantiomeric excess) of either R- or L- (6,5) SWNTs [162]. Additionally, pyrenes were observed to have a high interaction affinity to nanotubes when they increased in diameter. Pyrene, being planar, therefore, had a higher attraction to the increased planarity in the larger diameter tubes [165]. Therefore, planar systems

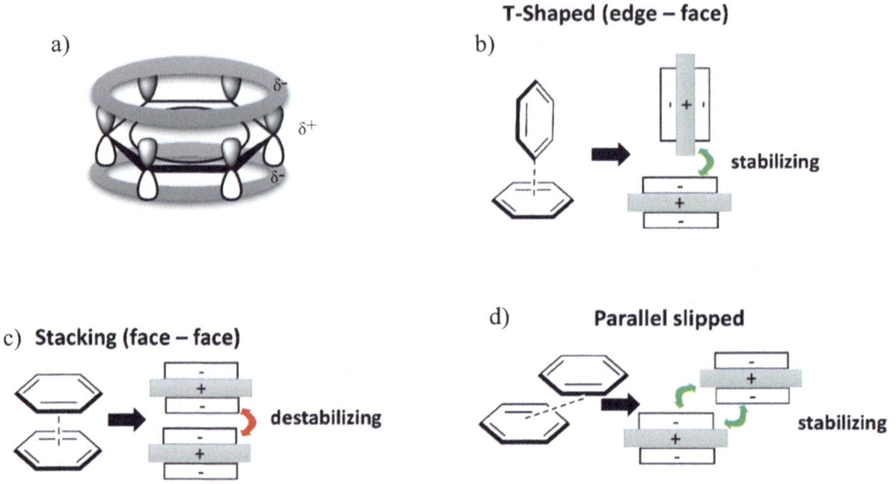

Fig. 1.13 Display for aromatic stacking (**a**) how the electrostatic charges are distributed in benzene, (**b**) illustration of T-shaped stacking, (**c**) illustration of face-stacking, (**d**) illustration of parallel stacking. Adapted from Martinez and Iverson [159] and Sinnokrot et al. [160]

have a higher affinity to similarly planar systems when discussing π-π stacking. As SWNTs have a curved or convex π-system, interactions with flat or planar compounds are mismatched. The convex sp^2 systems of the nanotubes require a bent or concave equivalent system for enhancing these π-π stacking interactions. A favorable configuration for concave-convex aromatic interactions can aid in the selectivity of the SWNTs [166]. Hence, the foundation for choosing a π-concave iptycene system in the ART, as its shape can be modified to complement different tube diameters and chiralities.

Additionally, these interactions can also occur with other polymers, small molecules, or even with other nanotubes. Unfortunately, the cause for one of the main problems with CNTs: bundling, is due to their ability to have non-covalent interactions across their long lengths with each other. The CNTs have a high propensity of form bundles, which negatively impacts their properties for FETs. Of course, avoiding this problem is a critical component of their applications into devices. High concentration bundles of SWNTs can typically be broken apart by high powered tip-sonication. Combining ART with sonication is an area to be explored, as this form of treatment could be used to effectively de-bundle the tubes once on the surface and thereby increasing the likelihood that individual SWNTs will bind with the ART iptycenes. The ART iptycene structure is shown in Fig. 1.14, where the core for SWNTs binding is made possible through the two triptycene components of this iptycene structure.

Fig. 1.14 Structure of ART molecule 1 used in this work. The phosphonate ester groups act as anchors to the substrate surface to retain the molecule. The benzene rings act as areas for π-π stacking with the SWNTs. The molecule also has a concave structure to interact with the cylindrical form of the CNTs

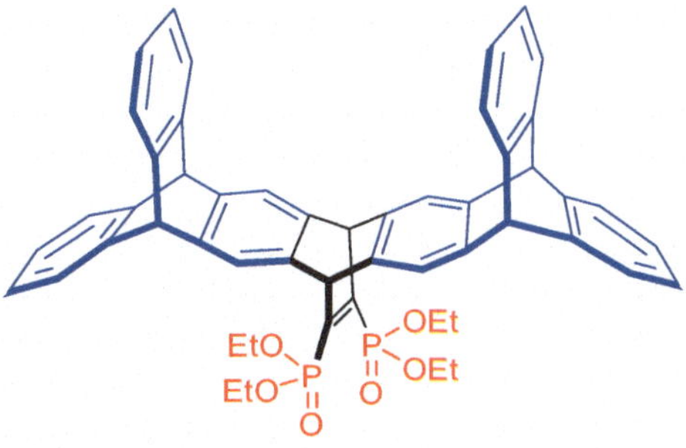

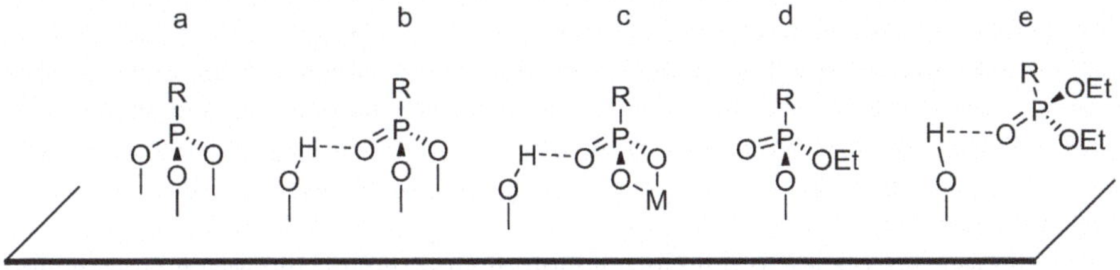

Fig. 1.15 Phosphonate ester binding with a metal oxide (M) surface. The –OEt group can be replaced with –OMe or –OH (phosphonic acid), in order of bond strength: (**a**) diative and two covalent bonds (strongest adhesion), (**b**) H-bonding through the carbonyl and two covalent bonds, (**c**) H-bonding and two covalent bonds coordinated on the same metal, (**d**) one covalent bond, (**e**) one H-bond (weakest adhesion). Adapted from Queffélec et al. [169]

1.2.2 Substrate Binding

The anchoring component is key to the ART molecule properly adhering to the surface, and without the iptycenes the nanotubes will not be aligned. The tethering component needs to be covalently bonded to the surface in order to prevent its removal when depositing the nanotubes and so the molecules position does not change once aligned and placed onto the substrate. The phosphonate group in molecule 1 (Fig. 1.14) was selected as it has been observed to bind well to a variety of metal oxide surfaces [167] and was readily accessible by using bis(diethoxyphosphoryl)acetylene [168]. The phosphonate functional group provides a point for either an H-bond or coordinate bond and two areas for covalent bonding through substitution with the ethoxy group [169]. Additionally, with two of these functional groups on the surface, six areas for bonding can emerge per iptycene molecule.

The different methods of binding with the metal oxide surface are shown in Fig. 1.15. The ethoxy group can be substituted for a methoxy or hydroxy group (phosphonic acid) [170]. The phosphonic acids are the most generally used,

Table 1.3 Binding table of different substrates to optimal adsorbate by Jadhav [174]

Surface type	Substrate	Adsorbates
Metal	Au	R-SH, R-S-S-R, R-S-R
	Ag	R-NH_2, R-NC, R-COOH
	Pt	R-NC, R-SH
	Pd, Cu, Hg	R-SH
Semiconductor	GaAs (III-V)	R-SH
	CdSe (II-VI)	R-SH
Oxide	Al_2O_3	R-COOH
	TiO_2	R-COOH, R-PO_3H
	ITO	R-COOH, R-SH, R-SiX_3
	SiO_2	R-SiX_3

especially in self-assembled monolayer structures [170–173]. The bonding occurs with the loss of the corresponding alcohol, and structure 15a is more likely in the presence of Lewis acid metal oxides. Structure 15b is more common on poor Lewis acid surfaces because the phosphoryl oxygen cannot efficiently bind, so it H-bonds instead. This result is because acidic surface sites can increase the electrophilicity of the P-atom, thereby helping with the condensation and hydrolysis of the POR groups on the surface [171]. Bonds can be chelated as in structure 15c or separated as in 15a and 15b.

Alternative anchoring components could involve silanes, phosphonic acids, and carboxylic acids [167]. As shown in Table 1.3, depending on the target surface, the anchoring component can be altered to be the most optimal tether to the surface [174]. Specifically, silanes have been demonstrated to have an efficient binding with SiO_2 surfaces [167, 174, 175], however, their addition to the iptycene backbone would be more synthetically elaborate. The phosphonic acid moiety has also been effectual as part of organic FETs fabrications [176]. Regardless, the most critical parameter is obtaining enough surface hydroxyls on the surface to ensure reactivity [170, 172]. The number of –OH groups provide the means for the formation of enough bonds such that the molecule **1** remains chemisorbed the surface during SWNTs deposition in the ART process.

1.2.3 Surface Oxidation and Cleaning

Acid "Piranha" oxidation is an ex-situ wet chemical method for removing trace organic impurities on surfaces as well as hydroxylate them. It is a solution of conc. H_2SO_4: H_2O_2 (30% aq.) that can range from 3:1 to 7:1 in ratio. The explosive nature is minimized by maintaining the hydrogen peroxide concentration below 50%. The exothermic reaction is shown in Scheme 1.1a. The exothermic reaction rapidly occurs to form peroxymonosulfuric (Caro's) acid [177]. The Piranha etch is also proficient at removing metals from surfaces and is used in various semiconducting

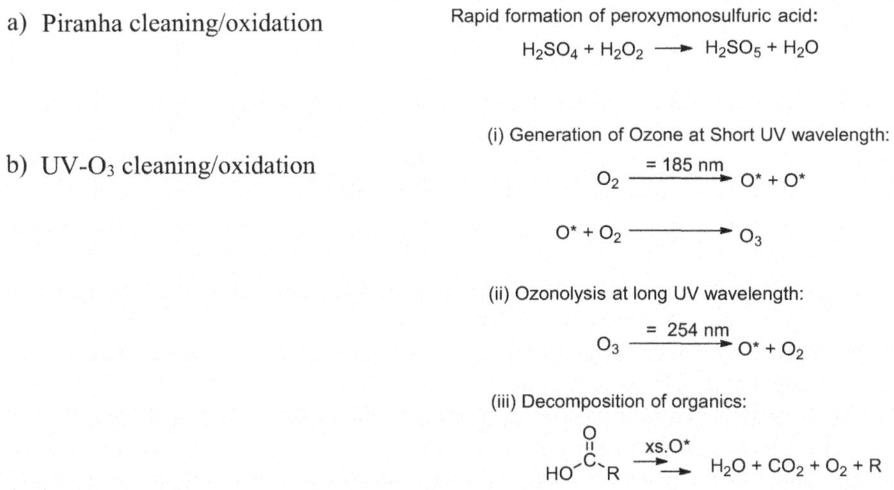

Scheme 1.1 (**a**) Piranha cleaning chemical process [177], (**b**) UV-Ozone chemical process decomposing a carboxylic acid as an example [178]

fabrication operations. It is also capable of oxidizing CNTs to incorporate carboxylic acid and hydroxyls along the length of the tube [179].

UV-Ozone (O_3) is alternative cleaning and oxidation method. This cleaning technology converts organic impurities to H_2O or CO_2. The highly oxidized states of inorganic contaminants can then be rinsed with water. Additionally, it has been shown to effectively create a fresh layer of Si-OH in under 20 s [180], and it has been extensively reviewed by Fink et al. [181]. Atomic oxygen is formed along with ozone by short wavelength (185 nm) UV-rays. Then, long-wavelength (255 nm) UV radiation decomposes the ozone to form more of the atomic oxygen for decomposing hydrocarbon impurities [178]. Essentially, the bond energy for several molecules can be broken if they are transcended by the energy per mole of the UV-rays (185 nm = 647 kJ/mol; 255 nm = 472 kJ/mol). The chemical processes are outlined in Scheme 1.1b. Step 1 generates the ozone, step ii indicates that the generated ozone is further dissociated to atomic oxygen. Simultaneously with 2, the impurities are also absorbing the UV rays and are creating free radicals, ions, and excited molecules. In step 3, the impurities are removed as simple volatile molecules by reacting with hydrocarbons and nitrogen-containing species. The contaminants then desorb from the surface [178, 181].

Both Piranha and UV-O_3 are capable of simultaneously cleaning and oxidizing the surface. However, the UV-O_3 technique introduces less hazardous parameters to ART, and therefore, the process merits investigation in this work.

1.2.4 Liquid Crystal Technology

Liquid crystals (LCs) are organic compounds that have both anisotropic and liquid properties [182]. As per their name, they are in between two phases of matter. A substantial interest in LCs technology is spurred by their ability for orientational and positional order that can be controlled by polarized light, alignment layers [183], electric and magnetic fields [182]. The LC mesophases can be achieved either through temperature variations or solvent conditions. At low temperature, the thermotropic LC is solid, and at too high heat, it becomes a regular liquid as per Fig. 1.16. If the LC has mesophases that are created through changes in heat, then it is classified as thermotropic [184]. If the phases are formed through changes in the solvent, affected by concentration and temperature, then is it considered a lyotropic LC. Some systems exhibit mesophase changes when exposed either to heat or solvent, or both and are considered as amphotropic LCs [185]. The lyotropic LCs are akin to micelles but with orientational order. They can form cylindrical, rod-like micelles or lamellar structures and are very common in biological systems [186]. Additionally, Lagerwall et al. have been able to utilize the micelle-rod alignment of lyotropic LCs toward SWNT alignment using magnetic fields below 1 T [187].

The mesophases of thermotropic LCs can be described as either smectic or nematic. The most common phase is nematic, in which case there is not a set positional order of the molecules, but they have alignment and, therefore, what can be described as long-range order (Fig. 1.16). The nematic phase is fluid like a liquid, but it can also be directionally controlled through external parameters such as magnetic fields or orientational templates. The other phase is smectic, and it occurs at lower temperatures than the nematic phase [186]. It has long-range order and alignment properties as the nematic, but it also has short-range positional order along one single

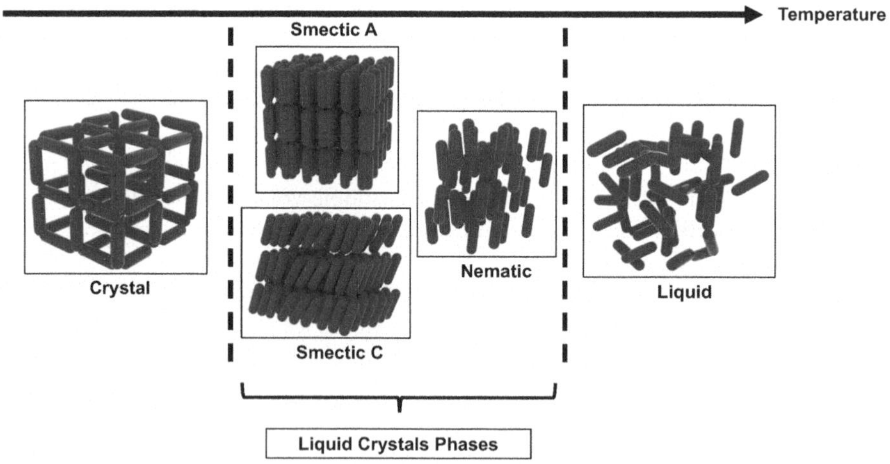

Fig. 1.16 Illustration of the solid crystal structure phase transition to liquid as temperature increases. The phase transitions of smectic and nematic for thermotropic LCs are shown

direction. The layers can glide over top of each other. The smectic phase has two sub-categories: A and C. Smectic A has horizontal layers, while in the smectic C phase, the rod-like molecules have a tilt to them.

Additionally, if a molecule is a chiral LC, it can also possess a chiral nematic phase or, rather a cholesteric phase. This phase has twisting within the layers [186]. For ART, the LC used is 4-cyano-4′-pentylbiphenyl (5CB), and it is in the nematic phase at ambient temperature, therefore allowing alignment at ~23 °C. Another liquid crystal, 4′-octyl-4-biphenylcarbonitrile (8CB), is in the smectic A category and also allows for room temperature alignment.

However, the LCs that dominate materials research and are used in electronic applications such as displays are of thermotropic nature [188]. This class of LCs can be further subdivided into calamitic (Fig. 1.17a) or discotic (Fig. 1.17d) structural types [189]. The calamitic structure is similar to a stick or rod. It consists of a hydrophobic tail, an aromatic, usually bi-phenyl unit, and a dipole group on the opposing end of the hydrocarbon chain. The tail limits crystallization, the stiff unit enables polarization (i.e., birefringence), and the dipole allows for electric field influence [184, 186]. The discotic LCs are akin to circles of aromatic rings surrounded by alkyl chains [190]. The discotic types, such as coronene derivatives (Fig. 1.17d), have shown useful charge carrier properties for device applications [191]. Either of these kinds of thermotropic LCs can provide controlled alignment and orientation.

For ART, the focus is on calamitic LCs; specifically, experiments are conducted with either the original 5CB (Fig. 1.17b) [168] or investigate 8CB (Fig. 1.17c) [192]. The work is inspired by findings on the triptycene molecule that can be aligned by 5CB [193]. Long and Swager have shown the T-like shape of the triptycene derivatives can incorporate the rod-like structure of 5CB within its structural matrix. The study concluded that when the 5CB is exposed to an aligning environment such as polyimide alignment layers, it orients itself with the template and then imparts that alignment on the molecules [193]. Hence, this form of alignment transfer with 5CB is one of the foundations for ART.

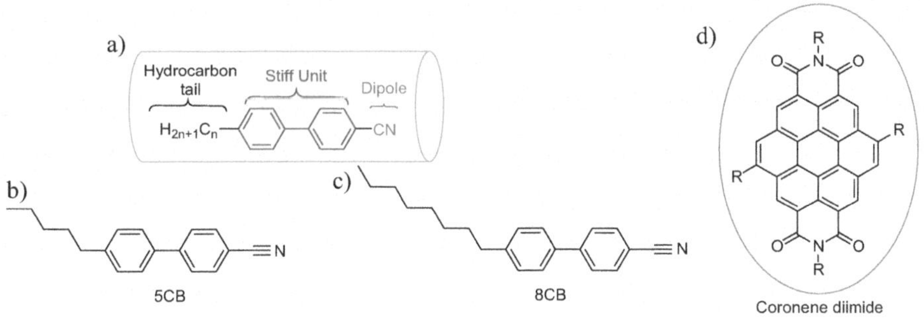

Fig. 1.17 (**a**) Structural backbone and components for the calamitic thermotropic LC, (**b**) structure of 5CB, (**c**) structure of 8CB, (**d**) example of a structure of a discotic thermotropic LC, coronene diimide

1.2.5 Synthesis of the CNT Tweezer in ART

The iptycene for the SWNT interactions (**1**) was synthesized by the following process (Scheme 1.2). Starting molecule **2** can be formed from *cis*-2-butene-1,4-diol with thionyl chloride and pyridine or be purchased directly. Compound **3** is made from the Diels-Alder addition reaction of anthracene with the synthesized *cis*-1,4-dichloro-2-butene over 2 days at 190°C in a 95% yield, and unreacted anthracene is recovered using flash chromatography purification. Note that either the cis or trans isomer of **2** can be used [168], as the positions of the chlorines do not matter due to the subsequent elimination reaction. An E2 elimination is then pursued using potassium *tert*-butoxide under argon conditions to form molecule **4** (70%). Then, the double triptycene-like structure molecule **5** (67%) is made through lithium-halogen exchange with 1,2,4,5-tetrabromobenzene and two equivalents of compound **4** via benzyne chemistry. This reaction is then followed by the treatment of **5** with 2,3-dichloro-5,6-dicyano-1,4-benzoquinone (DDQ) to form compound **6** in a 65%

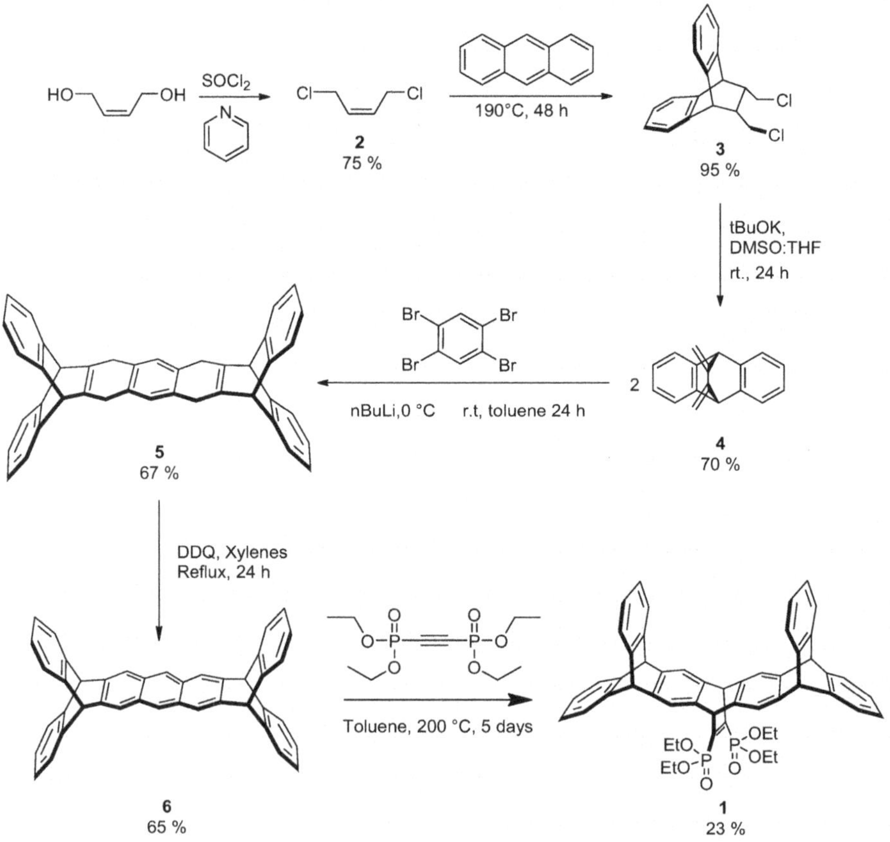

Scheme 1.2 Overall synthesis for the ART iptycene

yield. A Diels-Alder reaction of molecule **6** with bis(di-ethoxyphosphoryl)acetylene creates compound **1**, which is then the molecule used for alignment purposes, as shown in Fig. 1.14.

1.2.6 ART Process

The alignment relay technique (ART) is scalable and simple while showing selectivity for some SWNTs diameters over others [168]. The surface functionalization occurs with a molecule that is easily aligned, and it is this alignment of the molecules that continues over to the SWNTs. Primarily, the alignment relay technique utilizes an alignment molecule (**1**) (Fig. 1.18) that possesses an iptycene backbone paired with a phosphonate ester anchoring group [168]. The iptycene is integrated explicitly with ART to interact with carbon nanotubes through π-π stacking interactions favorably [194]. Additionally, **1** has a non-planar structure, which is advantageous for interacting with curved systems, such as the nanotubes [166]. The molecule itself is aligned on a surface by being solvated in a liquid crystal, 4-cyano-4′-pentylbiphenyl (5CB) in this case. A drop of the solution is placed on the surface of the polyimide alignment layer. This surface is the key component used for liquid crystal alignment [183]. The ART has been shown to work on silica, indium tin oxide (ITO) [168] and gold surfaces [195].

Fig. 1.18 Scheme outlining the alignment relay technique (ART) using iptycenes (blue, 1) and 5CB liquid crystal (green). The micro-aligned glass template (purple) controls the liquid crystal orientation, which in turn imparts the alignment onto the molecule. Only after placing the mixture of 1/5CB on the template is the mixture put in contact with Si/SiO_2 surface to covalently bind which is shown in previous work. Once the surface is functionalized, the LC is washed off and sample is placed in contact with nanotubes in solution for 24 h

As the LC aligns, it imparts its alignment towards the molecule. A surface capable of covalent binding with the phosphonate ester of molecule **1** is placed and compressed with the alignment layer. Excess of the **1**/5CB mixture is removed, and this results in a monolayer of **1**, where the triptycene backbone is in a prime position for interacting with CNTs (Fig. 1.18). Given the difficulty in directly imaging small molecules on a surface, the exact positioning of the iptycenes in the monolayer is as of yet unknown. It is also likely that the arrangement of the molecules is not as uniformly aligned as depicted in Fig. 1.18.

Evidence of the aromatic stacking between the iptycenes and SWNTs is shown by Atomic Force Microscopy (AFM), as without the functionalized surface, the nanotubes do not align [168]. A sample without any iptycenes was placed in the solution of SWNTs for the same deposition time as ART samples, but no alignment was observed. Further evidence is from a previously reported control experiment that had the iptycenes molecules but lacked the liquid crystal as the solvent, so there was no alignment relay of information. This trial also demonstrated no deposition nor alignment of the SWNTs [46]. The final structure after SWNTs application in Fig. 1.18 schematically represents SWNTs' alignment after standard ART.

Nonetheless, ART requires fine-tuning for eventual applicability in electronic devices. For instance, the alignment of the nanotubes may be sporadic across the sample surface, and the SWNTs themselves may deviate by size, length, and type. In addition to the interactions depicted in Fig. 1.12, it is also plausible that the SWNTs are interacting with the substrate surface directly in the space between neighboring iptycenes. Samples can also be plagued with CNT overlap, bundling, or partial to incomplete π-π stacking. Adjustments and modifications to explore the ART of SWNTs for FET applications are culminated in these pages.

1.3 Characterization Methods

Analysis of carbon nanotubes can be conducted by means of a wide assortment of surface characterization methods. Depending on the CNT sample, solutions can be characterized using UV-vis, photoluminescence, Raman, or IR spectroscopy. The former three are used for chiral characterization. The CNTs can be transmission electron microscopy (TEM) imaged, as well as with AFM or scanning electron microscopy (SEM) analyzed, with high accuracy. The microscopic methods are frequently used in combination with spectroscopic methods. AFM and SEM are the more widely used methods, where AFM tapping mode is preferred for minimal tube damage.

1.3.1 Atomic Force Microscopy

The AFM itself allows for the measurement and manipulation of atomic surfaces. Additionally, it does not need a clean conducting surface to provide a nice image, as is the case with scanning electron microscopy (SPM). Therefore, AFM has a wide variety of uses, from measuring insulators to biological samples such as live proteins. However, if conductive AFM (c-AFM) is needed, the tip can be coated with a conductive material, though usually, the tip is silicon or silicon nitride. A laser focuses on to the cantilever surface and is mirrored towards a 4- segmented photodiode (Fig. 1.19) [196].

A mechanical probe senses the material on the substrate surface, and electrical signals are generated with piezoelectric materials. The deflection that comes from the cantilever is measured via a series of these photodiodes that receive the laser signal from the tip of the cantilever surface. The c-AFM works in an equivalent manner, with the additional fact that it also concurrently measures the landscape of a material and the electric current that can flow at the point where the tip makes contact with the substrate surface [197]. The data for the topography comes from the deflection of the cantilever using the same laser/photodiode optical system as with conventional AFM. At the same time, the current is distinguished with a current-to-voltage preamplifier. Bright spots represent conductive elements on the sample surface, whereas dark areas show insulating or rather non-conductive segments [198, 199].

The working principle behind the AFM is associated with the spring constant (k) from Hooke's law ($F = -kX$, where X is the spring displacement, and F is applied force) that dictates the motion of the cantilever [199, 200]. When the cantilever is taken into close proximity to the sample surface, deflection occurs from all the interacting forces such as bonding, VdW, and mechanical contact forces [46]. Many other parameters may also be measured depending on the fitted probe. Nevertheless, if scanning occurs at a constant height, some tip damage may occur. Hence a feedback loop is designed to ensure constant force along the tip and the sample during the measurement, such that the distance between the tip and the sample stays constant.

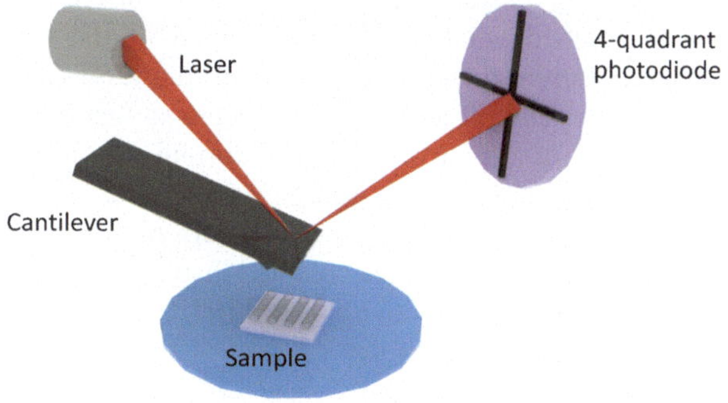

Fig. 1.19 The typical setup for the AFM. Example of approaching a CNT surface (not to scale)

The sample is placed on an XYZ stage, as shown in Fig. 1.19; the z-direction is chosen to keep the force constant, and the x/y directions are to help in the scanning of the sample. There are also three different working modes [198, 199]. One is the contact (static) approach that indicates the cantilever is moved across the surface through direct contact tip-to-sample, producing a deflection on the cantilever tip. The exact value of the deflection is directly measured and is used as signal feedback. When the cantilever nears the sample, the attractive forces tend to stick the two together; therefore, this method is done typically wherever a repulsive force is present [198, 199]. Another method is non-contact (dynamic); here, the cantilever oscillates externally with a frequency very near to its actual value but does not encounter the substrate surface. Information on the surface is obtained from comparing the external oscillations to the change in oscillation from the forces coming from the tip and material that are closely interacting. This mode is particularly useful for liquids, as the AFM wavers overhead the adsorbed liquefied layer so that both the fluid and surface are scanned [198, 199].

The last of the three is the most widely used and is the one utilized for experimentation in this report. The tapping (dynamic-contact) method consists of the tip of the cantilever being fixed to a piezo-electric material that dictates the motion of the cantilever to an up and down movement to values near its resonant frequency. As the tip nears the top of the sample, the amplitude of oscillation decreases due to the diverse interaction of forces on the cantilever. As the lever scans the surface material, the electronic feedback loops, moving the height to the correct position [198, 199]. Essentially, the AFM is a versatile tool to obtain height information and detailed micrographs of nanoscale features. It is, however, much slower to obtain images than with electron beam microscopes such as SEM and TEM. For c-AFM, this is always done under contact mode as a current is applied to the sample.

1.3.1.1 Conductive-Atomic Force Microscopy (c-AFM)

The c-AFM system utilizes a conductive cantilever to sense the surface current (I) of a sample. During c-AFM imaging, the conductive cantilever scans the substrate under contact mode to produce a topography map, and simultaneously, the electric current amongst the cantilever and the surface components is measured. This electric current is obtained through an electric current amplifier, whereas the topography is attained via the deflection signal of the lever [201]. Subsequently, the electrical current that goes through the cantilever can be as tiny as a pico-ampere, such that the current amplifier is selected so that the electrical noise is suppressed on the order of a femtoampere. The current data is compiled by the c-AFM [201]. Accordingly, as a metallic sample or metal area is placed into contact with the c-AFM tip, the tip can then collect the current based on the overall lateral conductivity of the metal. The images obtained display the high conductivity area as high up and bright on the conductivity scale [197, 202–204].

1.3.2 Scanning Electron Microscopy

Scientific evaluation involves thorough observations that consider all aspects of the form of material. However, such an analysis with a magnifying glass or optical microscope cannot account for details smaller than the wavelength of light. Therefore, any optical instruments using light are unable to identify nanoscale structures. Therein represents the use of an electron beam-based microscope like the scanning electron microscope (SEM), such that observing a structure down to several nanometers in scale becomes possible [205]. The primary portions of the SEM are the electronic console and the electron column (depicted in Fig. 1.20) [206, 207].

The working principle present involves an electron beam that scans across the substrate surface, and the subsequent electrons that are produced from the conducting sample are combined to create a micrograph [205]. This formation of images is reliant on signals that are produced from the specimen interactions with the electron beam. This connection can be broken down into two major types: elastic and inelastic scattering. For elastic interactions, these are caused by the deflection of any incident electrons by either the atomic nucleus' present or by the outer shell electrons of analogous energy on the substrate surface [205]. Elastic scattering is defined as insignificant energy being lost during a collision and by a sizeable directional shift of the scattered species [6].

If the elastically scattering of the electrons occurs at a degree of 90° or greater, they are called backscattered electrons (BSE) [205, 208]. Inelastic interactions

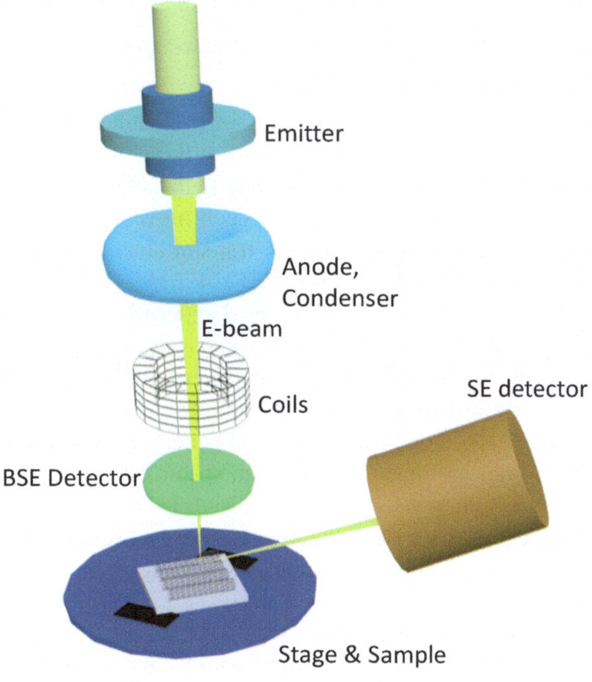

Fig. 1.20 Schematic diagram of a scanning electron microscope. Sample is of CNTs (not to scale)

happen from various scatterings amid the incident electrons and the electrons and atoms of the sample; this leads to the primary beam electron shifting a large amount of energy to that atom [209]. The quantity of energy lost is reliant upon whether or not the specimen electrons are excited alone or collectively. Another factor is the binding energy of the electron to the equivalent atom [209].

Hence, the generation of secondary electrons (SE2) comes from the excitation of the electrons while the sample atoms are being ionized. These electrons are described as having energies fewer than 50 eV [207, 208]. It is because of the SE2 that a high definition resolution of the topography is obtained. On the other hand, using BSE gives a better contrast that is dependent on the atomic number of the compounds on the surface, such that further topographical information is accessible [206].

Additionally to the BSE and SE2 signals, a variety of different signals are created when the electron beam is making contact with the substrate, including the emission of characteristic X-rays, Auger electrons, and cathodoluminescence [205]. For the electrons to travel effectively, a high-vacuum environment is required which minimizes electron scattering via air [207, 210, 211].

The technique is disadvantageous for imaging wet samples, and non-conducting samples must be sputter coated for effective charging and imaging. An important factor while performing SEM analysis of CNTs is to use a low voltage electron-beam source as to avoid damage to the CNTs [212]. Additionally, one form of SEM experimentation on SWNTs has been shown to be able to distinguish between semi-conducting and metallic SWNTs [213].

1.3.3 IR and Raman Spectroscopy

Infrared (IR) bands are a result of molecular vibrations causing a change in the dipole in a molecule. No IR band will occur if there is no alteration in the dipole moment. For instance, compounds such as O_2 and N_2 are not depicted in the IR, but CO_2 has a stretch where one oxygen moves opposingly to the other [214]. The IR data is also represented with % transmittance, such that 12% of light transmittance is indicative of 88% absorbance [214, 215].

Like IR, Raman spectroscopy also permits the recognition of different molecular components [216]. The Raman spectrum is a result of inelastically scattered light (see Fig. 1.21 that shows how the laser light is scattered from the surface), allowing for the identification of vibrational (phonon) states of compounds (Stokes or anti-Stokes) [216, 217]. Rayleigh is the general light scattering. However, the Stokes and anti-Stokes are inherent to Raman spectroscopy. The stokes scattering signal is due to phonon energy loss in the material. In contrast, anti-Stokes are a gain in energy, and these signals are indicative of energy excitations due to the laser light source. This property allows for the identification of materials that possess discrete energy bands, such as CNTs.

The Raman spectrum has sharp and strong signal peaks where the corresponding IR spectrum of the compound would have weak peaks, and vice versa, since IR

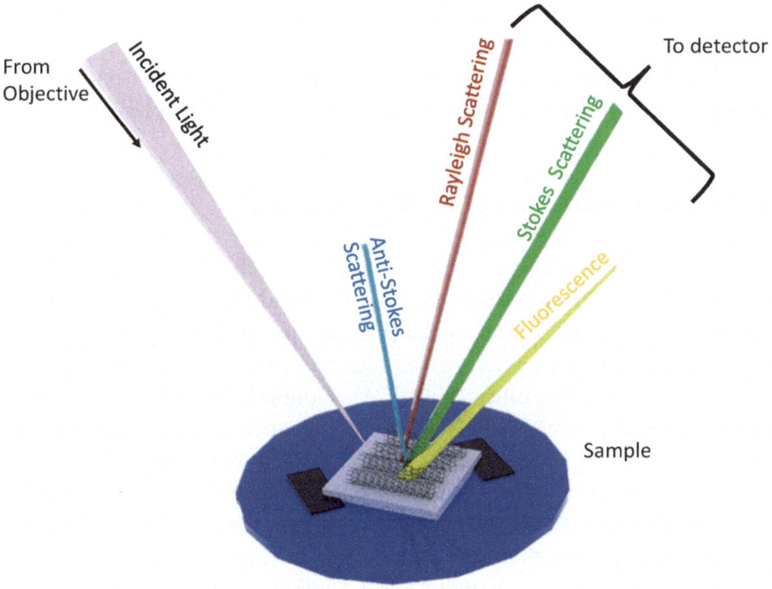

Fig. 1.21 Diagram categorizing the different light scatterings after the laser exposure on the surface for data acquisition from Raman spectroscopy. Adapted from Colthup [216]

direct absorptions are favored by dipole moments, and Raman scattering is favored by polarizability, which tends to be low in bonds with a strong dipole [216, 217]. The frequency position of groups in IR and Raman are identical since both are based on the same vibrational energy level difference. Raman spectroscopy is an asset for molecular fingerprinting as well as for observing changes in the bond structure [216, 217]. The IR is used in this work for characterization of the organic compounds, particularly for the identification of the very characteristic carboxylic acid carbonyl stretch of not only small organic molecules but also the SWNTs [218].

Raman has advantages versus other vibrational spectroscopy methods like Fourier Transform (FT)-IR and Near-IR (NIR) [216, 217]. The Raman effect happens when light is scattered off a sample as opposed to the light absorbed by a sample, and no permanent dipole is necessary for the molecule to be observed. It is also possible to identify the covalent character of molecules [216, 217]. Raman is also insensitive to aqueous absorption bands [216, 217]. However, Raman is inherently more expensive than IR and is not as efficient to identify key organic functional groups, such that the organic chemist uses IR for quick and cheap analysis of functional groups present in their samples alongside nuclear magnetic resonance for characterization rather than Raman [216, 217].

Non-resonant Raman spectroscopy utilizes any laser excitation for the scattering measurement, which can lead to low-intensity peaks. For resonant Raman, the laser wavelength is selected to be close to an electronic transition, usually very near the UV-visible absorption of the compound being analyzed. This selection of λ can result in significantly higher intensity peaks. Double resonant scattering is additionally possible for semiconductors, such as CNTs and graphite, and are only observable during special conditions. Duo phonon scattering methods—one phonon, and one of elastic scattering—make up double resonance Raman [219].

1.3.3.1 Raman Spectroscopy of Carbon Nanotubes

The Raman spectra of CNTs is essential in identifying their diameter and chiralities. The peaks of interest are the disorder band (D-band, 1350 cm^{-1}), graphene band (G-band, 1500–1600 cm^{-1}), and peaks present in radial breathing mode (RBM, 100–300 cm^{-1}, though some sources will cite up to 400 cm^{-1}), which can be correlated to the diameter of the tubes [220, 221]. Double resonant scatterings through tangential vibrations are responsible for the appearance of the D-band and G-band. In contrast, the RBM is a vibrational property akin to being the CNT's fingerprint [90]. The D-band is an indication of damage and deformities on the nanotube. The lower the D-band intensity, the higher the uniformity, and the fewer defects that are present in the sample. Semiconducting (s-) SWNTs tubes typically depict a G-band at approximately 1570 cm^{-1} (diameter) and 1590 cm^{-1} (length) that are relatively sharp with a strong signal depending on purity [222]. These peaks are shown in Fig. 1.22. The tangential modes are a term used to describe phonon bands of CNTs that are created from graphite's optical phonons, characteristic of two adjacent

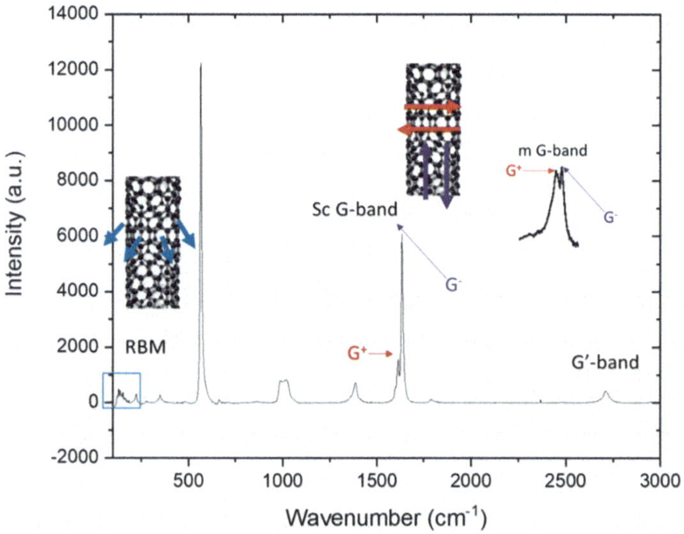

Fig. 1.22 Example of Raman spectrum of CNTs on silicon with distinct bands and peaks of interest and the reported difference in G-band for metallic(m) or semiconducting (sc) SWNTs. Adapted from Nanot et al. [223]

carbon atoms in out-of-phase displacement. In CNTs, they are radial (diameter) and longitudinal (length).

Moreover, the split peak structure of the G-band is unique to single-walled tubes. The G-band can identify metallic or semiconducting CNTs. As shown in Fig. 1.22, as the G^+ and G^- vibrations are nearly equal for m-SWNTs, but very far apart in intensity for sc-SWNTs. The G^+ band shifts up-field in the presence of electron acceptors and downward in the presence of electron donors, and is, therefore, an indicator of SWNT doping [90, 223]. The spectrum in Fig. 1.22 is identified as being semiconducting SWNTs, but not of a single chirality. If this were the spectrum of a single SWNT, there would only be one peak in the RBM region. Additionally, m-CNTs possess powerful electronic bands, which leads to an always resonant Raman spectrum regardless of the excitation energy being used. This observation means that real optical transitions at all energies are responsive to m-SWNTs [90].

The diameter of these tubes is slightly under 2 nm based on ω_{RBM}, where the relationship is described in Eq. (1.4). For archetypal SWNT bundles of diameter d_t, A = 234 cm^{-1}·nm and B = 10 cm^{-1} have been calculated [220]; where B is an upshifted value for ω_{RBM} due to the presence of tube–tube interactions [224]. For a single SWNT on silicon, A = 248 cm^{-1} and B = 0 has been used [96].

$$\omega_{RBM} = \frac{A}{d_t} + B\left[\mathrm{m}^{-1}\right] \tag{1.4}$$

The resonance scattering caused by different lasers gives us information about different excitation levels, as shown in the Kataura plot (Fig. 1.23) [220, 225]. The Kataura plot is primarily optical transition energies that have been measured in relation to the CNT diameter. The NT diameters can effectively lead to defining the specific chirality of the nanotubes, as has been shown in various other studies [92, 224, 226–228]. The nanotubes have efficiently disconnected ranges of transition energies in a given range of diameters, as those energies follow the $1/d_t$ trend. However, some tubes do fall above or below this trend. In this way, metallic tube transition energies are separated from the semiconducting ones. The m-SWNTs are not identifiable by luminescence, and single m-SWNTs chiralities can only be distinguishable using Raman [90]. Additionally, it is imperative to indicate that small fluctuations in A and B in Eq. (1.4) can give slightly different values. Nanotubes that are in the same family/branch have comparable physical properties, and environmental factors may be possible that can affect the RBM frequency to diameter relation, without altering the m- or sc-assignment [90].

Moreover, different kinds of SWNTs vibrate more intensely in the RBM region, depending on the laser strength used [221]. Both red and green laser light sources were used to obtain data for nanotube identification to create a more representative picture of all the SWNTs that can be present on the sample surface for the ART work. Raman spectroscopy allows for identifying the unique peak signals and assigning them to different chiral types. This correlation is possible based on the

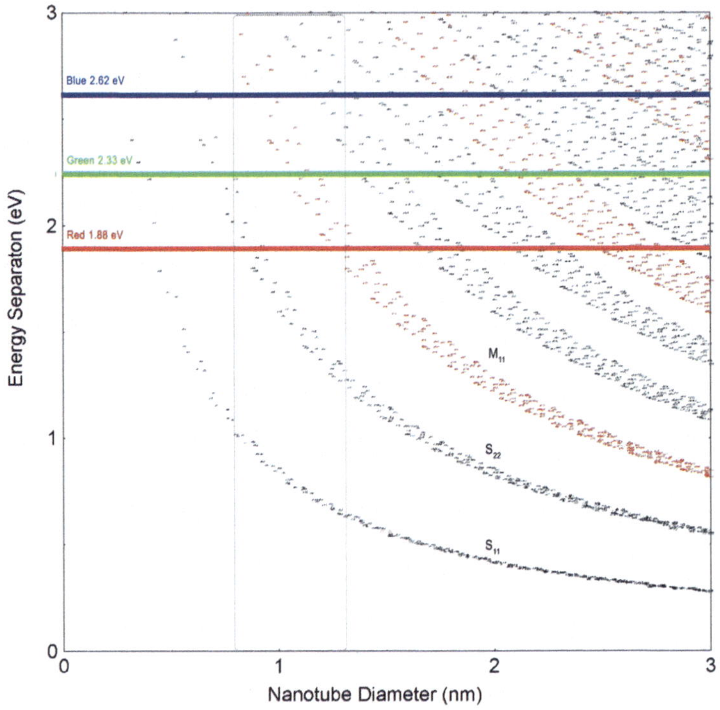

Fig. 1.23 Transition energies amongst VHSs in the valence and conduction bands. Black circles are for sc-SWNTs and red circles are for m-SWNTs. Laser energies are represented with colored horizontal lines, and vertical lines around the 1 nm tube diameter indicate CNTs of interest in this work. Plot adapted from the following website. http://www.photon.t.u-tokyo.ac.jp/maruyama/kataura/kataura.html

chiral dependence that unites the bandgap energy and the vibrational frequency in the RBM mode of the individual nanotubes within their nanotube branches or families. Raman-CNT chiral assignments should consider the various parameters of excitation energy, frequency, luminescence energy, and diameter, as well as any surfactant effects. The correlation between the optical transition energies (VHS, S_{11}, M_{11}, S_{22}) and vibrational properties (ω_{RBM}) equates to the ability to obtain nanotube chirality.

In ART, the nanotubes are individually separated, and therefore it is possible to characterize single SWNTs on the surface. For CNT bundles, the peak positions and energies can be affected by VdW interactions between the tubes themselves, resulting in shifts in the Kataura plot; the same happens when using different surfactants or substrate surfaces [229]. Additionally, obtaining information of all the nanotubes on the surface must involve many sample points; as a single Raman scan does not give a full idea of the kinds of SWNTs dispersed on the surface. Raman mapping, or rather, the collection of Raman spectrum take at a single point, helps formulate a representative picture of all the SWNTs on the surface. A point of disadvantage is also when scanning the distributed thin layer of CNTs, and it can be challenging to obtain a signal if the laser is not directly on the nanotube. If the laser is offset or not focused even slightly, the spectrum will not analyze the CNTs.

1.3.4 I-V Testing

Electrical testing typically occurs using a source measure unit (SMU) with a three-terminal device CNTFET. This arrangement includes the two metal S and D electrodes between the sample CNT and the probes. For testing, probes are brought into contact with the S- and D-electrodes, which are positioned similarly to Fig. 1.24. The third probe is capacitively connected to the CNTs using an insulating dielectric film and is defined as the transistor's gate electrode. This process is a form of testing FETs. The dielectric material is usually SiO_2, as it is easy to fabricate devices with and easy to acquire. Depending on the device, many CNTs can be connecting the S to D, as shown in Fig. 1.24. Alternatively, individual CNTs can be tested, or even bundles of SWNTs. Each system would have its individual I-V curve. These transport quantities include source/drain current (I_{DS}) data that is a function of the S-/D--voltage (V_{SD}) or the gate voltage (V_G). The V_{SD} is the voltage between the S and D electrodes to turn the device to the ON-state. Figure 1.24 is a rendition of a bottom gate (BG) CNTFET in this work.

In Fig. 1.25, a top-gate (TG) device is shown. For this TG CNTFET, the dielectric layer is placed above the patterned S- and D- electrodes and covers the sides and surface of the nanotubes. This setup is intended to eliminate any external

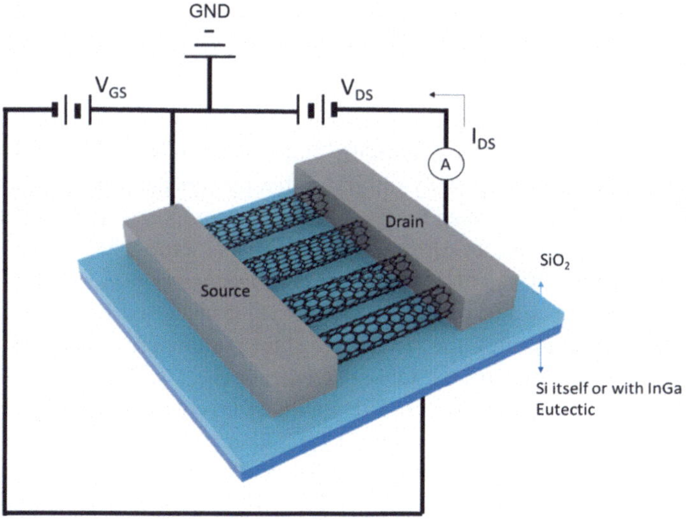

Fig. 1.24 Representation of a setup for electrical measurements of SWNTs in bottom-gate FETs. GND is ground

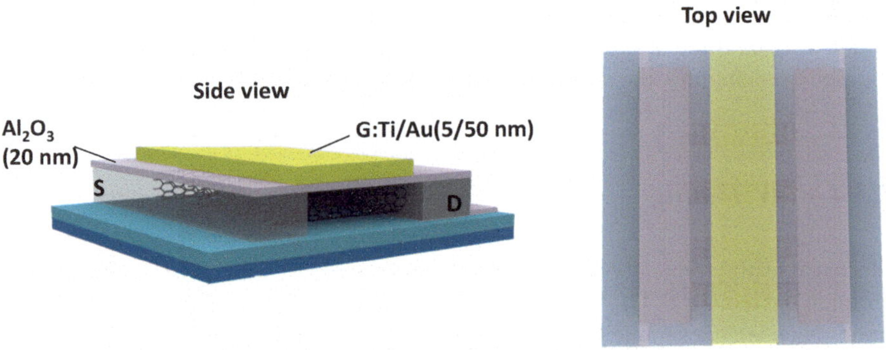

Fig. 1.25 Representation of SWNTs in top-gate FETs

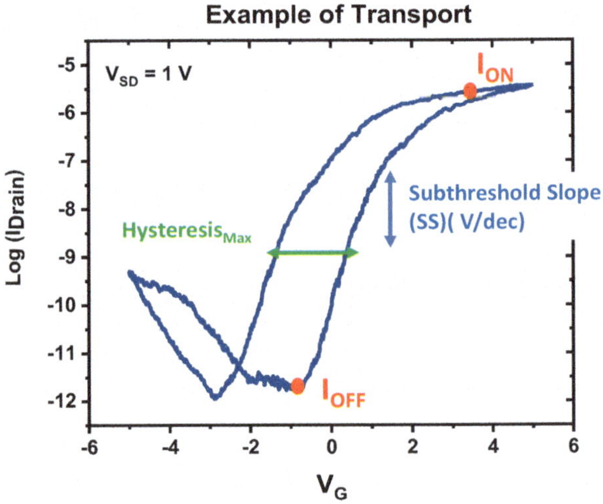

Fig. 1.26 Depiction of desirable transport data for CNTFETs with essential parameters that can be extrapolated from the curve

environmental factors that can potentially impede the CNTs electrical transport. Additionally, it has been previously observed that hysteresis, transconductance and subthreshold swing are improved for TG CNTFETs compared with their BG counterpart [230]. For testing the TG device, a third probe is setup that gently touches the surface of the gate electrode.

Both transport and output device properties are considered for the examination of CNTFETs formed through the ART. Important parameters considered are foremost the electron mobility (μ), the ON-state current (I_{ON}), and the ratio of ON-state current to current leakage (I_{ON}/I_{OFF}). The leakage current is defined by the current that is still flowing even in the OFF state (I_{OFF}). Other parameters considered are the transconductance (G_m) and the subthreshold slope (SS). The maximum hysteresis is also a further point of discussion. The SS, I_{ON}, I_{OFF} and hysteresis can be obtained through the analysis of the transport plot shown in Fig. 1.26. The transport figure depicts the drain current (I_D) over a function of V_G. The target is to also utilize a low

V_{SD} while having a high I_{ON}/I_{OFF} and μ. The μ value defines the ease at which the current carriers can travel across the channel. Some sources list the mobility as being irrelevant for nanoscale transistors that have channel lengths below 100 nm [9, 231]. In this work, the minimal channel length is 100 nm and therefore, the μ was calculated as the devices were on a larger scale.

The SS comes from the change in V_G over the change in the log value of I_D, as shown from the curve in Fig. 1.26 and is defined as shown from Eq. (1.5). The G_m comes from the linear curve of I_D vs. V_G (Eq. (1.6)). A high G_m is indicative of high device performance as it defines the change in current by the voltage. The SS characterizes how fast the device can transition between the I_{OFF} to I_{ON} state. Ideal values for MOSFETs are in the 60–70 mV/dec range. For example, a Si MOSFET with a ~22 nm channel size has I_{ON} of 1 mA/μm and an SS = 72 mV/dec where V_{SD} = 0.8–0.5 V, which has shown to be equivalent to a CNTFET with a 90 nm channel at V_{SD} = 0.4 V [232]. This example provides additional evidence to the improvement of transport characteristics that are possible with CNTs over Si for scaling potential in FETs.

$$SS = \frac{dV_G}{dI_{D\ \log}} \left[\mathrm{V/decade}\right] \tag{1.5}$$

$$G_m = \frac{dI_D}{dV_G}[\mathrm{S}] \tag{1.6}$$

For CNTFETs, the calculation for mobility is typically defined as shown in Eq. (1.7). The value is reliant on the length of the gap from the source and drain (L_{SD}) and the transconductance while being inversely proportional to the capacitance (C) of the system and the V_{SD}. Capacitance is the ability of the device to store charge and is targeted to be as low as possible. The capacitance is different for individual CNT devices and is defined in Eq. (1.8). For the tubes, 2π is multiplied by the constant of electric permittivity (ε_0) and relative permittivity (ε_r) for the dielectric layer used. The relative permittivity for Al_2O_3 is 10, and for SiO_2, it is 3.9; and these are the values used in this work. Equation (1.8) also incorporates the thickness of the oxide (t_{ox}) and the nanotube radius (r). For films of CNTs, the capacitance is defined by Eq. (1.9) and incorporates the width of the source-drain channel (W_{SD}) [233] as shown, although other definitions for film capacitance can also be utilized that take into consideration additional factors [234].

$$\mu = \frac{L_{SD}{}^2 G_m}{C V_{SD}} \left[\mathrm{cm2/Vs}\right] \tag{1.7}$$

$$C_{individual\,CNT} = \frac{2\pi\varepsilon_0\varepsilon_r}{\ln\left(2\frac{t_{ox}}{r}\right)} L_{SD}\left[\mathrm{F}\right] \tag{1.8}$$

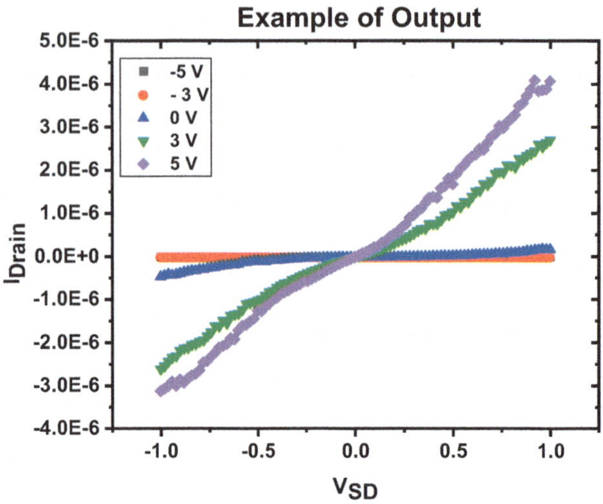

Fig. 1.27 Depiction of output data for CNTFETs

$$C_{CNT\,Film} = \varepsilon_0 \varepsilon_r W_{SD} \frac{L_{SD}}{t_{Ox}} [\mathrm{F}] \tag{1.9}$$

The second parameter for analysis is the output curve that is shown in Fig. 1.27. These curves showcase the different I_D possible for different V_G while keeping the V_{SD} constant. From this plot, it is possible to determine the ON-state resistance, as determined by the slope of the max V_G curve [235]. This parameter is important as it can dictate whether the device is under ballistic transport (~13.5 kΩ) or diffusive (200 kΩ) [236]. Ideally, the value would be as minimal as possible for the device to indicate effective electrical transport. The value of R_{ON} can be determined by using Eq. (1.10).

$$R_{ON} = \frac{dV_{ON}}{dI_{ON}} [\Omega] \tag{1.10}$$

1.3.5 Additional Techniques

Transmission electron microscopy (TEM) is an electron beam imaging method capable of identifying CNT chiralities [237]. The TEM operates by shining the e-beam through a very thin sample, and the interactions of the atoms in the material with the bombarding electrons are used to identify surface features such as optical diffraction patterns and morphology [238]. The electrons transform into the light to obtain the image. A condenser lens focuses the beam, and an objective lends then collects all the electrons after they interact and form an image on the substrate.

High-resolution (HR) TEM is often used in CNT studies and is capable of imaging their defects [75]. Transmitted and scattered beams work together to create an interference image in HRTEM that can have an image resolution on the scale of a crystal unit cell. In this case, all the electrons that have interacted with the sample are combined at one single point. When the electrons are propagated through the objective lens, they are modulated at very low angles that interfere with each other and then conform to a single point. The bright field images in HRTEM are a result of electron phase contrasts at high kV, whereas in TEM these images are the result of the electron waves amplitude contrast at lower kV. TEM images are analyzed using GMS 3 software.

It has also been demonstrated that CNT chiralities can also be selectively altered [239]. This work was possible by altering programmed bias pulses on the CNTs. Nucleation was promoted through the SW defects and (0,1) chiral dislocations on the circumference of the tube. The CNTs were transitioned from metallic to semiconducting that were experimentally confirmed by electron transport measurements. The process was shown to work on SWNT, DWNT, and even triple-walled CNTs. An exciting direction for future work would be to align individual CNTs regardless of chirality and then alter the chiralities to all sc-SWNTs [239].

However, HRTEM and TEM both require samples that can have electrons be transmitted through them (electron transparent materials). Therefore, regular silicon surface mounted CNTs are not possible to image. The sample has to be mounted on special SiO_2 wafers with windows sections having small pores to see through some regions of the CNTs The sample needs to also be able to withstand the high vacuum operating conditions [238].

Additional spectroscopic techniques that can identify nanotube chiralities are UV-vis absorption and photoluminescence. These methods are robust and useful for identifying the chiralities of CNTs when dispersed in a solution [240]. In UV-vis spectroscopy, the basic principle is to shine the broadband of light in the range of 200–2000 nm on a liquid dispersion and obtaining a measurement of the light that becomes transmitted. The absorption spectrum, which is a function of the peak intensity and the wavelength, is a result of the transmitted and incident light differences. The wavelength of high-intensity peaks can be correlated to the bandgap energy in semiconductors. For CNTs, the optical transitions are identified from a database of VHS of SWNTs. Because of absorption in the UV-vis light range, electrons that are present in the VHSs of the CNTs' valence band are augmented to the energy levels that are present in the conduction band ($S_{11} \rightarrow S_{22}$ in sc-SWNTs, M_{11} in m-SWNTs). The different energy transitions have been identified by Chernysheva et al. [241]. The CNTs are usually dispersed in a surfactant solution such as SDBS, centrifuged, and then characterized in the spectrometer [240].

Each peak output in the spectrum is associated with a single chirality, but similarly to Raman, the environment can alter the peak position as well as intensity. Factors that can affect the CNT peaks are mainly the quality of the dispersion, the sample purity as in the presence of organic and inorganic materials, and the surfactant used. For example, red-shifts are common when then SWNTs are exposed to organic dyes [242]. M-SWNTs are usually in the UV/Visible region, whereas the sc-SWNTs are in the NIR/Visible regions. The larger diameter tubes consist of a smaller energy gap and generally show a shift in absorbance towards longer wavelengths (lower energy) [242]. This technique is advantageous in identifying specific SWNTs types from a solution. It is not possible to use CNTs deposited on a solid sample.

Photoluminescence (PL) is also a technique capable of identifying CNT chiralities [243, 244]. The process is defined by photon emission from a substrate in the visible light range. The exposed sample can absorb the light source and incur photoexcitation. The luminescence or light energy released as the excited state photons relax back to the lower energy level is detected. As PL requires good optical emission properties in the sample, materials with bandgaps are practical candidates for characterization. This result is because the light energy that is emitted when electrons and holes recombine and are de-excited is equal to the bandgap energy. Emission is particularly useful and sensitive to surface atoms, and for SWNTs that is the entire structure [243, 245].

However, this also means that the SWNTs must be suspended and not in contact with a material to avoid PL quenching. Encapsulation of CNTs has also been shown to allow for luminescence detection. The PL signals are enhanced when the excitation energy is resonant with peaks in the DOS of the CNTs, which is ultimately a way of identifying the chirality [244]. The SWNTs types are typically observed in the NIR range [245]. Additionally, a combination of PL and Raman spectroscopy can be used to characterize the electronic and vibrational properties of suspended CNTs [246]. Stokes PL [244] and anti-Stokes PL [247] spectra have been obtained for SWNTs.

1.4 Concluding Remarks

The fundamental aspects of the project have been defined and described in this section and will be referred to throughout the remainder of the manuscript. The following Chaps. 2, 3, 4, and 5 will detail the materials, methods, results, and analysis. Chapter 6 summarizes the entirety of the research and elaborates on an outlook for ART. The experimental protocols are described in Chap. 7.

References

1. Cao, Q., Tersoff, J., Farmer, D.B., Zhu, Y., Han, S.-J.: Carbon nanotube transistors scaled to a 40-nanometer footprint. Science. **356**(6345), 1369 (2017)
2. Thompson, S.E., Parthasarathy, S.: Moore's law: the future of Si microelectronics. Mater Today. **9**(6), 20–25 (2006)
3. Rupp, K.: "48 years microprocessor trend data" plot until 2020, predictions 2021-2030. Philos Trans R Soc A. **378**, 20190061 (2020)

4. Javey, A., Guo, J., Wang, Q., Lundstrom, M., Dai, H.: Ballistic carbon nanotube field-effect transistors. Nature. **424**(6949), 654 (2003)
5. McEuen, P.L., Fuhrer, M.S., Hongkun, P.: Single-walled carbon nanotube electronics. IEEE Trans Nanotechnol. **99**(1), 78–85 (2002)
6. Thess, A., Lee, R., Nikolaev, P., Dai, H., Petit, P., Robert, J., Xu, C., Lee, Y.H., Kim, S.G., Rinzler, A.G., Colbert, D.T., Scuseria, G.E., Tománek, D., Fischer, J.E., Smalley, R.E.: Crystalline ropes of metallic carbon nanotubes. Science. **273**(5274), 483–487 (1996)
7. Baughman, R.H., Zakhidov, A.A., de Heer, W.A.: Carbon nanotubes--the route toward applications. Science. **297**(5582), 787–792 (2002)
8. Bunshi, F., Wenhai, H., Noriaki, E., Yuichi, K., Toshio, O.: Disassembling single-walled carbon nanotube bundles by dipole/dipole electrostatic interactions. Chem Lett. **34**(9), 1218–1219 (2005)
9. Franklin, A.D.: The road to carbon nanotube transistors. Nature. **498**(7455), 443–444 (2013)
10. Liu, L., Han, J., Xu, L., Zhou, J., Zhao, C., Ding, S., Shi, H., Xiao, M., Ding, L., Ma, Z., Jin, C., Zhang, Z., Peng, L.-M.: Aligned, high-density semiconducting carbon nanotube arrays for high-performance electronics. Science. **368**(6493), 850–856 (2020)
11. Iijima, S.: Helical microtubules of graphitic carbon. Nature. **354**(6348), 56–58 (1991)
12. Wang, X., Li, Q., Xie, J., Jin, Z., Wang, J., Li, Y., Jiang, K., Fan, S.: Fabrication of ultralong and electrically uniform single-walled carbon nanotubes on clean substrates. Nano Lett. **9**(9), 3137–3141 (2009)
13. Kaushik, B.K., Majumder, M.K.: Carbon nanotube: properties and applications. In: Carbon nanotube based VLSI interconnects, pp. 17–37. Springer, Cham (2015)
14. Hayashi, T., Kim, Y.A., Matoba, T., Esaka, M., Nishimura, K., Tsukada, T., Endo, M., Dresselhaus, M.S.: Smallest freestanding single-walled carbon nanotube. Nano Lett. **3**(7), 887–889 (2003)
15. Kasperski, A., Weibel, A., Datas, L., De Grave, E., Peigney, A., Laurent, C.: Large-diameter single-wall carbon nanotubes formed alongside small-diameter double-walled carbon nanotubes. J Phys Chem C. **119**(3), 1524–1535 (2015)
16. Truong, V., McMahon, P.J., Olsson-Jacques, C.L., Wilson, A.R., Mathys, G.I.: In multiwalled carbon nanotubes of 200 nm diameter and carbon micro-balloons. Int Conf Nanosci Nanotechnol. **2010**, 66–69 (2010)
17. Yin, L.-C., Saito, R., Dresselhaus, M.S.: The fermi level dependent electronic properties of the smallest (2,2) carbon nanotube. Nano Lett. **10**(9), 3290–3296 (2010)
18. Huang, S., Cai, X., Liu, J.: Growth of millimeter-long and horizontally aligned single-walled carbon nanotubes on flat substrates. J Am Chem Soc. **125**(19), 5636–5637 (2003)
19. Saito, R., Dresselhaus, G., Dresselhaus, M.S.: Physical properties of carbon nanotubes. World Scientific, Singapore (1998)
20. Lei, T., Pochorovski, I., Bao, Z.: Separation of semiconducting carbon nanotubes for flexible and stretchable electronics using polymer removable method. Acc Chem Res. **50**(4), 1096–1104 (2017)
21. Liu, Y., Dobrinsky, A., Yakobson, B.I.: Graphene edge from armchair to zigzag: the origins of nanotube chirality? Phys Rev Lett. **105**(23), 235502 (2010)
22. Zhang, M., Li, J.: Carbon nanotube in different shapes. Mater Today. **12**(6), 12–18 (2009)
23. Ao, G., Streit, J.K., Fagan, J.A., Zheng, M.: Differentiating left- and right-handed carbon nanotubes by DNA. J Am Chem Soc. **138**(51), 16677–16685 (2016)
24. Strano, M.S.: Sorting out left from right. Nat Nanotechnol. **2**(6), 340–341 (2007)
25. Damnjanović, M., Milošević, I., Vuković, T., Sredanović, R.: Full symmetry, optical activity, and potentials of single-wall and multiwall nanotubes. Phys Rev B. **60**(4), 2728–2739 (1999)
26. Cao, Q., Han, S.-J., Tersoff, J., Franklin, A.D., Zhu, Y., Zhang, Z., Tulevski, G.S., Tang, J., Haensch, W.: End-bonded contacts for carbon nanotube transistors with low, size-independent resistance. Science. **350**(6256), 68–72 (2015)
27. Bianco, S.: Carbon nanotubes: from research to applications. BoD–Books on Demand, Norderstedt (2011)

28. Zhang, J., Cui, J., Wang, X., Wang, W., Mei, X., Yi, P., Yang, X., He, X.: Recent progress in the preparation of horizontally ordered carbon nanotube assemblies from solution. Phys Status Solidi. **2018**, 1700719 (2018)
29. Baibarac, M., Matea, A., Ilie, M., Baltog, I., Magrez, A.: Anti-Stokes Raman spectroscopy as a method to identify metallic and mixed metallic/semiconducting configurations of multi-walled carbon nanotubes. Anal Methods. **7**(15), 6225–6230 (2015)
30. Duong, B., Seraphin, S., Wang, L., Peng, Y., Xin, H.: Production of predominantly semiconducting double-walled carbon nanotubes. Carbon. **49**(11), 3512–3521 (2011)
31. Wen, Q., Qian, W., Nie, J., Cao, A., Ning, G., Wang, Y., Hu, L., Zhang, Q., Huang, J., Wei, F.: 100 mm long, semiconducting triple-walled carbon nanotubes. Adv Mater. **22**(16), 1867–1871 (2010)
32. Lu, J.P., Han, J.: Carbon nanotubes and nanotube-based nano devices. Int J High Speed Electron Syst. **9**(1), 101–123 (1998)
33. Berber, S., et al.: Unusually high thermal conductivity of carbon nanotubes. Phys. Rev. Lett. **84**(20), 4613–4616 (2000)
34. Wang, X.-D., et al: Synthesis of carbon nanotubes by catalytic chemical vapor deposition. Perspectives on carbon nanotubes. H. E.-D. Saleh, IntechOpen: 2. Page 2 (2019)
35. Spear, K. E.; Dismukes, J. P.. John Wiley & Sons: 1994; Vol. 25. p 4.
36. Hirama, K. et al.. Jpn J Appl Phys 2012, 51, 090112
37. Tombros, N., et al.: Quantized conductance of a suspended graphene nanoconstriction. Nat Phy **7**(9), 697–700 (2011)
38. Lee, C., Wei, X., Kysar, J. W., Hone, J.: Science **321**(5887), 385–388 (2008)
39. Duryat, R. S.: Graphene Nanoribbons (GNRs) for future interconnect. IOP Conference Series: Mater Sci Eng 131: 012018 (2016)
40. Li, T. C., Lu, S.-P.: Quantum conductance of graphene nanoribbons with edge defects. Physl Rev B **77**(8):085408 (2008)
41. Hoffmann, S., et al.: Fracture strength and Young's modulus of ZnO nanowires. Nanotechnology **18**(20), 205503 (2007)
42. Yuldashev, S. U., et al.: Thermal conductivity of ZnO single nanowire. J Nanosci Nanotechnol. **16**(2), 1592–1595 (2016)
43. Lu, J. G.: Fundamental Properties of Zinc Oxide Nanowires. Encyclopedia of Nanotechnology. B. Bhushan. Dordrecht, Springer Netherlands 919–927 (2012)
44. Lee, C. J., et al.: Field emission from well-aligned zinc oxide nanowires grown at low temperature. Appl Phys Lett. **81**(19), 3648–3650 (2002)
45. Saito, Y., Uemura, S.: Field emission from carbon nanotubes and its application to electron sources. Carbon. **38**(2), 169–182 (2000)
46. Yue, G.Z., Qiu, Q., Gao, B., Cheng, Y., Zhang, J., Shimoda, H., Chang, S., Lu, J.P., Zhou, O.: Generation of continuous and pulsed diagnostic imaging X-ray radiation using a carbon-nanotube-based field-emission cathode. Appl Phys Lett. **81**(2), 355–357 (2002)
47. Milne, W.I., Teo, K.B.K., Minoux, E., Groening, O., Gangloff, L., Hudanski, L., Schnell, J.P., Dieumegard, D., Peauger, F., Bu, I.Y.Y., Bell, M.S., Legagneux, P., Hasko, G., Amaratunga, G.A.J.: Aligned carbon nanotubes/fibers for applications in vacuum microwave amplifiers. J Vac Sci Technol B. **24**(1), 345–348 (2006)
48. Qiu, C., Zhang, Z., Xiao, M., Yang, Y., Zhong, D., Peng, L.-M.: Scaling carbon nanotube complementary transistors to 5-nm gate lengths. Science. **355**(6322), 271 (2017)
49. Hashim, U., Low, F.W., Liu, W.-W.: Precise alignment of individual single-walled carbon nanotube using dielectrophoresis method for development and fabrication of pH sensor. J Nanomater. **2013**, 139–139 (2013)
50. Cole, M.T., Cientanni, V., Milne, W.I.: Horizontal carbon nanotube alignment. Nanoscale. **8**(35), 15836–15844 (2016)
51. Eatemadi, A., Daraee, H., Karimkhanloo, H., Kouhi, M., Zarghami, N., Akbarzadeh, A., Abasi, M., Hanifehpour, Y., Joo, S.: Carbon nanotubes: properties, synthesis, purification, and medical applications. Nanoscale Res Lett. **9**(1), 393 (2014)

52. Khan, W., Sharma, R., Saini, P.: Carbon nanotube-based polymer composites: synthesis, properties and applications. In: Carbon nanotubes current progress of their polymer composites. IntechOpen, London (2016)
53. Szabó, A., Perri, C., Csató, A., Giordano, G., Vuono, D., Nagy, J.B.: Synthesis methods of carbon nanotubes and related materials. Mater Ther. **3**(5), 3092–3140 (2010)
54. Sengupta, J.: Carbon nanotube fabrication at industrial scale: opportunities and challenges. In: Handbook of nanomaterials for industrial applications, pp. 172–194. Elsevier, London (2018)
55. Plaza, E., Briceño-Fuenmayor, H., Arévalo, J., Atencio, R., Corredor, L.: Electric field effect in the growth of carbon nanotubes. J Nanopart Res. **17**, 6 (2015)
56. Azam, M.A., Manaf, N.S.A., Talib, E., Bistamam, M.S.A.: Aligned carbon nanotube from catalytic chemical vapor deposition technique for energy storage device: a review. Ionics. **19**(11), 1455–1476 (2013)
57. Baker, R.: Nucleation and growth of carbon deposits from the nickel catalyzed decomposition of acetylene. J Catal. **26**(1), 51–62 (1972)
58. Tessonnier, J.-P., Su, D.S.: Recent progress on the growth mechanism of carbon nanotubes: a review. ChemSusChem. **4**(7), 824–847 (2011)
59. Page, A.J., Chandrakumar, K.R.S., Irle, S., Morokuma, K.: SWNT nucleation from carbon-coated SiO2 nanoparticles via a vapor–solid–solid mechanism. J Am Chem Soc. **133**(3), 621–628 (2011)
60. Kumar, M., Ando, Y.: Chemical vapor deposition of carbon nanotubes: a review on growth mechanism and mass production. J Nanosci Nanotechnol. **10**(6), 3739–3758 (2010)
61. Tessonnier, J. P., Su, D. S.: Recent progress on the growth mechanism of carbon nanotubes: a review. Chem Sus Chem. **18**;4(7):824-47. https://doi.org/10.1002/cssc.201100175. Epub 2011 Jul 5. PMID: 21732543 (2011)
62. Cheng, H.M., Li, F., Su, G., Pan, H.Y., He, L.L., Sun, X., Dresselhaus, M.S.: Large-scale and low-cost synthesis of single-walled carbon nanotubes by the catalytic pyrolysis of hydrocarbons. Appl Phys Lett. **72**(25), 3282–3284 (1998)
63. Wang, J., Shen, B., Lan, M., Kang, D., Wu, C.: Carbon nanotubes (CNTs) production from catalytic pyrolysis of waste plastics: the influence of catalyst and reaction pressure. Catal Today. **351**, 50–57 (2019)
64. Joselevich, E., Dai, H., Liu, J., Hata, K., Windle, A.H.: Carbon nanotube synthesis and organization. In: Carbon nanotubes, pp. 101–165. Springer, New York (2007)
65. Arora, N., Sharma, N.N.: Arc discharge synthesis of carbon nanotubes: comprehensive review. Diam Relat Mater. **50**, 135–150 (2014)
66. Das, R., Shahnavaz, Z., Ali, M.E., Islam, M.M., Abd Hamid, S.B.: Can we optimize arc discharge and laser ablation for well-controlled carbon nanotube synthesis? Nanoscale Res Lett. **11**, 1 (2016)
67. Kalam, A., Mieno, T.: Pressure and discharge current dependence of production rate of single-walled carbon nanotubes by the bipolar pulsed arc-discharge method. Fullerenes. **26**(7), 458–464 (2018)
68. Kataura, H., Kumazawa, Y., Maniwa, Y., Ohtsuka, Y., Sen, R., Suzuki, S., Achiba, Y.: Diameter control of single-walled carbon nanotubes. Carbon. **38**(11-12), 1691–1697 (2000)
69. Scott, C.D., Arepalli, S., Nikolaev, P., Smalley, R.E.: Growth mechanisms for single-wall carbon nanotubes in a laser-ablation process. Appl Phys A Mater Sci Process. **72**(5), 573–580 (2001)
70. Chrzanowska, J., Hoffman, J., Małolepszy, A., Mazurkiewicz, M., Kowalewski, T.A., Szymanski, Z., Stobinski, L.: Synthesis of carbon nanotubes by the laser ablation method: effect of laser wavelength. Phys Status Solidi. **252**(8), 1860–1867 (2015)
71. Sanchez-Valencia, J.R., Dienel, T., Gröning, O., Shorubalko, I., Mueller, A., Jansen, M., Amsharov, K., Ruffieux, P., Fasel, R.: Controlled synthesis of single-chirality carbon nanotubes. Nature. **512**(7512), 61–64 (2014)
72. Wang, J., Zhuang, G., Huang, Q., Xiao, Y., Zhou, Y., Liu, H., Du, P.: Precise synthesis and photophysical properties of a small chiral carbon nanotube segment: cyclo[7]paraphenylene-2,6-naphthylene. Chem Commun. **55**(64), 9456–9459 (2019)

73. Collins, P.G.: Defects and disorder in carbon nanotubes. In: Oxford handbook of nanoscience and technology volume 2: materials: structures, properties and characterization techniques, pp. 156–184. Oxford University Press, Oxford (2009)
74. Li, Y., Maruyama, S.: Single-walled carbon nanotubes: preparation, properties and applications. Springer, New York (2019)
75. Rahmandoust, M., Ayatollahi, M.R.: Carbon nanotubes. In: Characterization of carbon nanotube based composites under consideration of defects, pp. 5–63. Springer, New York (2016)
76. Diarra, M., Amara, H., Bichara, C., Ducastelle, F.: Role of defect healing on the chirality of single-wall carbon nanotubes. Phys Rev B. **85**, 24 (2012)
77. Ajayan, P.M., Ravikumar, V., Charlier, J.C.: Surface reconstructions and dimensional changes in single-walled carbon nanotubes. Phys Rev Lett. **81**(7), 1437–1440 (1998)
78. Saxena, R.K., Williams, W., McGee, J.K., Daniels, M.J., Boykin, E., Ian Gilmour, M.: Enhanced in vitro and in vivo toxicity of poly-dispersed acid-functionalized single-wall carbon nanotubes. Nanotoxicology. **1**(4), 291–300 (2007)
79. Bekyarova, E., Davis, M., Burch, T., Itkis, M.E., Zhao, B., Sunshine, S., Haddon, R.C.: Chemically functionalized single-walled carbon nanotubes as ammonia sensors. J Phys Chem B. **108**(51), 19717–19720 (2004)
80. Ebbesen, T.W., Takada, T.: Topological and SP3 defect structures in nanotubes. Carbon. **33**(7), 973–978 (1995)
81. Patel, A.M., Gosai, N., Joshi, A.Y.: A review on defects in carbon nanotubes. Appl Mech Mater. **2015**, 145–150 (2015)
82. Joshi, A.Y., Sharma, S.C., Harsha, S.P.: Dynamic analysis of a clamped wavy single walled carbon nanotube based nanomechanical sensors. J Nanotechnol Eng Med. **1**, 3 (2010)
83. Cao, Y., Cong, S., Cao, X., Wu, F., Liu, Q., Amer, M.R., Zhou, C.: Review of electronics based on single-walled carbon nanotubes. In: Li, Y., Maruyama, S. (eds.) Single-walled carbon nanotubes: preparation, properties and applications, pp. 189–224. Springer, Cham (2019)
84. Meyyappan, M.: Carbon nanotubes: science and applications. CRC Press, Boca Raton (2004)
85. Park, J.Y., Chapter 1: band structure and electron transport physics of one-dimensional SWNTs (2009)
86. Sato, K., Saito, R., Jiang, J., Dresselhaus, G., Dresselhaus, M.: Discontinuity in the family pattern of single-wall carbon nanotubes. Phys Rev B. **76**(19), 195446 (2007)
87. Charlier, J.-C., Blase, X., Roche, S.: Electronic and transport properties of nanotubes. Rev Mod Phys. **79**(2), 677–732 (2007)
88. Zheng, M.: Sorting carbon nanotubes. In: Single-walled carbon nanotubes, pp. 129–164. Springer, Cham (2019)
89. Lu, X., Chen, Z.: Curved Pi-conjugation, aromaticity, and the related chemistry of small fullerenes (<C60) and single-walled carbon nanotubes. Chem Rev. **105**(10), 3643–3696 (2005)
90. Thomsen, C., Reich, S.: Raman scattering in carbon nanotubes, pp. 115–234. Springer, Berlin (2007)
91. Hagen, A., Hertel, T.: Quantitative analysis of optical spectra from individual single-wall carbon nanotubes. Nano Lett. **3**(3), 383–388 (2003)
92. Jorio, A., Araujo, P.T., Doorn, S.K., Maruyama, S., Chacham, H., Pimenta, M.A.: The Kataura plot over broad energy and diameter ranges. Phys Status Solidi. **243**(13), 3117–3121 (2006)
93. Liu, Z., Tabakman, S., Welsher, K., Dai, H.: Carbon nanotubes in biology and medicine: in vitro and in vivo detection, imaging and drug delivery. Nano Res. **2**(2), 85–120 (2009)
94. Miyata, Y., Kawai, T., Miyamoto, Y., Yanagi, K., Maniwa, Y., Kataura, H.: Chirality-dependent combustion of single-walled carbon nanotubes. J Phys Chem C. **111**(27), 9671–9677 (2007)
95. Jorio, A., Santos, A., Ribeiro, H., Fantini, C., Souza, M., Vieira, J., Furtado, C., Jiang, J., Saito, R., Balzano, L.: Quantifying carbon-nanotube species with resonance Raman scattering. Phys Rev B. **72**(7), 075207 (2005)
96. Jorio, A., Saito, R., Hafner, J., Lieber, C., Hunter, D., McClure, T., Dresselhaus, G., Dresselhaus, M.: Structural (n, m) determination of isolated single-wall carbon nanotubes by resonant Raman scattering. Phys Rev Lett. **86**(6), 1118 (2001)

97. Franklin, A.D., Koswatta, S.O., Farmer, D., Tulevski, G.S., Smith, J.T., Miyazoe, H., Haensch, W.: Scalable and fully self-aligned n-type carbon nanotube transistors with gate-all-around. In: International electron devices meeting, pp. 4.5.1–4.5.4. IEEE, Piscataway (2012)
98. Perebeinos, V., Tersoff, J., Avouris, P.: Mobility in semiconducting carbon nanotubes at finite carrier density. Nano Lett. **6**(2), 205–208 (2006)
99. Singh, A., Khosla, M., Raj, B.: Design and analysis of electrostatic doped Schottky barrier CNTFET based low power SRAM. AEU. **80**, 67–72 (2017)
100. Guo, J., Datta, S., Lundstrom, M.: A numerical study of scaling issues for Schottky-barrier carbon nanotube transistors. IEEE Trans Electron Dev. **51**(2), 172–177 (2004)
101. Prakash, A., Ilatikhameneh, H., Wu, P., Appenzeller, J.: Understanding contact gating in Schottky barrier transistors from 2D channels. Sci Rep. **7**, 1 (2017)
102. Avouris, P., Appenzeller, J., Martel, R., Wind, S.J.: Carbon nanotube electronics. Proc IEEE. **9**(11), 1772–1784 (2003)
103. Derycke, V., Martel, R., Appenzeller, J., Avouris, P.: Carbon nanotube inter- and intramolecular logic gates. Nano Lett. **1**(9), 453–456 (2001)
104. Kong, J., Franklin, N.R., Zhou, C., Chapline, M.G., Peng, S., Cho, K., Dai, H.: Nanotube molecular wires as chemical sensors. Science. **287**(5453), 622 (2000)
105. Tans, S.J., Verschueren, A.R.M., Dekker, C.: Room-temperature transistor based on a single carbon nanotube. Nature. **393**, 49 (1998)
106. Qiu, C., Zhang, Z., Zhong, D., Si, J., Yang, Y., Peng, L.-M.: Carbon nanotube feedback-gate field-effect transistor: suppressing current leakage and increasing on/off ratio. ACS Nano. **9**(1), 969–977 (2015)
107. Liu, L., Zhao, C., Ding, L., Peng, L., Zhang, Z.: Drain-engineered carbon-nanotube-film field-effect transistors with high performance and ultra-low current leakage. Nano Res. **13**, 7 (2019)
108. Swapna, P., Babu, K.K., Rambabu, B., Rao, Y.S.: Ambipolar CNTFET: basic characterization and effect of high dielectric material, pp. 1–4. IEEE, Piscataway (2011)
109. Phaedon, A., Zhihong, C., Vasili, P.: Carbon-based electronics. Nat Nanotechnol. **2**(10), 605 (2007)
110. Yamada, T.: Nanoelectronics applications. In: Carbon nanotubes, pp. 223–262. CRC Press, Boca Raton (2004)
111. Li, F., et al.: Characterization of single-wall carbon nanotubes by N2 adsorption. Carbon. **42**(12), 2375–2383 (2004)
112. Zhang, Z., et al.: Self-aligned ballistic n-Type single-walled carbon nanotube field-Effect Transistors with Adjustable Threshold Voltage. Nano Lett. **8**(11), 3696–3701 (2008)
113. Zhang, X., et al.: Fabrication of hundreds of field effect transistors on a single carbon nanotube for basic studies and molecular devices. J. Vac. Sci. Technol. B. Nanotechnology and Microelectronics: Materials, Processing, Measurement, and Phenomena **31**(6), 06FI01 (2013)
114. Kim, J. W., et al.: Atomic layer deposition of ruthenium on plasma-treated vertically aligned carbon nanotubes for high-performance ultracapacitors. Nanotechnology **25**(43), 435404 (2014)
115. Cao, Q., et al.: Carbon nanotube transistors scaled to a 40- nanometer footprint. Science **356**(6345), 1369
116. Arnold, M.S., Green, A.A., Hulvat, J.F., Stupp, S.I., Hersam, M.C.: Sorting carbon nanotubes by electronic structure using density differentiation. Nat Nanotechnol. **1**(1), 60–65 (2006)
117. Ghosh, S., Bachilo, S.M., Weisman, R.B.: Advanced sorting of single-walled carbon nanotubes by nonlinear density-gradient ultracentrifugation. Nat Nanotechnol. **5**, 443 (2010)
118. Moshammer, K., Hennrich, F., Kappes, M.M.: Selective suspension in aqueous sodium dodecyl sulfate according to electronic structure type allows simple separation of metallic from semiconducting single-walled carbon nanotubes. Nano Res. **2**(8), 599–606 (2009)
119. Cui, J., Su, W., Yang, D., Li, S., Wei, X., Zhou, N., Zhou, W., Xie, S., Kataura, H., Liu, H.: Mass production of high-purity semiconducting carbon nanotubes by hydrochloric acid assisted gel chromatography. ACS Appl Nano Mater. **2**(1), 343–350 (2019)

120. Wang, G., Wei, X., Tanaka, T., Kataura, H.: Diameter-selective separation of semiconducting single-walled carbon nanotubes in large diameter range. Phys Status Solidi. **254**(11), 1700294 (2017)
121. Wei, X., Tanaka, T., Akizuki, N., Miyauchi, Y., Matsuda, K., Ohfuchi, M., Kataura, H.: Single-chirality separation and optical properties of (5,4) single-wall carbon nanotubes. J Phys Chem C. **120**(19), 10705–10710 (2016)
122. Yomogida, Y., Tanaka, T., Zhang, M., Yudasaka, M., Wei, X., Kataura, H.: Industrial-scale separation of high-purity single-chirality single-wall carbon nanotubes for biological imaging. Nat Commun. **7**(1), 12056 (2016)
123. Liu, H., Tanaka, T., Kataura, H.: Optical isomer separation of single-chirality carbon nanotubes using gel column chromatography. Nano Lett. **14**(11), 6237–6243 (2014)
124. Khripin, C.Y., Fagan, J.A., Zheng, M.: Spontaneous partition of carbon nanotubes in polymer-modified aqueous phases. J Am Chem Soc. **135**(18), 6822–6825 (2013)
125. Subbaiyan, N.K., Cambré, S., Parra-Vasquez, A.N.G., Hároz, E.H., Doorn, S.K., Duque, J.G.: Role of surfactants and salt in aqueous two-phase separation of carbon nanotubes toward simple chirality isolation. ACS Nano. **8**(2), 1619–1628 (2014)
126. Krupke, R., Hennrich, F., Löhneysen, H.V., Kappes, M.M.: Separation of metallic from semiconducting single-walled carbon nanotubes. Science. **301**(5631), 344–347 (2003)
127. Lutz, T., Donovan, K.J.: Macroscopic scale separation of metallic and semiconducting nanotubes by dielectrophoresis. Carbon. **43**(12), 2508–2513 (2005)
128. Xiao, Z., Elike, J., Reynolds, A., Moten, R., Zhao, X.: The fabrication of carbon nanotube electronic circuits with dielectrophoresis. Microelectron Eng. **164**, 123–127 (2016)
129. Park, J., Han, Y.R., Park, M., Jun, C.-H., Ju, S.-Y.: Coral reef-like functionalized self-assembled monolayers for network formation of carbon nanotube with diameter selectivity. Carbon. **161**, 599–611 (2020)
130. Ren, Z., Lan, Y., Wang, Y.: Technologies to achieve carbon nanotube alignment. In: Aligned carbon nanotubes: physics, concepts, fabrication and devices, pp. 111–156. Springer, Berlin (2013)
131. Ajayan, P.M., Stephan, O., Colliex, C., Trauth, D.: Aligned carbon nanotube arrays formed by cutting a polymer resin-nanotube composite. Science. **265**, 1212 (1994)
132. Ma, Y., Wang, B., Wu, Y., Huang, Y., Chen, Y.: The production of horizontally aligned single-walled carbon nanotubes. Carbon. **49**(13), 4098–4110 (2011)
133. Rao, S.G., Huang, L., Setyawan, W., Hong, S.: Large-scale assembly of carbon nanotubes. Nature. **425**(6953), 36–37 (2003)
134. Tulevski, G.S., Hannon, J., Afzali, A., Chen, Z., Avouris, P., Kagan, C.R.: Chemically assisted directed assembly of carbon nanotubes for the fabrication of large-scale device arrays. J Am Chem Soc. **129**(39), 11964–11968 (2007)
135. Sharma, R., Strano, M.S.: Centerline placement and alignment of anisotropic nanotubes in high aspect ratio cylindrical droplets of nanometer diameter. Adv Mater. **21**(1), 60–65 (2009)
136. Velázquez, M.M., Alejo, T., López-Díaz, D., Martín-García, B., Merchán, M.D.: Langmuir-blodgett methodology: a versatile technique to build 2D material films. In: Two-dimensional materials, p. 21. Langmuir, London (2016)
137. Joo, Y., Brady, G.J., Arnold, M.S., Gopalan, P.: Dose-controlled, floating evaporative self-assembly and alignment of semiconducting carbon nanotubes from organic solvents. Langmuir. **30**(12), 3460–3466 (2014)
138. Li, X., Zhang, L., Wang, X., Shimoyama, I., Sun, X., Seo, W.-S., Dai, H.: Langmuir–blodgett assembly of densely aligned single-walled carbon nanotubes from bulk materials. J Am Chem Soc. **129**(16), 4890–4891 (2007)
139. Jinkins, K.R., Chan, J., Brady, G.J., Gronski, K.K., Gopalan, P., Evensen, H.T., Berson, A., Arnold, M.S.: Nanotube alignment mechanism in floating evaporative self-assembly. Langmuir. **33**(46), 13407–13414 (2017)
140. Chai, Z., Seo, J., Abbasi, S.A., Busnaina, A.: Assembly of highly aligned carbon nanotubes using an electro-fluidic assembly process. ACS Nano. **12**(12), 12315–12323 (2018)

141. Cao, Q., Han, S.-J., Tulevski, G.S., Zhu, Y., Lu, D.D., Haensch, W.: Arrays of single-walled carbon nanotubes with full surface coverage for high-performance electronics. Nat Nanotechnol. **8**(3), 180–186 (2013)
142. Vijayaraghavan, A., Blatt, S., Weissenberger, D., Oron-Carl, M., Hennrich, F., Gerthsen, D., Hahn, H., Krupke, R.: Ultra-large-scale directed assembly of single-walled carbon nanotube devices. Nano Lett. **7**(6), 1556–1560 (2007)
143. Zhang, J., Cui, J., Cheng, Y., Wang, W., He, X., Mei, X.: A molecular dynamics study on self-assembly of single-walled carbon nanotubes: from molecular morphology and binding energy. Adv Mater Interfaces. **6**(19), 1900983 (2019)
144. Fujii, S., Tanaka, T., Nishiyama, S., Kataura, H.: High performance thin-film transistors using moderately aligned semiconducting single-wall carbon nanotubes. Phys Status Solidi. **248**(11), 2692–2696 (2011)
145. Goh, G.L., Agarwala, S., Yeong, W.Y.: Directed and on-demand alignment of carbon nanotube: a review toward 3D printing of electronics. Adv Mater Interfaces. **6**(4), 1801318 (2019)
146. Ishigami, N., Ago, H., Imamoto, K., Tsuji, M., Iakoubovskii, K., Minami, N.: Crystal plane dependent growth of aligned single-walled carbon nanotubes on sapphire. J Am Chem Soc. **130**(30), 9918–9924 (2008)
147. Kang, S.J., Kocabas, C., Ozel, T., Shim, M., Pimparkar, N., Alam, M.A., Rotkin, S.V., Rogers, J.A.: High-performance electronics using dense, perfectly aligned arrays of single-walled carbon nanotubes. Nat Nanotechnol. **2**(4), 230–236 (2007)
148. Li, P., Zhang, J.: Preparation of horizontal single-walled carbon nanotubes arrays. In: Single-walled carbon nanotubes, pp. 69–98. Springer, New York (2019)
149. Ibrahim, I., Gemming, T., Weber, W.M., Mikolajick, T., Liu, Z., Rümmeli, M.H.: Current progress in the chemical vapor deposition of type-selected horizontally aligned single-walled carbon nanotubes. ACS Nano. **10**(8), 7248–7266 (2016)
150. Engel, M., Small, J.P., Steiner, M., Freitag, M., Green, A.A., Hersam, M.C., Avouris, P.: Thin film nanotube transistors based on self-assembled, aligned, semiconducting carbon nanotube arrays. ACS Nano. **2**(12), 2445–2452 (2008)
151. Park, S., Lee, H.W., Wang, H., Selvarasah, S., Dokmeci, M.R., Park, Y.J., Cha, S.N., Kim, J.M., Bao, Z.: Highly effective separation of semiconducting carbon nanotubes verified via short-channel devices fabricated using dip-pen nanolithography. ACS Nano. **6**(3), 2487–2496 (2012)
152. Shekhar, S., Stokes, P., Khondaker, S.I.: Ultra-high density alignment of carbon nanotubes array by dielectrophoresis. arXiv preprint arXiv:1101.2237 (2011)
153. Einarsson, E., Murakami, Y., Kadowaki, M., Maruyama, S.: Growth dynamics of vertically aligned single-walled carbon nanotubes from in situ measurements. Carbon. **46**(6), 923–930 (2008)
154. LeMieux, M.C., Roberts, M., Barman, S., Jin, Y.W., Kim, J.M., Bao, Z.: Self-sorted, aligned nanotube networks for thin-film transistors. Science. **321**(5885), 101 (2008)
155. Si, J., Zhong, D., Xu, H., Xiao, M., Yu, C., Zhang, Z., Peng, L.-M.: Scalable preparation of high-density semiconducting carbon nanotube arrays for high-performance field-effect transistors. ACS Nano. **12**(1), 627–634 (2018)
156. Du, F., Felts, J.R., Xie, X., Song, J., Li, Y., Rosenberger, M.R., Islam, A.E., Jin, S.H., Dunham, S.N., Zhang, C., Wilson, W.L., Huang, Y., King, W.P., Rogers, J.A.: Laser-induced nanoscale thermocapillary flow for purification of aligned arrays of single-walled carbon nanotubes. ACS Nano. **8**(12), 12641–12649 (2014)
157. Li, J., Liu, K., Liang, S., Zhou, W., Pierce, M., Wang, F., Peng, L., Liu, J.: Growth of high-density-aligned and semiconducting-enriched single-walled carbon nanotubes: decoupling the conflict between density and selectivity. ACS Nano. **8**(1), 554–562 (2014)
158. Zhou, W., Zhan, S., Ding, L., Liu, J.: General rules for selective growth of enriched semiconducting single walled carbon nanotubes with water vapor as in situ etchant. J Am Chem Soc. **134**(34), 14019–14026 (2012)
159. Martinez, C.R., Iverson, B.L.: Rethinking the term "pi-stacking". Chem Sci. **3**(7), 2191 (2012)

160. Sinnokrot, M.O., Valeev, E.F., Sherrill, C.D.: Estimates of the ab initio limit for π–π interactions: the benzene dimer. J Am Chem Soc. **124**(36), 10887–10893 (2002)
161. De Juan, A., Mar Bernal, M., Pérez, E.M.: Optimization and insights into the mechanism of formation of mechanically interlocked derivatives of single-walled carbon nanotubes. ChemPlusChem. **80**(7), 1153–1157 (2015)
162. Wang, F., Matsuda, K., Rahman, A.F.M.M., Peng, X., Kimura, T., Komatsu, N.: Simultaneous discrimination of handedness and diameter of single-walled carbon nanotubes (SWNTs) with chiral diporphyrin nanotweezers leading to enrichment of a single enantiomer of (6,5)-SWNTs. J Am Chem Soc. **132**(31), 10876–10881 (2010)
163. Hammershøj, P., Bomans, P.H.H., Lakshminarayanan, R., Fock, J., Jensen, S.H., Jespersen, T.S., Brock-Nannestad, T., Hassenkam, T., Nygård, J., Sommerdijk, N.A.J.M., Kilså, K., Bjørnholm, T., Christensen, J.B.: A triptycene-based approach to solubilising carbon nanotubes and C60. Chemistry. **18**(28), 8716–8723 (2012)
164. Nish, A., Hwang, J.-Y., Doig, J., Nicholas, R.J.: Highly selective dispersion of single-walled carbon nanotubes using aromatic polymers. Nat Nanotechnol. **2**(10), 640–646 (2007)
165. Stepanian, S.G., Karachevtsev, V.A., Glamazda, A.Y., Dettlaff-Weglikowska, U., Adamowicz, L.: Combined Raman scattering and ab initio investigation of the interaction between pyrene and carbon SWNT. Mol Phys. **101**(16), 2609–2614 (2003)
166. Selmani, S., Schipper, D.J.: π-Concave hosts for curved carbon nanomaterials. Chemistry. **25**(27), 6673–6692 (2019)
167. Pujari, S.P., Scheres, L., Marcelis, A.T.M., Zuilhof, H.: Covalent surface modification of oxide surfaces. Angew Chem Int Ed. **53**(25), 6322–6356 (2014)
168. Selmani, S., Schipper, D.: Orientation control of molecularly functionalized surfaces applied to the simultaneous alignment and sorting of carbon nanotubes. Angew Chem Int Ed. **130**(9), 2423–2427 (2018)
169. Queffélec, C., Petit, M., Janvier, P., Knight, D.A., Bujoli, B.: Surface modification using phosphonic acids and esters. Chem Rev. **112**(7), 3777–3807 (2012)
170. Guerrero, G., Mutin, P.H., Vioux, A.: Anchoring of phosphonate and phosphinate coupling molecules on titania particles. Chem Mater. **13**(11), 4367–4373 (2001)
171. Mutin, P.H., Guerrero, G., Vioux, A.: Hybrid materials from organophosphorus coupling molecules. J Mater Chem. **15**(35-36), 3761 (2005)
172. Hotchkiss, P.J.: The design, synthesis, and use of phosphonic acids for the surface modification of metal oxides. Georgia Institute of Technology, Atlanta (2008)
173. Hotchkiss, P.J., Jones, S.C., Paniagua, S.A., Sharma, A., Kippelen, B., Armstrong, N.R., Marder, S.R.: The modification of indium tin oxide with phosphonic acids: mechanism of binding, tuning of surface properties, and potential for use in organic electronic applications. Acc Chem Res. **45**(3), 337–346 (2012)
174. Jadhav, S.A.: Self-assembled monolayers (SAMs) of carboxylic acids: an overview. Cent Eur J Chem. **9**(3), 369–378 (2011)
175. Lessel, M., McGraw, J.D., Bäumchen, O., Jacobs, K.: Nucleated dewetting in supported ultra-thin liquid films with hydrodynamic slip. Soft Matter. **13**(27), 4756–4760 (2017)
176. Ma, H., Acton, O., Hutchins, D.O., Cernetic, N., Jen, A.K.-Y.: Multifunctional phosphonic acid self-assembled monolayers on metal oxides as dielectrics, interface modification layers and semiconductors for low-voltage high-performance organic field-effect transistors. Phys Chem Chem Phys. **14**(41), 14110–14126 (2012)
177. Jones, C.W.: Applications of hydrogen peroxide and derivatives. Royal Society of Chemistry, London (2007)
178. Kohli, R.: UV-ozone cleaning for removal of surface contaminants. In: Developments in surface contamination and cleaning: applications of cleaning techniques, vol. 8, pp. 355–390. Elsevier, London (2019)
179. Wei, C., Akinwolemiwa, B., Wang, Q., Guan, L., Xia, L., Hu, D., Tang, B., Yu, L., Chen, G.Z.: Mechano-Fenton–Piranha oxidation of carbon nanotubes for energy application. Adv Sustainable Syst. **3**(11), 1900065 (2019)

180. Tosaka, A., Nonaka, H., Ichimura, S., Nishiguchi, T.: Reaction analysis of initial oxidation of silicon by UV-light-excited ozone and the application to rapid and uniform SiO2 film growth. J Appl Phys. **101**(3), 034909 (2007)
181. Fink, C.K., Nakamura, K., Ichimura, S., Jenkins, S.J.: Silicon oxidation by ozone. J Phys Condens Matter. **21**(18), 183001 (2009)
182. Outram, B.: Liquid crystals, pp. 2–5. IOP Publishing, Bristol (2018)
183. Van Aerle, N.A.J.M., Tol, A.J.W.: Molecular orientation in rubbed polyimide alignment layers used for liquid-crystal displays. Macromolecules. **27**(22), 6520–6526 (1994)
184. Andrienko, D.: Introduction to liquid crystals. J Mol Liq. **267**, 520–541 (2018)
185. Kato, T.: Self-assembly of phase-segregated liquid crystal structures. Science. **295**(5564), 2414 (2002)
186. Bruschi, M.L.: Strategies to modify the drug release from pharmaceutical systems. Woodhead Publishing, Sawston (2015)
187. Lagerwall, J., Scalia, G., Haluska, M., Dettlaff-Weglikowska, U., Roth, S., Giesselmann, F.: Nanotube alignment using lyotropic liquid crystals. Adv Mater. **19**(3), 359–364 (2007)
188. Kawamoto, H.: The history of liquid-crystal displays. Proc IEEE. **90**(4), 460–500 (2002)
189. Hoogboom, J., Elemans, J.A.A.W., Rowan, A.E., Rasing, T.H.M., Nolte, R.J.M.: The development of self-assembled liquid crystal display alignment layers. Phil Trans R Soc A. **2007**(365), 1553–1576 (1855)
190. Wöhrle, T., Wurzbach, I., Kirres, J., Kostidou, A., Kapernaum, N., Litterscheidt, J., Haenle, J.C., Staffeld, P., Baro, A., Giesselmann, F., Laschat, S.: Discot Liq Cryst. **116**(3), 1139–1241 (2016)
191. An, Z., Yu, J., Domercq, B., Jones, S.C., Barlow, S., Kippelen, B., Marder, S.R.: Room-temperature discotic liquid-crystalline coronene diimides exhibiting high charge-carrier mobility in air. J Mater Chem. **19**(37), 6688 (2009)
192. Selmani, S., Hawes, G.F., Schipper, D.J.: Liquid-crystal phase optimization using the alignment relay technique for the deposition of single-walled carbon nanotubes. ACS Appl Nano Mater. **3**(3), 2118–2122 (2020)
193. Long, T.M., Swager, T.M.: Minimization of free volume: alignment of triptycenes in liquid crystals and stretched polymers. Adv Mater. **13**(8), 601–604 (2001)
194. Tournus, F., Latil, S., Heggie, M., Charlier, J.-C.: π-stacking interaction between carbon nanotubes and organic molecules. Phys Rev B. **72**(7), 075431 (2005)
195. Kulak, M.R., Schipper, D.J.: Carbon nanotube alignment and sorting: attempting a sulfur moiety as anchoring component. Phosphorus. **194**(7), 760–763 (2019)
196. Meyer, E.: Atomic force microscopy. Prog Surf Sci. **41**(1), 3–49 (1992)
197. Giessibl, F.J.: Advances in atomic force microscopy. Rev Mod Phys. **75**(3), 949 (2003)
198. Leng, Y.: Materials characterization: introduction to microscopic and spectroscopic methods. Wiley, New York (2009)
199. Zhang, L., Atomic force microscopy CHEM 750 course notes 2017.
200. Eaton, P., West, P.: Atomic force microscopy, pp. 5–30. Oxford University Press, Oxford (2010)
201. Lanza, M., Porti, M., Nafría, M., Aymerich, X., Whittaker, E., Hamilton, B.: Note: Electrical resolution during conductive atomic force microscopy measurements under different environmental conditions and contact forces. Rev Sci Instrum. **81**(10), 106110 (2010)
202. Murrell, M., Welland, M., O'Shea, S., Wong, T., Barnes, J., McKinnon, A., Heyns, M., Verhaverbeke, S.: Spatially resolved electrical measurements of SiO2 gate oxides using atomic force microscopy. Appl Phys Lett. **62**(7), 786–788 (1993)
203. Al-Jassim, M.: Conductive Atomic Force Microscopy. https://www.nrel.gov/materials-science/conductive-atomic.html
204. Brandon, D., Kaplan, W.D.: Microstructural characterization of materials. Wiley, New York (2013)
205. Joy, D.C.: Scanning electron microscopy. Wiley, New York (2006)
206. Heinig, N.: Microscopy: light and electron. CHEM750 course notes 2017, Lecture 1
207. Egerton, R.F.: Physical principles of electron microscopy, pp. 60–70. Springer, New York (2005)

208. Aharinejad, S., Lametschwandtner, A.: Fundamentals of scanning electron microscopy. In: Microvascular corrosion casting in scanning electron microscopy, pp. 44–51. Springer, New York (1992)
209. Ltd, L.E.M.: Software control LEO 15XX manual. Zeiss, Inc., Thornwood (2005)
210. Goldstein, J.I., Newbury, D.E., Echlin, P., Joy, D.C., Lyman, C.E., Lifshin, E., Sawyer, L., Michael, J.R.: Special topics in electron beam X-ray microanalysis. In: Scanning electron microscopy and X-ray microanalysis, pp. 453–536. Springer, New York (2003)
211. Apkarian, R.P.: The fine structure of fenestrated adrenocortical capillaries revealed by in-lens field-emission scanning electron microscopy and scanning transmission electron microscopy. Scanning. **19**(5), 361–367 (1997)
212. Choi, J.H., Lee, J., Moon, S.M., Kim, Y.-T., Park, H., Lee, C.Y.: A low-energy electron beam does not damage single-walled carbon nanotubes and graphene. J Phys Chem Lett. **7**(22), 4739–4743 (2016)
213. Li, D., Wei, Y., Zhang, J., Wang, J., Lin, Y., Liu, P., Fan, S., Jiang, K.: Direct discrimination between semiconducting and metallic single-walled carbon nanotubes with high spatial resolution by SEM. Nano Res. **10**(6), 1896–1902 (2017)
214. Silverstein, R.M., Webster, F.X., Kiemle, D.J., Bryce, D.L.: Spectrometric identification of organic compounds. Wiley, New York (2014)
215. Guiliano, M., Asia, L., Onoratini, G., Mille, G.: Applications of diamond crystal ATR FTIR spectroscopy to the characterization of ambers. Spectrochim Acta A. **67**(5), 1407–1411 (2007)
216. Colthup, N.: Introduction to infrared and Raman spectroscopy. Elsevier, New York (2012)
217. Rahman, A.: Raman, IR and UV-Vis spectroscopy. CHEM 750 Course Notes 2017, Lecture 4
218. Peng, H., Alemany, L.B., Margrave, J.L., Khabashesku, V.N.: Sidewall carboxylic acid functionalization of single-walled carbon nanotubes. J Am Chem Soc. **125**(49), 15174–15182 (2003)
219. Dresselhaus, M.S., Dresselhaus, G., Saito, R., Jorio, A.: Raman spectroscopy of carbon nanotubes. Phys Rep. **409**(2), 47–99 (2005)
220. Dresselhaus, M., Dresselhaus, G., Jorio, A., Souza Filho, A., Saito, R.: Raman spectroscopy on isolated single wall carbon nanotubes. Carbon. **40**(12), 2043–2061 (2002)
221. Dresselhaus, M.S., Jorio, A., Hofmann, M., Dresselhaus, G., Saito, R.: Perspectives on carbon nanotubes and graphene Raman spectroscopy. Nano Lett. **10**(3), 751–758 (2010)
222. Rao, C.N.R., Govindaraj, A.: Nanotubes and nanowires. RSC Publishing, London (2011)
223. Nanot, S., Thompson, N.A., Kim, J.-H., Wang, X., Rice, W.D., Hároz, E.H., Ganesan, Y., Pint, C.L., Kono, J.: Single-walled carbon nanotubes. In: Vajtai, R. (ed.) Springer handbook of nanomaterials, pp. 105–146. Springer, Berlin (2013)
224. Jorio, A., Souza Filho, A., Dresselhaus, G., Dresselhaus, M., Swan, A., Ünlü, M., Goldberg, B., Pimenta, M., Hafner, J., Lieber, C.: G-band resonant Raman study of 62 isolated single-wall carbon nanotubes. Phys Rev B. **65**(15), 155412 (2002)
225. Costa, S., Borowiak-Palen, E., Kruszynska, M., Bachmatiuk, A., Kalenczuk, R.: Characterization of carbon nanotubes by Raman spectroscopy. Mater Sci. **26**(2), 433–441 (2008)
226. Weisman, R.B., Bachilo, S.M.: Dependence of optical transition energies on structure for single-walled carbon nanotubes in aqueous suspension: an empirical kataura plot. Nano Lett. **3**(9), 1235–1238 (2003)
227. Pfeiffer, R., Kuzmany, H., Kramberger, C., Schaman, C., Pichler, T., Kataura, H., Achiba, Y., Kürti, J., Zólyomi, V.: Unusual high degree of unperturbed environment in the interior of single-wall carbon nanotubes. Phys Rev Lett. **90**(22), 225501 (2003)
228. Al-Zanganawee, J., Katona, A., Moise, C., Bojin, D., Enachescu, M.: Krypton gas for high quality single wall carbon nanotubes synthesis by KrF excimer laser ablation. J Nanomater. **16**(1), 404 (2015)
229. Thomsen, C., Reich, S., Goni, A., Jantoljak, H., Rafailov, P., Loa, I., Syassen, K., Journet, C., Bernier, P.: Intermolecular interaction in carbon nanotube ropes. Phys Status Solidi. **215**(1), 435–441 (1999)

230. Yang, M.H., Teo, K.B.K., Gangloff, L., Milne, W.I., Hasko, D.G., Robert, Y., Legagneux, P.: Advantages of top-gate, high-k dielectric carbon nanotube field-effect transistors. Appl Phys Lett. **88**(11), 113507 (2006)
231. Franklin, A.D.: Nanomaterials in transistors: from high-performance to thin-film applications. Science. **349**(6249), 2750 (2015)
232. Peng, L.-M., Zhang, Z., Wang, S.: Carbon nanotube electronics: recent advances. Mater Today. **17**(9), 433–442 (2014)
233. Snow, E.S., Novak, J.P., Campbell, P.M., Park, D.: Random networks of carbon nanotubes as an electronic material. Appl Phys Lett. **82**(13), 2145–2147 (2003)
234. Trivedi, K., Yuk, H., Floresca, H.C., Kim, M.J., Hu, W.: Quantum confinement induced performance enhancement in sub-5-nm lithographic Si nanowire transistors. Nano Lett. **11**(4), 1412–1417 (2011)
235. Lundstrom, M., Lecture 1: review of MOSFET fundamentals. 2008.
236. Lundstrom, M.: Notes on the Ballistic MOSFET. EEc. **1000**, 2 (2005)
237. Zhu, H., Suenaga, K., Hashimoto, A., Urita, K., Iijima, S.: Structural identification of single and double-walled carbon nanotubes by high-resolution transmission electron microscopy. Chem Phys Lett. **412**(1-3), 116–120 (2005)
238. Thomas, S., Thomas, R., Zachariah, A.K., Kumar, R.: Microscopy methods in nanomaterials characterization, vol. 1. Elsevier, New York (2017)
239. Tang, D.-M., Kvashnin, D.G., Cretu, O., Nemoto, Y., Uesugi, F., Takeguchi, M., Zhou, X., Hsia, F.-C., Liu, C., Sorokin, P.B., Kawamoto, N., Mitome, M., Cheng, H.-M., Golberg, D., Bando, Y.: Chirality transitions and transport properties of individual few-walled carbon nanotubes as revealed by in situ TEM probing. Ultramicroscopy. **194**, 108–116 (2018)
240. Njuguna, J., Vanli, O.A., Liang, R.: A review of spectral methods for dispersion characterization of carbon nanotubes in aqueous suspensions. J Spectrosc. **2015**, 1–11 (2015)
241. Chernysheva, M., Rozhin, A., Fedotov, Y., Mou, C., Arif, R., Kobtsev, S.M., Dianov, E.M., Turitsyn, S.K.: Carbon nanotubes for ultrafast fibre lasers. Nano. **6**(1), 1–30 (2017)
242. Ryabenko, A.G., Dorofeeva, T.V., Zvereva, G.I.: UV–VIS–NIR spectroscopy study of sensitivity of single-wall carbon nanotubes to chemical processing and Van-der-Waals SWNT/SWNT interaction. Verification of the SWNT content measurements by absorption spectroscopy. Carbon. **42**(8-9), 1523–1535 (2004)
243. Lefebvre, J., Homma, Y., Finnie, P.: Bright band gap photoluminescence from unprocessed single-walled carbon nanotubes. Phys Rev Lett. **90**(21), 217401 (2003)
244. Bachilo, S.M.: Structure-assigned optical spectra of single-walled carbon nanotubes. Science. **298**(5602), 2361–2366 (2002)
245. Shiraki, T., Shiraishi, T., Juhász, G., Nakashima, N.: Emergence of new red-shifted carbon nanotube photoluminescence based on proximal doped-site design. Sci Rep. **6**(1), 28393 (2016)
246. Chen, G., Davis, R.C., Futaba, D.N., Sakurai, S., Kobashi, K., Yumura, M., Hata, K.: A sweet spot for highly efficient growth of vertically aligned single-walled carbon nanotube forests enabling their unique structures and properties. Nanoscale. **8**(1), 162–171 (2016)
247. Akizuki, N., Aota, S., Mouri, S., Matsuda, K., Miyauchi, Y.: Efficient near-infrared up-conversion photoluminescence in carbon nanotubes. Nat Commun. **6**(1), 8920 (2015)

Chapter 2
Modifications to the ART Procedure

Abstract The alignment relay technique (ART) was designed as an approach to order and select for single-walled carbon nanotubes simultaneously. Based on the original proof-of-concept, the general motivation was to investigate if the method could be optimized, focusing on SWNTs' alignment and selectivity. According to the standard procedure for ART, as described in the introduction chapter (Sect. 1.2.6), the discussion that follows refers to three approaches to optimize the ART procedure guided by these goals: (1) To improve the alignment of the SWNTs, and (2) to enhance the density of the SWNTs on the surface. For the first modification, simple benchtop sonication was added after the deposition to the NTs to eliminate nanotube bundles and remove loosely bound tubes from the substrate (Sect. 2.1). Layer-by-layer and Langmuir-Blodgett treatments were also attempted on the samples (Sect. 2.2). For the third modification, filler molecules were added to account for the gaps in between the already aligned iptycenes to increase π–π stacking interactions with the SWNTs, such that more of them would adhere to the surface (Sect. 2.3). Hence, Chap. 2 focuses on simple add-ons to ART to achieve goals I and II.

Keywords Alignment · Sonication · Selectivity · Raman · Radial breathing mode · Carbon nanotubes · Single-walled · Semiconducting · Layer-by-layer · Filler · Phosphonate

2.1 Post-carbon Nanotube Deposition: Sonication

Here, sonication was tested after a 24 h nanotube deposition period to improve nanotubes' alignment (Fig. 2.1). Following previous exploits [1], the hypothesis was to observe if it was possible to agitate away unaligned tubes from the surface with simple sonication. Theoretically, this process should allow only the tubes with strong interactions with the iptycenes to remain on the surface. The incorporation of sonication, an easy step to add to the ART, was speculated to offer a highly aligned surface with the re-distribution of un-aligned or stacked SWNTs. Therefore, the proposed methodology could simultaneously provide a higher degree of

M. Snowdon, *Methods of the Alignment-Relay Technique for Nanosystems*, Springer Theses, https://doi.org/10.1007/978-3-030-84413-4_2

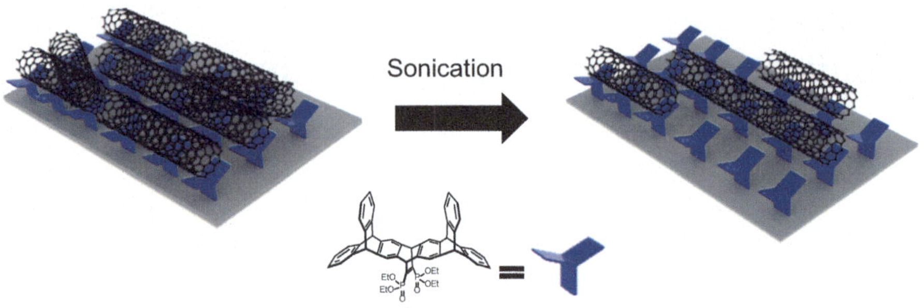

Fig. 2.1 Scheme outlining the proposed result of this work adapting ART with additional post-SWNTs deposition sonication, estimating that the nanotubes sit in the concave structure of the iptycenes (molecule 1)

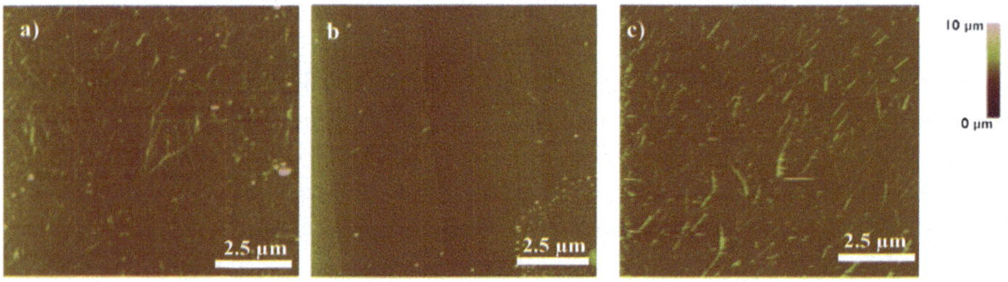

Fig. 2.2 An ART functionalized silicon sample with (**a**) immediate 40-min sonication then 24 h nanotube deposition (**b**) 40-min sonication after 24 h of nanotube deposition (**c**) alternating 10 min for 40-min sonication after 24 h nanotube deposition

orientational control than previously reported. Additionally, the sc-SWNTs diameters and chiralities were characterized.

2.1.1 Results and Discussion

2.1.1.1 Sonication Solvent and Time

When using the ART, the functionalized sample was placed in a solution of SWNTs for approximately 1–2 days. This timeframe allows the SWNTs to deposit on the surface and interact with the iptycenes (**1**). The sample was placed in a test tube in the center of a benchtop sonicator bath. First, it was confirmed that the sonication should occur after allowing the carbon nanotubes deposit on the sample, rather than as soon as the functionalized surface was placed in the SWNTs solution. The nanotube solution was a mixture of sodium cholate and sodium dodecyl sulfate (SDS), 1% wt/volume with 90% pure sc-SWNTs, purchased from Nanointegris. Figure 2.2 shows that the results of sonicating the sample within 5 min of being placed in the

SWNTs solution are shown (Fig. 2.2a). This result was compared to sonication after letting the sample rest in the nanotube solution for 24 h (Fig. 2.2b).

The samples were each sonicated for 40 min. The 40 min was arbitrarily chosen as a preliminary test to observe the sonication treatment effects on the ART sample. It was observed that with 40 min of continuous sonication, the water bath would gradually heat up. It was unknown if the temperature affected the ART samples, and the temperature was not a parameter investigated here. To maintain room temperature for all the samples and be consistent, the time frames were distributed in 10-min intervals. A sample was also attempted in increments of 10 min sonication and allowing the water bath cool down for 1 h, for a total time of 40 min of sonication (Fig. 2.2c). From this initial testing, it was observed that sonication should occur after the nanotubes have had time to deposit on the aligned surface to have any effect on agitating the tubes into a set orientation.

The preliminary results for applying sonication are qualitatively assessed through the SWNTs' alignment in the AFM micrographs from Fig. 2.2. The alignment and distribution of SWNTs did not seem to improve from the standard ART method when sonicating at the start of the CNTs' deposition (Fig. 2.2a). When applying continuous 40-min sonication after allowing the SWNTs to deposit as usual, very few CNTs remained on the surface (Fig. 2.2b). However, when applying sonication in 10-min increments, an increase in SWNTs alignment was observed (Fig. 2.2c). The alignment and distribution of the SWNTs appeared more uniform across the sample surface over the other two trials. Therefore, the shorter bursts of sonication times were chosen for further study.

The experiments were then continued based on the alternating sonication time model in the SWNTs solvent. The resulting AFM data is shown in Fig. 2.3. Here, a sample is sonicated for 10 min only (Fig. 2.3a), two 10 min sonication times with a 1 h pause to allow for re-deposition of the SWNTs (Fig. 2.3b), five 10 min sonication cycles with a 1 h pause for SWNTs re-deposition in between each session (Fig. 2.3c), and six 10 min sonication times with a 1 h pause for SWNTs re-deposition in between each session (Fig. 2.3d). The more prolonged the sonication and manipulation of the sample, the more considerable the increase in layers of unidentifiable particles on the sample surface. These layers of unidentified particles are defined as roughness in this work. Although the tubes appear qualitatively aligned based on the AFM images, fewer were on the surface. By a total of 60 min—even with pausing the treatment and keeping the sample in the SWNTs solution—a limited number of tubes remained on the substrate. The results from 8 × 10 min (Fig. 2.3d) appear to have equivalent outcomes to sonicating for 40 continuous minutes (Fig. 2.2b)—in that very few nanotubes remain on the surface.

Additionally, the sample size is shown (scale size) for Fig. 2.3c, d is double the size of Fig. 2.3a, b, demonstrating how much further apart the nanotubes are from each other. Specifically, in Fig. 2.3c, there appears to be a high degree of aligned tubes, but they are very sparsely distributed on the surface. The concept of re-exposing the sample to allow the SWNTs to redistribute did not prove functional nor improve the ART when applying 10-min increments. Subsequent experiments then continued with lower sonication times.

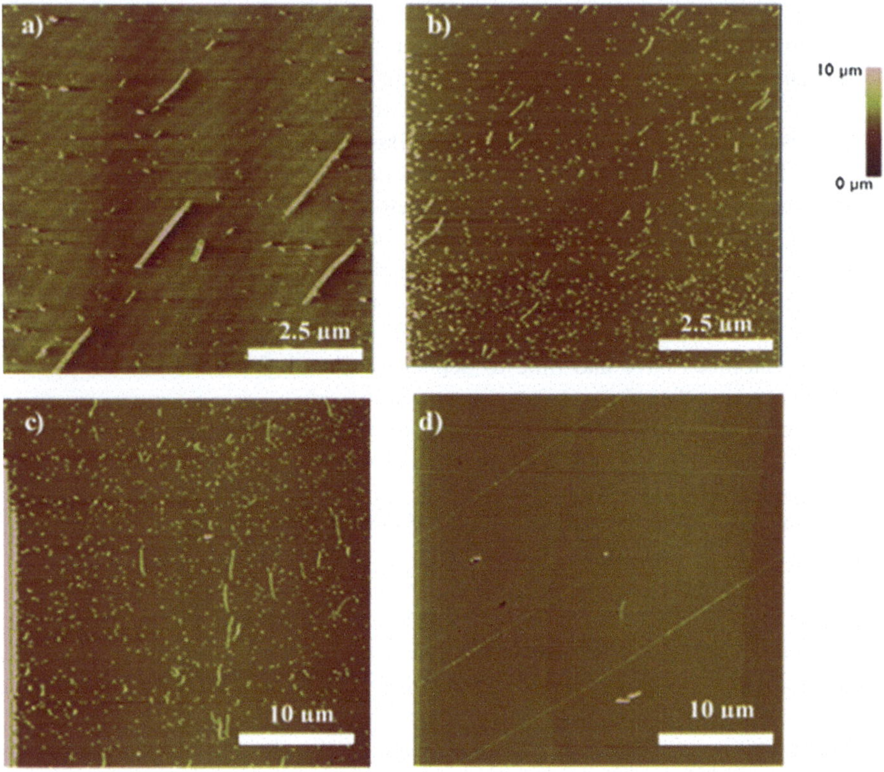

Fig. 2.3 (**a**) 10-min of sonication, (**b**) 2 × 10min of sonication with 1-h pause in-between at room temperature, (**c**) 5 × 10min of sonication with 1-h pause, (**d**) 8 × 10min of sonication with 1-h pause in between treatments

Next, sonication experiments were conducted by placing the ART-treated silicon surface in 10 mL of various solvents for shorter time frames. The time frames were separated into 5-, 10-, and 15- min sonication times, pausing for 1 h every 5 min to allow for SWNTs re-distribution. The 5 min was chosen as it was a method to identify what occurs at the halfway point between the non-sonicated samples to the 10 min sonicated sample (Fig. 2.3a). It was also interesting to determine if the 10 min continuous sonication would be different from separating into two sonication segments with a 1 h pause in between the treatments.

Additionally, because of the increase in particle impurities observed in Fig. 2.3, it was estimated that changing the solvent would allow for a cleaner sample surface. The SWNTs solution contains surfactants such as SDS that could be wrapping around the SWNTs' structure and interfering with the iptycene interactions for alignment. It was hypothesized that by changing the solvent for sonication, it would enhance aromatic stacking of the SWNTs with the ART surface by rinsing away surfactant molecules. Therefore, the sample was tested in water. It was also deliberated if the SWNTs not bound to the surface yet still in solution were causing interference for the already secured SWNTs on the surface. As a test of the surfactant solution without the presence of other SWNTs, a sample was placed in a 1% sodium dodecylbenzene sulfonate (SDBS) solution. Thus, three samples were compared: one with SWNTs in surfactant, one with surfactant and no SWNTs, and one with neither SWNTs nor surfactant (i.e., water).

The one sample was sonicated in water (Milli-Q, Fig. 2.4a) and another in the 1% SDBS (Fig. 2.4b) for 5 min. This information was compared to a sample sonicated for 5 min in the SWNTs solution (Fig. 2.4c). From these AFM images in Fig. 2.4, it appeared that the water seems to affect the integrity of the surface as various spots appear. In the presence of SDBS, it is observed that the sonication treatment increases the tendency of SWNTs to form bundles, something CNTs already have a propensity to do [2]. Perhaps this is because the added benzene ring in the SDBS structure increases the π–π stacking interactions of the nanotubes with the surfactant [3–5]. Nevertheless, sonication in the original SWNTs solution appears to distribute the CNTs more evenly in solely 5 min.

The three samples of 5 min Milli-Q, SDBS, and SWNTs were then again sonicated in their respective solvents for an additional 5 min after performing AFM (approximately 1 h). The pause also allowed for the water bath to remain at room

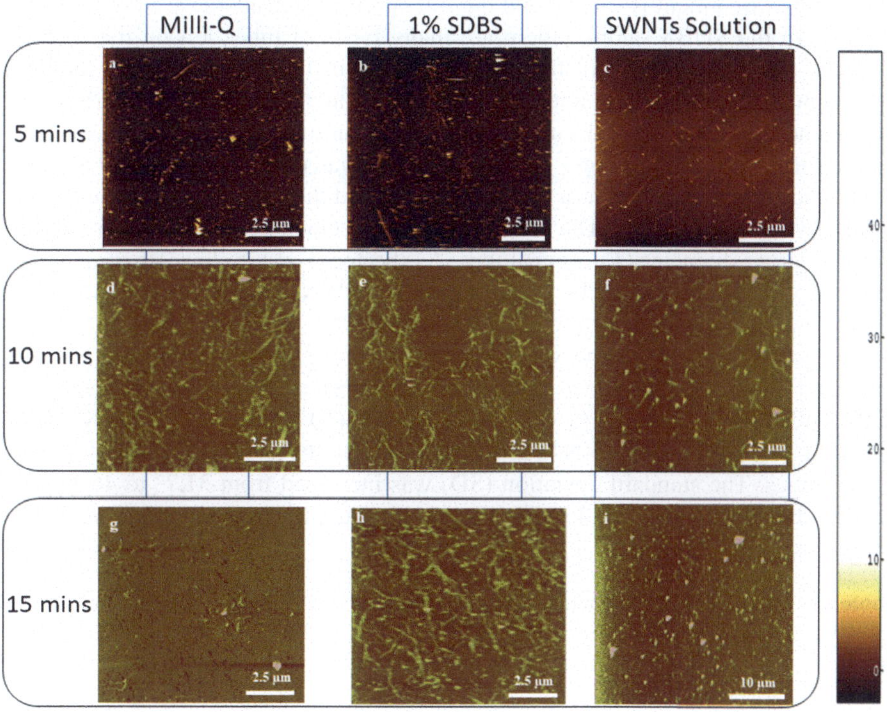

Fig. 2.4 Gradually increasing sonication of samples of Milli Q, 1% SDBS and original SWNT solution in 5-min increments from first row: 5-min, second row:10 min, third row: 15 min. Height scale bar in nanometers. (**a**) 5 mins MilliQ, (**b**) 5 mins in 1% aqueous SDBS, (**c**) 5 mins in SWNTs solution, (**d**) 10 mins in MilliQ, (**e**) 10 mins in 1% aqueous SDBS, (**f**) 10 mins in SWNTs solution, (**g**) 15 mins in MilliQ, (**h**) 15 mins in 1% aqueous SDBS, (**i**) 15 mins in SWNTs solution.

temperature. Additionally, the total time of sonication was 10 min, and it could be compared to the sample that was sonicated for 10 min continuously. It was also observed if the additional treatment had any influence on the surface (Fig. 2.4d–f). An increase in SWNTs was observed on all the samples; however, there was no alignment. Comparing the two-by-5-min treatments (Fig. 2.4) to the continuous 10-min (Fig. 2.3a), there were more SWNTs, but they were randomly distributed on the surface. The samples were again re-solvated and sonicated for an additional 5 min (15 min total sonication time in Fig. 2.4g–i).

Further analysis was conducted to affirm the qualitative observations. The average tube lengths are tabulated in Table 2.1 for each treatment. From the data, the lengths of the CNTs stay approximately the same. The degrees away from average orientation are calculated for each sample at each time frame in Fig. 2.5. An improved alignment of SWNTs is characterized by many tubes with minimal degrees away from average. The Raman RBM regions are reported for these samples in Fig. 2.6.

The Raman spectra showed that not all the original SWNTs peaks were present on the surface (refer to Fig. 1.23 in the introduction for the dropcast SWNTs sample). These results were all performed at a single laser wavelength (red laser, $\lambda = 632$ nm). In Fig. 2.6, the Raman RBM region for the three samples after a total of 15 min of treatment is shown, and the full spectra are available at the end of this chapter. In the SDBS sample, there are many types of tubes left on the surface, whereas there were fewer in the sample treated in the original CNTs solution. Although the water sample and SDBS samples did not exhibit any increases in alignment, the SDBS treated sample seems to retain many different types of tubes on the surface (diameter ranges 1.6–1.4 nm). The water sample had barely any singly aligned tubes on the surface as per the AFM, and the CNT signals in the Raman were of very low intensity. The SWNTs solution seemed to have a greater inclination for larger diameter (1.45 nm) tubes. As the focus was to have alignment and selectivity, it appeared that the sonication in the SWNTs solution itself was most effective.

The sample treated in the water was the longest, but also the least SWNTs remained on the surface. The average density of tubes on the surface changed from 0.20 SWNTs/μm^2 (5 min) to 0.38 SWNTs/μm^2 (10 min) to 0.02 SWNTs/μm^2 (15 min). The increase in SWNTs at 10-min is due to the SWNTs bundles, not individual tubes. The standard deviation (SD) was increased from 31.7° to 46.3° then remained about the same at 45.3° in the 15-min sample. For the sample treated in

Table 2.1 Average lengths in 5 min increments of Si samples with SWNTs sonicated in different solvents

Surfactant	SWNTs Solution	Milli-Q	1% wt/vol SDBS
Time sonicated (min)		avg length (μm)	
5	1.006	1.234	1.035
10	1.275	1.169	0.977
15	0.996	1.627	0.840

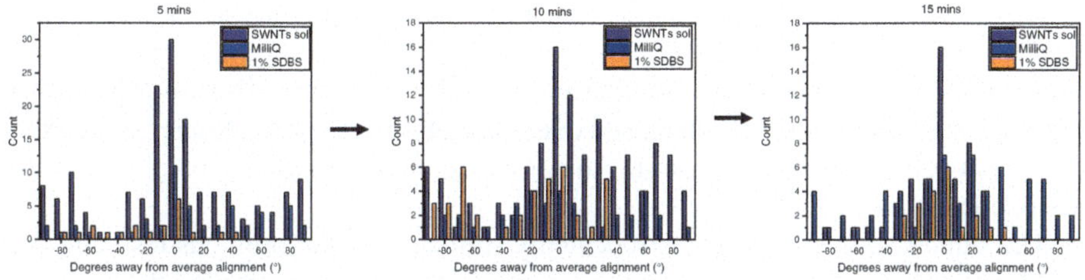

Fig. 2.5 Gradually increasing sonication of samples of Milli Q, 1% SDBS and original SWNT solution in 5-min, 10-min, and 15-min increments. The histograms represent the count of carbon nanotubes over the degrees away from average alignment

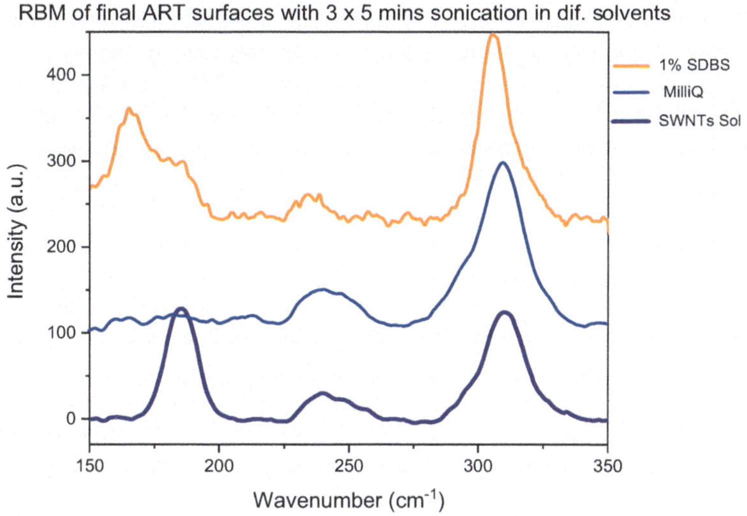

Fig. 2.6 The final ART substrates' RBM region after being treated for 3 × 5-min sonications in different solvents

SDBS, the density increased from 0.20 SWNTs/μm^2 to 0.40 SWNTs/μm^2 to 0.52 SWNTs/μm^2 by 15 min. There was a similar increase in bundling or stacking within the SWNTs. This result could be due to the additional aromatic ring on SDBS, increasing additional π-stacking between the surfactant and the SWNTs themselves. The SD also increased gradually in this sample, starting from 31.7° to 39.0°, and finally 45° for the counted SWNTs.

For the sample that remained in the SWNTs solution, then density increased from 0.28/μm^2 (SD = 38.3°) at 5-min to 0.55/μm^2 (SD = 48.13°) at 10 min to 0.73/μm^2 (SD = 58.3°) by 15-min. It was therefore beneficial for the SWNTs' density to maintain the SWNTs in the original surfactant solution from the data. Additionally, the tubes that remained on the surface after multiple sonication cycles have increased curvature throughout the SWNTs structure. It was also observed that all the samples became increasingly rough, with many large unknown particle impurities adhering to the surface the more the sample was treated. As per the alignment data (Fig. 2.5), the best distribution of alignment appeared in the SWNTs solution sample that had been agitated for 5 min; or, in other terms, sonicated one time. The SWNTs in this sample also did not have the increased curvature observed for the longer sonication times. This observation demonstrates that it was suitable to sonicate only once, with no time intervals.

The same slide was taken and repeatedly sonicated in the previous test, and the AFM was collected in between the treatments. It was also attempted to determine if simply sonicating for a set time frame in water would affect the disappearance or bundling of SWNTs on the surface. Pure CNTs are hydrophobic; however, the surfactants present in the solution could be removed in water as SDS and sodium cholate are readily dispersed in aqueous media. Therefore, it was further speculated that the water could potentially remove any surfactant molecules that interfere with the π–π stacking interactions of the nanotubes with the iptycene alignment layer [6]. It was tested once more if a more significant number of SWNTs on the surface could be obtained. Five samples were sonicated for 5-, 10-, 15-, 25- and 35-min, respectively, without any breaks in between, unlike in the first trial. Continuous treatment was done to minimize the previous contamination effects observed on the sample surface. No additional treatment of samples followed the initial sonochemical procedure in water.

The SWNT lengths and overall area density are expressed in Table 2.2. Figure 2.7 demonstrates the AFM micrographs, where (Fig. 2.7a) was before any sonication inflicted onto the sample surface. The subsequent samples shown are incremental with the time of sonication from least to most times of agitation. Following the AFM data, Raman spectra revealed a weak graphene (G)-band, which indicated that very few tubes remained on the surface after treatment (see Appendix Chapter 2). Based on these results, exposing ART aligned samples to water is not beneficial to increase SWNTs density or orientation.

Furthermore, other organic solvents were investigated and compared with the positive results from the sample sonicated for 5 min in the original SWNTs solution. These were a mix of organic polar and non-polar solvents. Dispersion studies with various organic solvents have been conducted with pristine multi-walled NTs and gave a general idea of CNT solubility [7]. The samples were tested in 1-min and 5-min sonication time frames to observe any changes. Samples were imaged before

Table 2.2 Overall lengths and density of samples treated with water at various times

	Length (μm)		# CNTs/μm^2 (area density)	
Sonication time (min) and solvent	Before	After	Before	After
5 H_2O	0.8	0.9	0.2	0.2
10 H_2O	1.7	0.9	0.25	0.2
15 H_2O	2.0	1.4	0.3	0.05
25 H_2O	1.6	0.9	0.1	0.05
35 H_2O	1.4	N/A	0.1	0

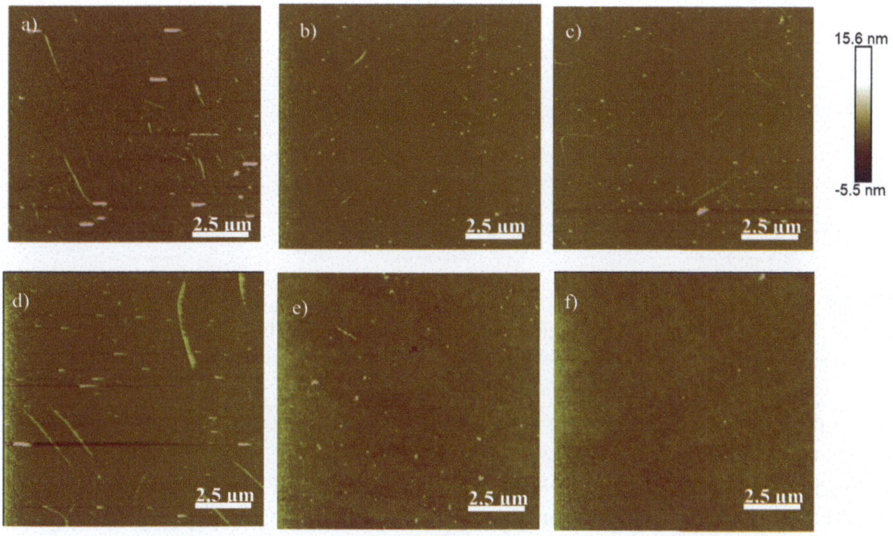

Fig. 2.7 (**a**) No sonication (**b**) 5 min with water sonication (**c**) 10 min water sonication (**d**) 15 min water sonication (**e**) 25 min water sonication (**f**) 35 min water sonication

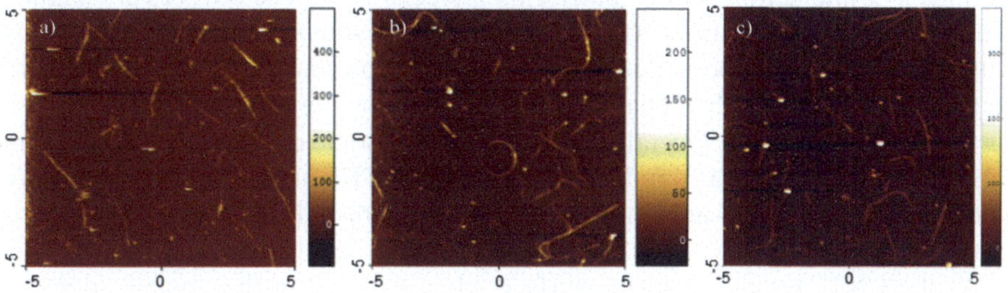

Fig. 2.8 1 min sonication of the samples in (**a**) toluene, (**b**) THF (**c**) DMSO. The y/x-range is 10 µm and the height scale bar is in angstroms

any sonication took place to ensure the samples properly represented the ART. First, the 1 min sonicated samples are shown in Fig. 2.8. The 1-min trials were done to determine if exposure to the organic solvent would dissolve away and eliminate all the tubes from the surface. The solvents were dimethyl sulfoxide (DMSO), tetrahydrofuran (THF), and toluene.

Figure 2.8a shows the appearance of the sample after 1-min of toluene sonication. There is some alignment of SWNTs, and the Raman spectrum appears very similar to ART-treated samples (Fig. 2.9a). In Fig. 2.8b, the sample treated with THF was observed to have many curved SWNTs. The AFM results matched what was observed in the Raman spectrum of this THF treated sample. As in, the D-band of the sample in Fig. 2.9b had significantly increased, indicating greater disorder in the SWNTs structure. Additionally, an appearance of bands between 1200–1600 cm^{-1}

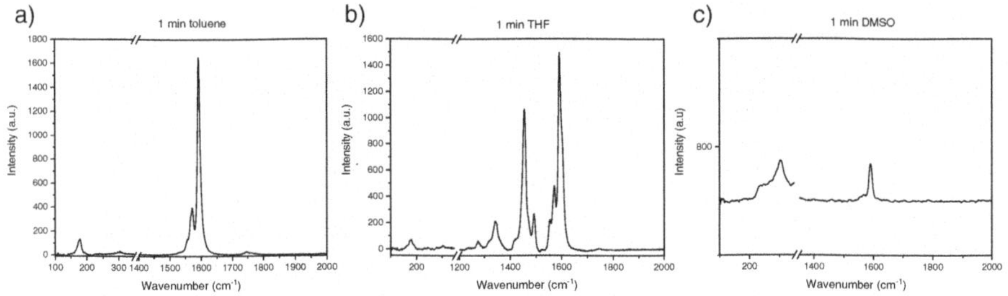

Fig. 2.9 Raman spectra of the 1 min sonicated samples at 632 nm for (**a**) toluene, (**b**) THF, (**c**) DMSO. The data shows the RBM regions and the G-band signals. In (**c**), only the molecule and Si-peaks are apparent in the RBM region, and no nanotube chiralities are identifiable

also occurred, perhaps indicating effects to the functionalized surface as these peaks are not characteristic of CNTs.

In Fig. 2.8c, the sample was sonicated in DMSO, and it also appeared to have curved SWNTs. Although the corresponding Raman of the sample (Fig. 2.9c) gives a very weak G-band signal, equating to an even weaker RBM signal, making it difficult to identify any tube types. As SWNTs usually are curvy [8, 9], the increase in curvature being observed in the nanotubes could be due to the solvents interfering with the π–π interactions of the underlying iptycene layer. The SWNTs are aligned and straightened when using ART due to the underlying sp^2 stacking interactions with molecule **1**. It is possible that some of their intermolecular interactions are being interfered with when changing the solvent, leading to this increase in tube curvature on the surface. Overall, this data provides some evidence towards certain solvents' selectivity towards individual tubes as the THF and toluene samples have a peak at 176 cm^{-1} (1.50 nm diameter tubes). In contrast, the DMSO sample seems to have completely removed this kind of tube and has a small number of curved tubes.

As the CNTs remained on the surface when exposed to the organic solvents, 5-min benchtop sonication was then tested. The solvents screened for the 5 min were DMSO, toluene, THF, acetonitrile (ACN), chloroform, and acetone. The AFM images are shown in Fig. 2.10a–f. From these images, the acetone (Fig. 2.10e) treated sample has the longest sc-SWNTs, nearing 4 μm, but they were very dispersed on the surface. There appeared to be a high degree of curvature present on the tubes that remained on the surface for all the organic solvent treated tubes. The alignment of SWNTs is only visible in Fig. 2.10e (acetone) and Fig. 2.10f (chloroform). The toluene, THF, DMSO, and ACN were not observed to be effective. The DMSO and THF seem to wash away the tubes from the surface. The toluene sample appeared the most contaminated by tiny nanoscale particles—similar to the 1 min sample (Fig. 2.8a) but with more unknown nanodots on the surface. As this solvent is the only one with a benzene-ring and the one where these particles are observed, it could be the solvent interacting with the iptycene monolayer through additional sp^2 stacking interactions. These findings dictate that 5 min in chloroform sonication could also be somewhat favorable for sc-SWNT alignment (Fig. 2.10f), but overall,

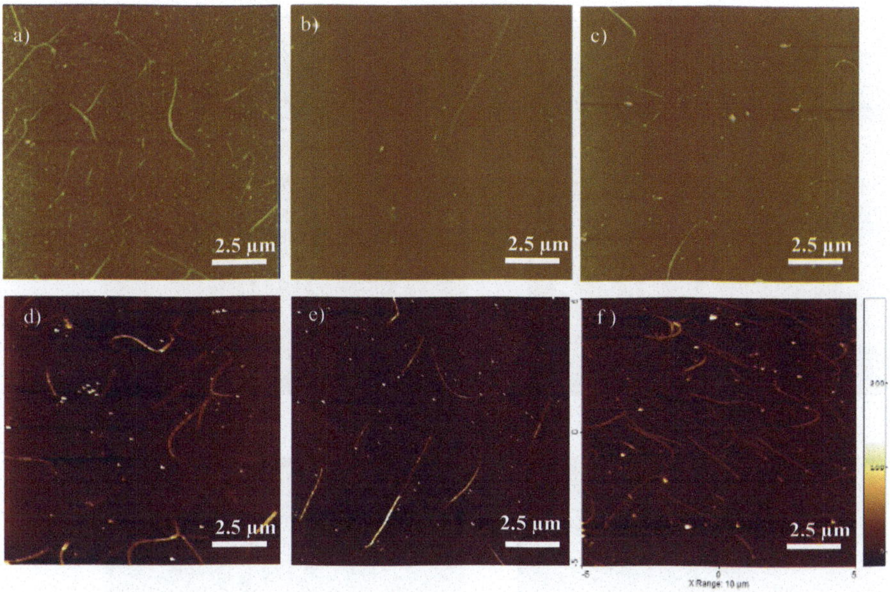

Fig. 2.10 AFM micrographs (**a**) 5 min toluene, (**b**) 5 min THF, (**c**) 5 min DMSO, (**d**) 5 min ACN, (**e**) 5 min acetone, (**f**) chloroform. Y/x-range is 10 µm for all the images

not as beneficial as merely keeping the sample in the original SWNTs depositing solution.

The expected goal of obtaining increased alignment and density of the SWNTs with minimal impurities was not achieved. Instead, the organic solvents increased the curvature of the SWNTs or decreased the density of SWNTs on the surface. Therefore, sonicating the sample in the original SWNTs solution was used for the continued screening of an optimal sonication time for SWNTs' alignment. Hence, the following experiments occurred without the removal of the sample from the depositing SWNTs medium.

The AFM results from previous treatments of sonication in the SWNTs solution are compiled in Fig. 2.11. The control sample micrograph in Fig. 2.11a was subjected to no sonication and compared with previous exploits' data [1]. In Fig. 2.11b, the sample underwent 5 min of sonication after the 24 h deposition period. Fig. 2.11c, d are the AFM micrographs of samples that underwent 10 min and 40 min of continuous post-nanotube deposition sonication, respectively.

In Fig. 2.11a, the control sample exposed to no sonication showed aligned tubes with some overlapping across the surface and various lengths. The image in Fig. 11b reveals a more even distribution of nanotubes on the surface with less stacking of tubes and evenly distributed lengths. In Fig. 2.11c, the alignment of the SWNTs remains and appears very precise, though it does appear as though the count of tubes diminishes. The AFM tip quality affected the appearance of the nanotubes in Fig. 2.11c. The height measurements were ~3 nm (shown in Fig. A.1), which are

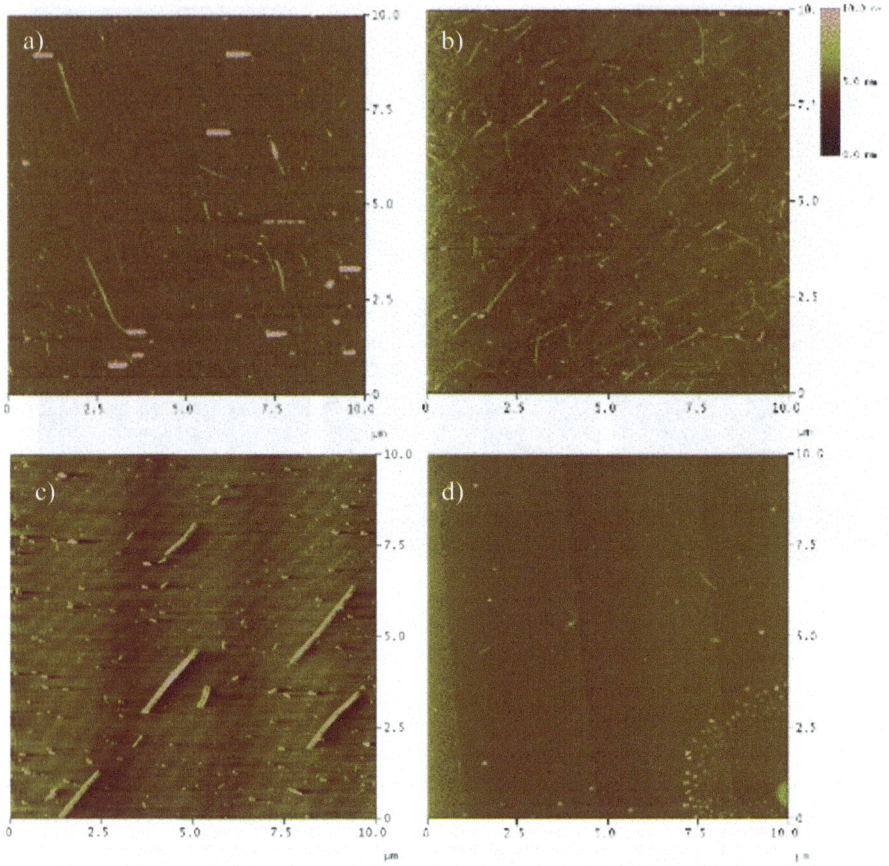

Fig. 2.11 (**a**) No-sonication—control, (**b**) 5-min sonication, (**c**) 10-min sonication, (**d**) 40-min continuous sonication. Sonication was applied after nanotube deposition

reasonable for SWNTs atop the iptycene monolayer. Samples treated with 40 min of continuous sonication (Fig. 2.11d) showed two tubes across five 20 × 20 μm AFM images. Overall, these qualitative results confirm the hypothesis that sonication can remove or displace un-aligned tubes present on the ART surface. These results indicate that sonication should be kept to a maximum of 10 min to ensure the tubes' integrity on the surface and a constant temperature in the water bath.

Afterward, investigations for an optimal time frame of sample sonication were then distributed to 1 min intervals, up to 10 min. In Table 2.3, the change in the density of nanotubes is correlated with sonication times. The data is compiled from at least ten different AFM images taken per sample. Analysis from both Fig. 2.11 and Table 2.3 indicates fewer SWNTs on the surface the more the sample is subjected to sonication. The results with increasing agitation from the micrographs in Fig. 2.11 depict overall more aligned SWNTs with respect to one another. Fewer tubes appear to be stacked on top of each other over time as well. Mainly, sonication

Table 2.3 Sonication time correlated with the density of the nanotubes on the silicon surface

Time sonication (min)	Density (μm^2)
0	0.25
1	0.21
2	0.17
3	0.14
4	0.12
5	0.13
6	0.11
7	0.10
8	0.09
9	0.10
10	0.09
40	0.01
10 × 2	0.1
10 × 3	0.05
10 × 4	0.01/3 μm^2
10 × 5	0.09/3 μm^2
10 × 6	0.07/3 μm^2
10 × 7	0.01/9 μm^2
10 × 8	0.01/9 μm^2

appears to remove loosely bound tubes that are not efficiently adhered to the surface through π–π interactions. The sound wave agitation seems to be enough to remove tubes with insufficient binding from the surface, leaving only the more aligned tubes. This observation is likely because the more aligned tubes are the ones that have better adherence to the layer of molecule **1**. The density results from the alternating 10 min sonication treatments shown from Fig. 2.3 are also included.

2.1.1.2 Orientation and Length Analysis

To further analyze the alignment of the samples, the relative orientation of each tube was obtained. From thc alignment, the standard deviation (SD) of the nanotubes was calculated. Fig. 2.12a shows that the SD of SWNTs orientation on a SiO_2 ART surface is graphed as per the sonication intervals from Table 2.3, from the 1- to 10-min test. The plot depicts the SD of the alignment of SWNTs for each different sonication time. As the graph in Fig. 2.12a shows, the dispersion in the nanotubes' orientation on the surface deviates less with increasing sonication time. The association that the alignment increased with sonication time correlates with Table 2.3, as there are fewer tubes on the surface. Still, the ones that remain are the aligned ones as per the observations in Fig. 2.11. Overall, these results demonstrate that SWNTs will remain on the surface if efficiently interacting with the ART molecules.

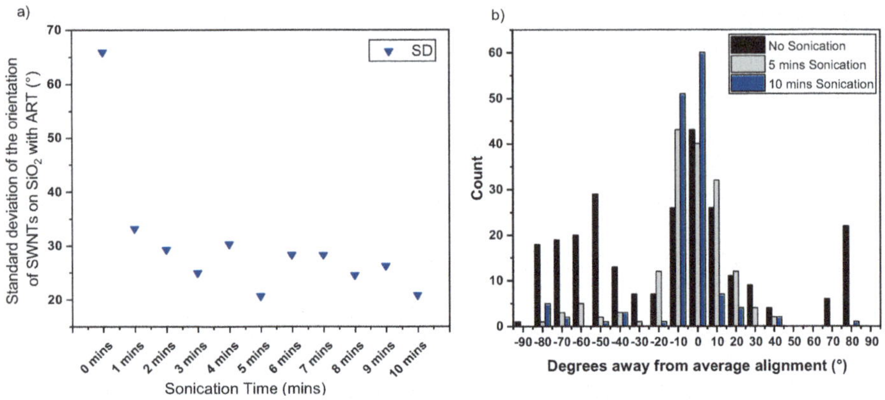

Fig. 2.12 (**a**) Standard deviation tabulated from all the orientations of each respective sample of different sonication time treatment for comparing the alignment of the samples (**b**) degrees away from average alignment for all three-time frames being compared. Represented here, 0° = 120°, and subsequent degrees outward are all the average orientation deviations

Based on the findings shown in Fig. 2.12a, the overall degrees away from average alignment for the control, the 5- and 10-min sonicated samples are tabulated to create the histogram in Fig. 2.12b. These samples were chosen as they demonstrated the overall trends of increasing alignment with decreasing SWNTs count. From Fig. 2.12b, it is apparent that there is an increase in apparent alignment with increasing sonication time.

The SD data (Fig. 2.12a) represents 68% of the counted nanotubes being within a degree of orientation. The control (0 min sonication) sample is akin to a non-Gaussian distribution, as shown in the black bars of Fig. 2.12b, hence a relatively large SD value was obtained. The mean absolute deviation, commonly used for non-Gaussian distributions, was 46.5° for the control sample, which is still higher than the SD for subsequent samples. The deviation in orientation decreases with increasing sonication, which is indicative of increased alignment. However, the number of SWNTs on the surface after 10-min of sonication decreases significantly compared to the control. Thus, the more the sample is sonicated, the more un-adhered tubes are displaced off the surface, leaving tubes that are only well-aligned with proper stacking interactions as can be designated by the connection between Fig. 2.11, Table 2.3 and Fig. 2.12.

From Fig. 2.12b, the percentage of aligned versus non-aligned SWNTs was calculated. Based on the range of orientation of nanotubes from −10° to +10°, 36.4% of nanotubes in the control sample possess a high degree of alignment. By 5 min of sonication, this amount increases to 76.7%, as some tubes are redistributed or selectively removed. By 10 min of sonication, 86.8% of SWNTs are aligned on the ART-treated surface. Hence, with sonication, the carbon nanotubes' alignment increased by two-fold with 5 min of sonication compared to the control non-sonicated sample, and over two-fold by 10 min of sonication.

The SWNT lengths on the surface were additionally analyzed and are shown in Fig. 2.13, where the nanotube length in microns is distributed with the corresponding count. The average length of tubes for the control, 5 min, and 10 min samples are shown in Table 2.4. It was observed that with sonication, the presence of shorter length tubes diminishes, and by 10 min, SWNTs with lengths below 1 μm vanish from the surface. A possible explanation of this effect is that longer tubes provide a more robust network of π–π stacking interactions with the layer of molecule **1** such that they can withstand longer sonication times and remain on the surface. This distinction between shorter and longer nanotubes indicates that sonication treatment can remove tubes that are not well adhered to the surface.

It was observed that the longer the nanotube, the stronger the interactions with the ART sample. Therefore, it is speculated that the shorter tubes have fewer interactions with the aligned molecules and are removed from the substrate. At 5 min of sonication, the tubes are majorly around the lengths of 1.42 μm, and by 10 min, the average increases to 1.78 μm SWNTs on the surface. There is approximately a 1 μm length difference between the control and 10-min sonication (Table 2.4). This change in average nanotube lengths further establishes the effects of sonication on the nanotube-bound surface. The tabulated length data depicted in Fig. 2.13 and Table 2.4 shows that the longer the sample is sonicated equates to a surface area with longer tubes.

In conjunction with the findings in Fig. 2.12 and Table 2.3, the control sample has less uniform alignment with a wider distribution of tube lengths and sizes. Sonicating for up to 10 min continuously results in the loss of tubes; however, all the

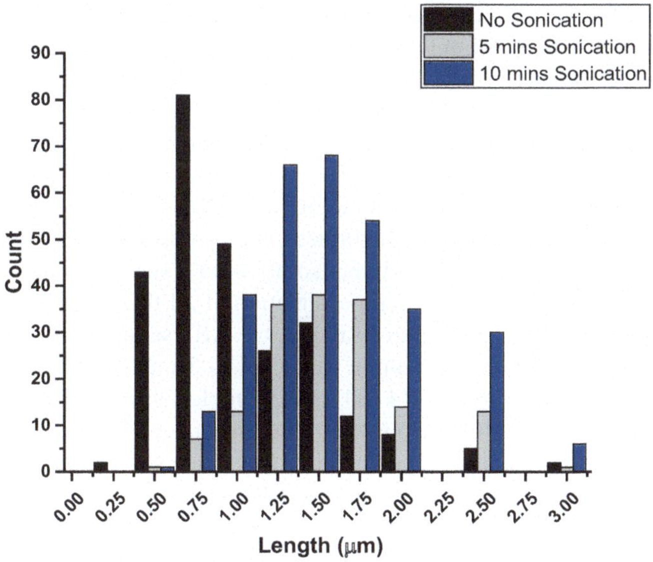

Fig. 2.13 Representation comparing the lengths of SWNTs on ART surface on control (no sonication), 5-min, and 10-min sonication samples

Table 2.4 Sonication time of the samples with average lengths

Sonication time (min)	Avg. lengths (μm)
0	0.73
5	1.42
10	1.78

tubes that remain on the surface have a higher degree of alignment and are observed to be the more extended length SWNTs. For improvement on the original non-sonicated ART sample on silica, it appears as though 5-min sonication is the best-balanced outcome between nanotube count, alignment, and length of SWNTs. The treatment was repeated on multiple occasions to verify the data was constant (Appendix Chapter 2).

2.1.1.3 Selectivity and Chirality Assessment

The results demonstrated that sonication modifications can benefit the alignment of carbon nanotubes in ART. It was also essential to characterize the tubes' selectivity that remained on the surface, and if ART with sonication shows selectivity to what SWNT types remain on the surface. Assessment of the SWNTs' chiralities was analyzed using the Raman shift RBM (ω_{RBM}) and nanotube diameter (d_t) [10–14]. It needed to be addressed if metallic (m-) and semiconducting (sc-) SWNTs remain on the surface. Does sonication provide us with more aligned NTs of the sc-SWNT variety, or is there no favoritism between removing sc- vs. m-SWNTs?

To answer this, two laser frequencies were used ($\lambda = 532$ nm and 632 nm) with Raman mapping. A 30 × 30 μm Raman mapping region was used for the control and 5-min samples, and a 15 × 15 μm mapped area for the 10-min sonication. The 10-min sonicated sample, due to having decreased density of the tubes, needed a higher amount of laser accumulations for obtaining proper peak signals. The SWNTs depicted a G-band between 1570 cm^{-1} and 1590 cm^{-1} that is relatively sharp following the literature [15]. The solution of SWNTs used was dropcast from the bottle onto a Piranha cleaned Si surface. The dropcast sample was created to compare the tubes present in the original solution on a surface to the ART-treated ones. Therefore, sample 1 (S1) is a representation of all the possible CNTs on the surface. Sample 2 (S2) is the control sample, the ART-sample with no sonication exposure. Sample 3 (S3) is the sample subjected to 5 min of sonication after the ART nanotube deposition. Sample 4 (S4) is the sample subjected to 10 min of sonication after the ART nanotube deposition. The associated Raman characterization for all the samples in both red and green laser frequencies is shown in Fig. 2.14.

The non-weighted Lorentzian fits of the RBM peaks for the SWNTs are tabulated in the Appendix of this chapter. The chiralities were determined through

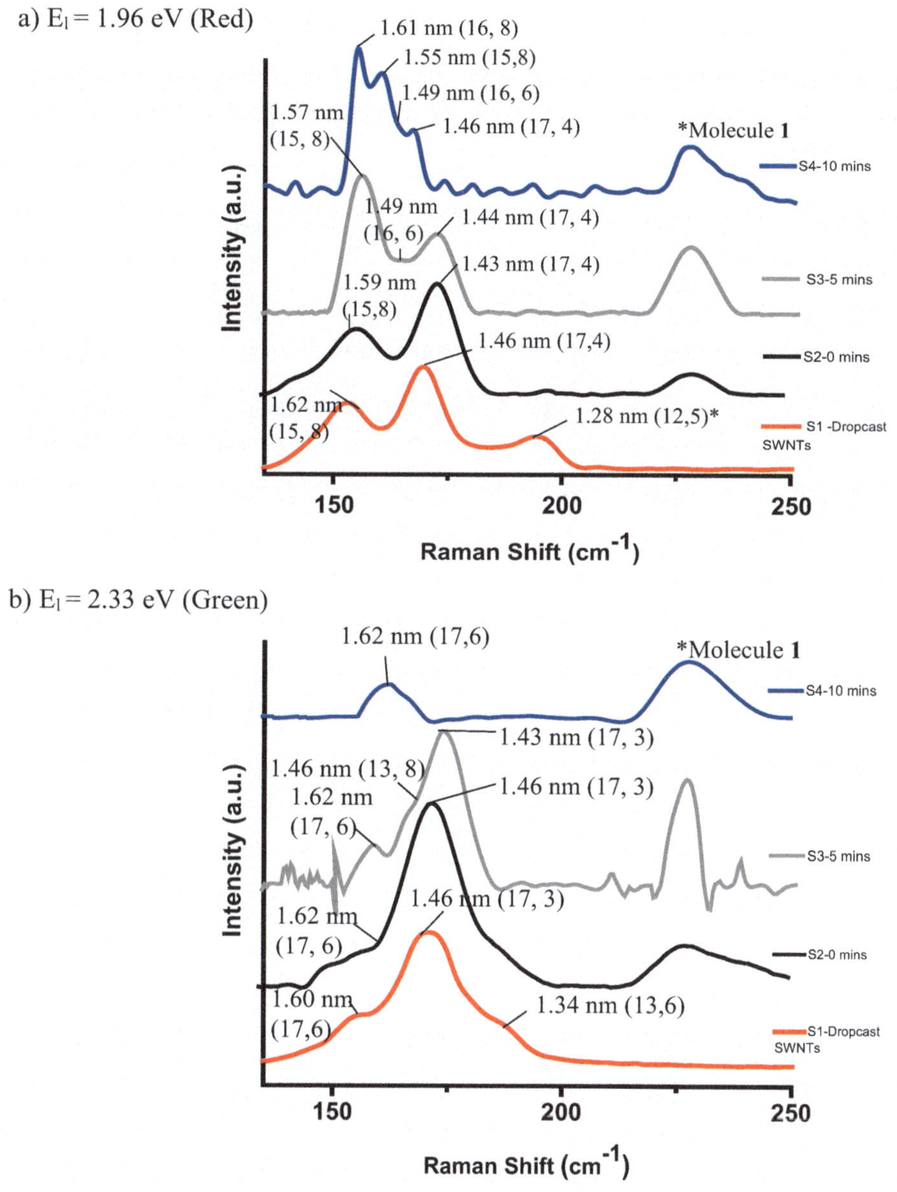

Fig. 2.14 Average Raman mapping of sonicated and non-sonicated sample (30 × 30 μm) at (**a**) red laser excitation energy and (**b**) green laser excitation energy correlated alongside the dropcasted 90% pure sc-SWNTs on silicon wafer. Peak at 233 cm-1 is the organic layer and a peak at 303 cm-1 was observed that resonates from the silicon and is used for calibration. S1 is the dropcast solution of SWNTs evaporate on silicon. S2 is the control ART sample. S3 is an ART sample with 5 min of sonication post-deposition. S4 is an ART sample with 10 min of sonication post-deposition. Diameters and chiralities for the deconvoluted peaks are indicated. *could also be (12,5) chiral nanotube

Kataura plot analysis by matching the energy separation data available from the Maruyama group website.[1] The Kataura plot is generated using γ_o = 2.9 eV, C–C bond length = 0.144 nm, and ω_{RBM} = 248/diameter. All RBMs equate to the nanotubes under dry conditions. Table 2.5 was used to identify the different metallicities of the SWNTs.

Table 2.5 Assessment of SWCNT chiralities using kataura plot as stated above

Sample laser source and sonication	Fitted RBM peak (cm^{-1})	Diameter (nm)	Chirality regions (n,m)	Closest chiral association (n,m)	Metallicity
Red dropcast	152.75	1.62	(19,5) (10,3) (15,8)	(15,8)	SC
	169.44	1.46	(17,4) (16,8)	(17,4)	SC
	194.27	1.28	(9,9) (12,6) (12,5)	(12,6) or (12,5)[a]	M or SC[a]
Green dropcast	154.98	1.60	(15,8) (17,6) (17,6)	(17,6)	SC
	184.82	1.34	(11,9) (12,7) (13,6)	(13,6)	SC
	170.74	1.45	(14,7) (18,1) (11,10) (17,3)	(17,3)	SC
Red no sonic	156.45	1.59	(12,11) (15,8)	(15,8)	SC
	173.40	1.43	(11,10) (17,4)	(17,4)	SC
	150.60	1.65	(17,6) (10,3)	(10,3)	SC
Green no sonic	153.07	1.62	(17,6)	(17,6)	SC
	172.66	1.44	(11,10) (14,7) (18,1) (17,3)	(17,3)	SC
Red 5 min	158.28	1.57	(18,2) (15,8) (13,11)	(15,8)	SC
	166.74	1.49	(16,6) (15,8)	(16,6)	SC
	171.86	1.44	(17,4)	(17,4)	SC
Green 5 min	158.42	1.57	(17,6)	(17,6)	SC
	169.63	1.46	(11,10) (14,6) (14,7)	(13,8)[a]	SC[a]
	175.19	1.42	(17,3)	(17,3)	SC[a]
Red 10 min	153.88	1.61	(16,8) (13,11) (10,3)	(16, 8)	SC
	160.04	1.55	(15,8)	(15,8)	SC
	165.05	1.50	(15,8) (16,6)	(16,6)	SC
	169.00	1.47	(17,4)	(17,4)	SC
Green 10 min	153.00	1.62	(17,6)	(17,6)	SC

M metal, *SC* semiconductor

[a]No value close on the Maruyama group website, so identified using work by Weisman and Bachilo [36]

[1] Kataura-Plot for Resonant Raman http://www.photon.t.u-tokyo.ac.jp/~maruyama/kataura/kataura.html.

The Raman spectra in Fig. 2.14a (red laser, $E_l = 1.96$ eV) and Fig. 2.14b (green laser, $E_l = 1.96$ eV) reveal carbon nanotubes' presence in the 150–200 cm^{-1} range, corresponding to SWNTs in the diameter range of 1.25–1.65 nm. The lack of a peak at 218 cm^{-1} indicates the absence of SWNTs bundles [16]. The organic molecule appears as a peak at approx. 233 cm^{-1}, and grows in relative signal intensity with decreasing nanotube density. Additionally, using pilot Kataura analysis based on $E_L \sim E_{ii}$ from the resonance condition [13] with constant $\kappa = 2.22$ [14, 17, 18], following previous assignment methodologies [10, 11], specific nanotube chiralities can be estimated. The diameters of the SWNTs are also calculated, based on the ω_{RBM}-diameter relation equation [10]. Mainly, from $d_t = 1.28$–1.62 nm found in the dropcast sample, ART alone is selective for nanotubes in the diameter range from 1.40–1.62 nm.

Firstly, the dropcast sample in both the red and green frequencies contain at least three prominent peaks that are very broad and overlapping. Large broadness between peaks indicates other nanotubes that do not have as strong a Raman intensity due to the number of these tubes or not vibrating within the laser frequency strongly enough. In the dropcast SWNTs, diameters for tubes of 1.62, 1.46, and 1.28 nm are present in the red laser region. Similarly, the highest signal SWNTs in the green region have diameters within the same range: 1.60, 1.46, and 1.34 nm. However, even though the diameters for tubes may be similar, they can still have different metallicities, as their conductive properties are ultimately dependent on their chirality.

In Fig. 2.14a, the dropcast sample has prominent peaks for semiconducting tubes (15,8) and (17,4), and what could either be a (12,6) metallic or (12,5) semiconducting SWNTs at the 194 cm^{-1} position [19]. The smaller diameter SWNTs are removed when subjected to ART, so regardless if it was an sc- or m-SWNTs, it is selectively detached and therefore no need to further investigate its true conductivity. These findings correspond with previous results [1]. The diameter tubes of 1.57 nm (15,8) and 1.44 nm (17,4) remain on the surface. The dominant intensity peak is the 1.44 nm SWNTs in both the control and dropcast samples.

Upon treatment with simple sonication for 5-min, the same peaks remain, however, in opposing ratios. The peak around the 1.57 nm region (15,8) has increased intensity over the 1.44 nm peak. With Lorentzian peak fitting, a point in between the two dominant peaks can be revealed. This observation leads to the assessment that there are underlying tubes beneath the more dominant and intensely Raman resonating SWNTs. For the 5-min plot in Fig. 2.14a, the presence of a nanotube peak with a diameter of 1.49 nm can be deconvoluted to a (16,6) sc-SWNTs. This tube was likely present in the dropcast and control sample but did not have a strong enough signal for characterization.

In turn, the 10-min sonicated sample shows a smaller, sharper peak in the red wavelength. This result is likely related to the previous observation that there are fewer SWNTs on the surface. The diameters are all within the same range as the previous samples (±0.2 nm) and correspond to the same chiralities of nanotubes. Now a distinction can be made between the (15,8) tube that corresponds with the 1.55 nm diameter peak, the (16,6) sc-SWNTs that resembles more with the 1.50 nm diameter tube, and a 1.60 nm (16, 8) sc-SWNTs. The (17,4) tube peak is still present on the surface. Hence, with ART and sonication, the red laser frequency spectra represent that there is selectivity for larger diameter tubes. Certain chiralities appear to have a high relative intensity in the Raman spectrum throughout the sonication process, such as the (16,8) tube. In contrast, other peak intensities decrease with more exposure to sonication, such as the (17,4) SWNTs. Over time, the decrease in signal intensity indicates that there are fewer smaller diameter tubes present on the sample surface.

For the higher energy green laser light, the spectra are all shown in Fig. 2.14b. For the dropcast sample, similar diameters are found to the red laser sample, but the peaks are closer together, increasing broadness. The larger RBM smaller diameter peak at 1.34 nm is no longer present after ART functionalization, much like Fig. 2.14a, and pertains to the (13,6) sc-CNT. Furthermore, in Fig. 2.14b, the center peak of 1.46 nm diameter becomes more intense and represents the (17,3) tube. The smaller tube possesses a 1.60 nm diameter and can be determined to be the (17,6) SWNTs. Furthermore, the control at green laser frequency shows only two chiralities of tubes: (17,3) and (17,6) to be present after ART. These are different from the (15,8) and (17,4) peaks that remain on the control red laser Raman in Fig. 2.14a, and this observation is due to different SWNTs being more Raman active at different laser intensities [11–14, 17, 20].

With sonication, it is possible to differentiate that there are both 1.46 nm (17,3) and 1.43 nm (13,8) tubes and that the larger diameter tube at 1.62 nm (17,6) remains and grows to be more intense. The peak of 1.43 nm was likely present throughout but could not be deconvoluted from being too broad of a peak and close to another signal of similar diameter (1.46 nm). By 10 min of sonication, however, only one peak is distinguishable, and it is the peak at 1.62 nm diameter, correlating with the (17,6) tube, although the signal-to-noise was quite high for the last green laser sample due to the loss of density of tubes. The red laser source was much more efficient for the use of chiral assessments for the nanotubes. However, both laser strengths show that larger diameter peaks are preferable and that other SWNTs chiralities can be observed on the surface using deconvolution and sonication. The sc-SWNTs identified here have also been confirmed to be in the 90% sc-SWNTs mixture by Nanointegris, Inc.

From the characterization results regarding Fig. 2.14, there are no longer any RBM intensities in the red past 175 cm [1], and respectively, fewer nanotube peaks in the green spectra in about the same Raman shift range. Mainly, from d_t = 1.28–1.62 nm for the dropcast sample, ART alone is selective for nanotubes in the diameter range from 1.43–1.65 nm. With sonication, the diameter range of the nanotubes is still around 1.4–1.6 nm. Nevertheless, more nanotube types are distinguishable in the red laser spectrum versus the control due to fewer tubes on the surface (Fig. 2.14a). By 10 min of sonication, very few tubes remain that are green laser light active (Fig. 2.14b), showing only nanotubes in the diameter range around 1.6 nm. Therefore, longer sonication times result in fewer tubes on the surface, allowing for the analysis of a broader range of SWNTs over the sample surface due to less overlapping relative intensities.

Also, this correlates with the longer tube lengths observed with AFM images represented in Fig. 2.13. It seems that nanotubes that could not previously be detected now have an observable Raman intensity, as other tubes are agitated off the sample's surface. From the data in Fig. 2.14, ART with sonication appears to be further selective for a more restrictive set of sc-SWNTs. There is selectivity for chiral sc-SWNTs of a large diameter using sonication.

The observation that the nanotubes are exclusively selective for the larger diameter tubes between 1.4–1.6 nm in diameter is limited as there are no larger diameter tubes past the 1.6 nm range. We hypothesize that the smaller tubes around 1.2 nm in diameter are removed as they may bind less efficiently to the concave structure of the iptycenes as their larger counterparts. However, it is interesting to note that the vertical distance in the AFM images (see Appendix) exhibits an increase in overall height in nanotubes. This observation is likely due to the remaining surfactant molecules that deposit on top of the sample's surface. This observation is interesting for future device applicability and is to be studied in our future work.

2.1.1.4 Additional Analysis

As an added test, if only sonication is applied to the sample without ART, no nanotubes and no alignment are present on the surface, as shown in the Helium Ion Microscope (HIM) image Fig. 2.15a. To add, the propensity for SWNTs to bundle is very likely, and ART helps to negate that inherent bundling. In Fig. 2.15b, a sample is treated with 5-min of sonication after being Piranha cleaned and

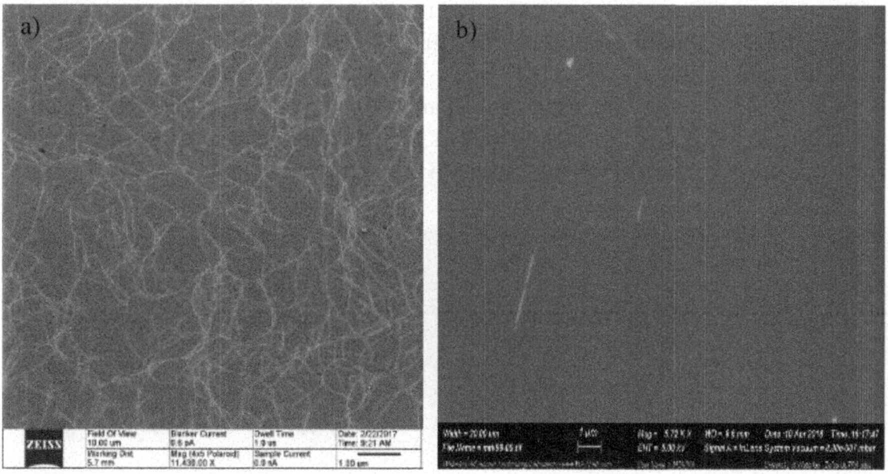

Fig. 2.15 (**a**) HIM resolution image of SWNTSs on bare silicon without any aligned organic molecule present on the surface, 10 × 10 μm image. (**b**) The SEM In-lens 5 kV image at 20 × 20 μm of a silicon surface that has not been functionalized with the ART but does have 5-min sonication applied to it after a 24 h SWNTSs deposition period. These were the only tubes to appear on the entire substrate surface (3 × 10 mm)

submerged in the sc-SWNTs solution for two days. There are barely any nanotubes that remain on the surface. Therefore, sonication must be combined with the aligned iptycenes to influence the orientation of the CNTs.

Additionally, SEM was explored as a faster way to identify alignment. Figure 2.16a is an unfunctionalized silicon surface with NTs (dropcast, then rinsed with MilliQ), whereas Fig. 2.16b contains the organic layer. Both are In-Lens images using 5 kV, so there could be some damage to the SWNTs, as only the 1 kV strength has been determined not to damage the tubes [21]. The sample without organic layer appeared more randomly disordered and dirty (Fig. 2.16a), and the one with the organic layer demonstrated some degree of orientation present as well as possess a higher tube density, indicative that ART-treatment increases the likelihood of not only NT alignment but also the number of NTs on the surface.

For the nanotubes, without any gold coating, it is somewhat surprising the level of detail and the ease of use of the SEM towards imaging these types of samples. Even with an organic monolayer present on the surface, there remains some degree of conductivity between the 90% semiconducting CNTs, although it is not possible to dictate which are sc-SWNTs vs. m-SWNTs (Fig. 2.16a, b). The lengths of the tubes are determined to be from 700 nm to 1.5 μm using ImageJ software to obtain the lengths. The SEM was quick and efficient in obtaining the position of the tubes on the surface. However, it was observed that the resolution of a single SWNT was challenging to obtain, and the sample surface appeared to darken very quickly after taking a single image. It also took much longer to adjust the focus due to the charging (darkening) effect, potentially damaging the tubes. Additionally, the tubes' diameter was not obtained (require AFM or tilting the SEM sample to extrapolate an angle that gives the z-data).

Therefore, AFM was continuously used for alignment analysis for the ART samples and to obtain high-resolution micrographs with a rough estimate of height (diameter).

It was also attempted to observe any changes in conductivities using conductive (c-)-AFM and SEM. For the c-AFM images, two different scan sizes are depicted in Fig. 2.17 where (a) and (c) are the conductivity micrographs, and (b) and (d) are the

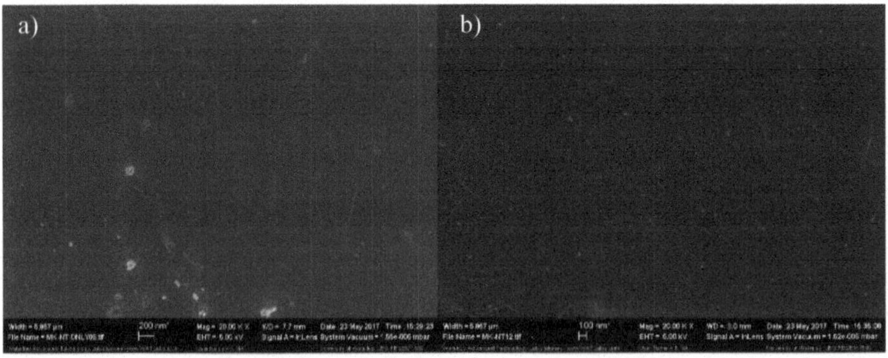

Fig. 2.16 In-lens images of carbon nanotubes on (**a**) silicon surface, (**b**) ART treated silicon surface

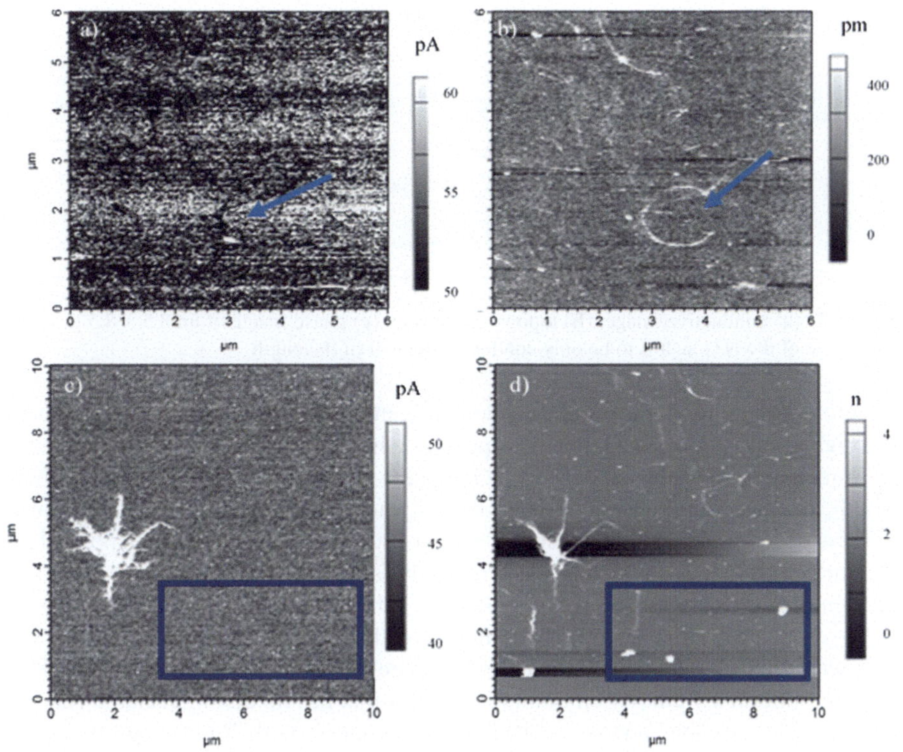

Fig. 2.17 6 × 6 µm image of (**a**) conductivity and (**b**) height image of NTs on silicon. The 10 × 10 µm image of (**c**) conductive image and (**d**) height image of NTs on silicon

topography images. The brighter the color, the more conductive the specimen is on the substrate surface; thus, the dark regions are insulating. The same is observed about the topography images, in that the brighter areas are higher in topography that the darker regions. Some particles seem to be high in terms of height but do not appear to possess conductivity and hence result in darker regions in the c-AFM image.

From the information in Fig. 2.17, the curved nanotube, which is apparent in Fig. 2.17b as white, appears as a dark region in Fig. 2.17a only where it is straight. This observation indicates that the conductive properties of the nanotubes are only present when the tube possesses a straight orientation. If it is curved, it becomes less conducting such that it appears as an insulator. This observation is also depicted in the literature [22, 23]. It is one of the reasons these NTs need to be aligned in a specific orientation for transistor applications because if the NTs curve unpredictably, they can no longer be used for efficient electrical transport [24]. Additionally, in Fig. 2.17d, there are some round particles present on the sample surface that are very bright, indicating they are very high on the surface. The same spots are not present in Fig. 2.17c and Fig. 2.17d, as represented by the blue boxes, and are

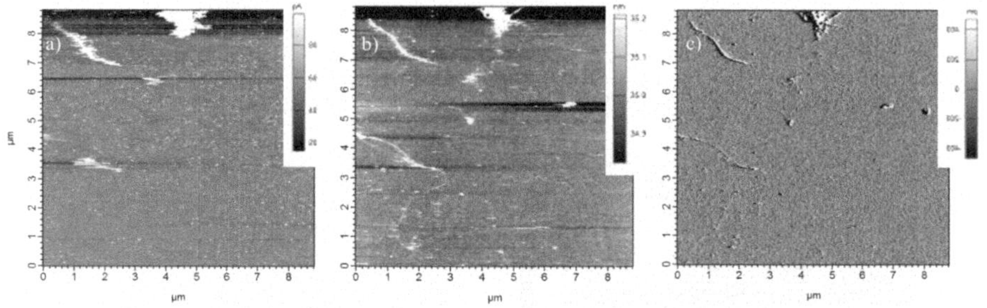

Fig. 2.18 (**a**) conductive image, (**b**) topography image, (**c**) phase image of an 8.5 × 8.5 μm area where one of the NTs seems to be only conductive on half of its length

therefore non-conducting, insulating contaminants [25]. Some tubes also appear to be brighter, indicating that some NTs possess a higher conductivity than others.

Similarly, some tubes appear to be conductive only in certain portions of their length, and then their brightness seems to dissipate, as is shown in Fig. 2.18. Because of these observations, and the fact that a mix of sc- and m- SWNTs are deposited on the surface, it makes sense that the brighter NTs would be metallic tubes whereas the semi-bright may be the semi-conductive tubes, though this would have to be confirmed with Raman spectroscopy that provides a higher degree of analysis for differentiating sc- and m-SWNTs [11]. Nevertheless, it is challenging to go back to a single spot on a 1 cm × 1 cm unmarked wafer that has no distinguishable characteristics. Perhaps this sort of c-AFM, in combination with Raman spectroscopy, could be revisited with ART on marked wafers for precise conductivity and chirality assessments.

2.1.2 Summary of Sonication Experiments

In summary, sonication can improve the surface orientation and selectivity of SWNTs in conjunction with ART. Five minutes sonication proved to be most effective in terms of increased alignment and minimal loss of tubes on the surface. Selectivity is shown for larger diameter semiconducting carbon nanotubes, as the (13,6) and smaller tubes under approximately 1.3 nm in diameter were removed immediately open sonication exposure. By 5 min of sonication, this amount increases to 76.7%, as some tubes are redistributed or selectively removed. No change in the solvent is necessary for sonication to be effective with ART. With prolonged sonication, fewer tubes remain on the surface, and red laser frequencies had higher intensity RBM peaks.

With the result of fewer tubes on the surface, it also appears as though the π–π stacking interactions between the iptycenes and the CNTs are weakened, and therefore, can be removed with prolonged agitation. In particular, the organic solvents add curvature to the tubes, which additionally impacts the π stacking. The Raman analysis demonstrates the selectivity between ART and the addition of sonication for SWNTs on the surface of silicon. The sonication step was shown to select for longer nanotubes lengths greater than 1 micron, also possessing wider diameters. Overall, it is possible to affirm that larger diameter tubes remain on the surface after sonication, and smaller diameter nanotubes are agitated away from the surface. More mixtures of less pure SWNTs should be tested in ART for further developing the selectivity of the method. This work lays the foundation for future simple alterations that can be done to ART to increase its effectiveness in simultaneous nanotube alignment and selectivity for FET applications. These results are promptly beneficial for the simultaneous alignment and sorting of single-walled carbon nanotubes. However, because the density of the SWNTs on the surface was not improved, changes were then explored before the deposition of nanotubes on the surface (pre-deposition).

2.2 Post-carbon Nanotube Deposition: Layer-by-Layer (LBL) Treatment

A form of the Layer-by-layer (LBL) dipping technique was also attempted as it was hypothesized this treatment could increase the density of SWNTs on the surface through repetitive exposure of the nanotubes to the functionalized surface [26, 27]. First, the solution of nanotubes was identified to have a slightly negative charge of −28.98 mV through zeta-potential analysis (measurements were averaged from 5 different trials, shown in Appendix Chapter 2). After having determined the solution was of a negative charge, a polyethylenimine (PEI) aqueous solution (2 mg/mL) was made as a positive counterpart. An ART functionalized silicon surface was placed in the positive solution for 10 min, rinsed with MilliQ water, then placed into the original SWNTs solution for 10 min, as represented in Fig. 2.19.

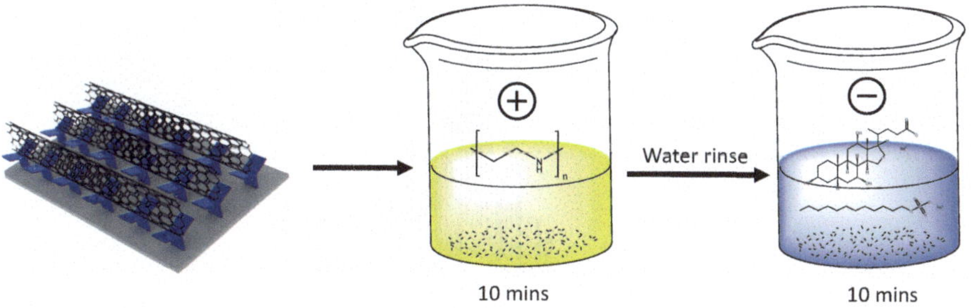

Fig. 2.19 Illustration of a functionalized surface being treated with two differently charged solutions. The positive is an aqueous PEI dispersion, and the negative is an aqueous surfactant dispersion of 90% sc-SWNTs

2.2.1 Results and Discussion

The resulting AFM micrographs are shown in Fig. 2.20 alongside the alignment data. The sample was an SiO_2 wafer with ART using **1/5CB** with 5-min of sonication (Fig. 2.20a). After a single treatment, the surface had additional impurities that increased surface roughness, and there was no increase in SWNTs alignment (Fig. 2.20b). At the recommended five total treatments, the trend continued, and the samples did not exhibit increased alignment of the SWNTs (Fig. 2.20c). The alignment data is shown in Fig. 2.20d, comparing the degrees away from average alignment for all the samples. A new solution of **1/5CB** was used along with newly purchased SWNTs.

The number of nanotubes (count data) shown in Fig. 2.20d is representative of all the tubes from eight AFM images of the untreated sample, five images from the one round LBL, and two from the five rounds of LBL. The density of SWNTs on the surface after one LBL treatment was 0.35 NTs/μm^2 (SD = 42.2°, 43.2% SWNTs 0 ± 10°), and increased to 0.76 NTs/μm^2 (SD = 51.7°, 40.3% SWNTs 0 ± 10°) by the end of five cycles. No LBL treatment resulted in a density of 0.25 NTs/μm^2

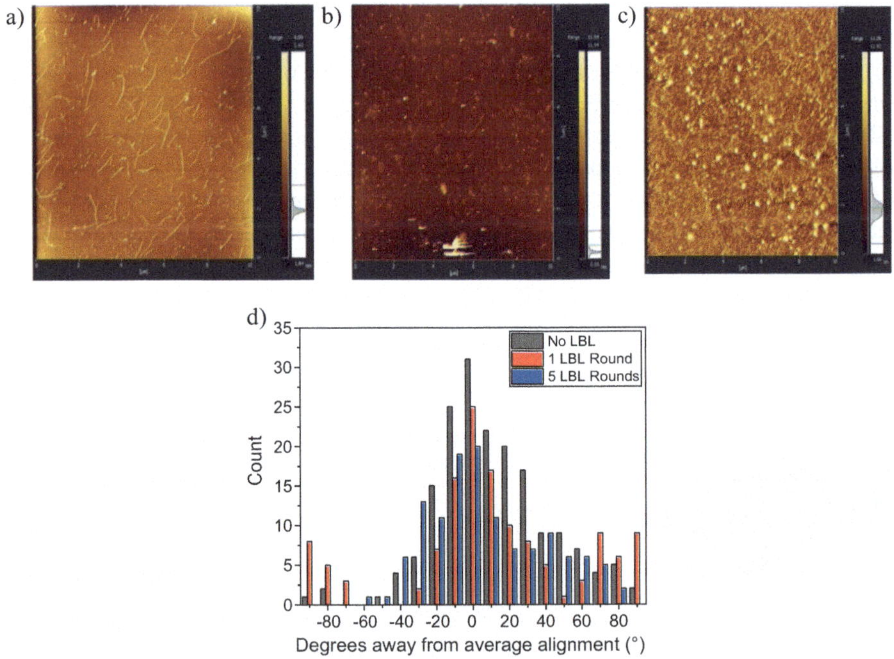

Fig. 2.20 Results from applying LBL treatment to ART samples, AFM micrographs of (**a**) ART silicon surface without any LBL treatment, but had sonication 5-min, (**b**) with one cycle of dipping into the positive and negative solutions, (**c**) repeating LBL cycle for a total of five times. All the AFM images are of the same sample. (**d**) Histogram showing the degrees away from the average alignment of all the samples together

(SD = 30°, 55.7% SWNTs 0 ± 10°, 5 min sonicated). These numbers indicate that although there were fewer AFM images to gather data from the 5 LBL sample, the number of nanotubes on the surface is still very high. Although the LBL method did not increase the alignment, it did increase the number of CNTs present on the surface. The surface roughness also significantly increased, which made it increasingly challenging to obtain AFM data without the tip breaking or smearing the images. The SWNTs in Fig. 2.20c are also much higher in height (z-data), indicating SWNTs stacking on top of themselves. Therefore, LBL treatment with ART results in increased density of tubes on the surface but decreases their uniformity and alignment.

2.2.2 Langmuir-Blodgett Attempt

An effort was also made to incorporate the Langmuir-Blodgett (LB) method in combination with the ART in order to obtain a higher concentration of aligned CNTs on the surface. The LB method for deposition has been popular for the formation of aligned CNTs on surfaces [28–30]; hence it was deemed an interesting direction to complement ART. Both an already finalized ART sample with CNTs and a silicon wafer with only the iptycenes were attempted. The samples were cleaned by exposure to $O_{2(g)}$ for 15 min, UV for 5 min, and $N_{2(g)}$ for 15 min. The sample was then dipped into a thin layer of 1% concentrated sc-SWNTs (95% purity) in chloroform with Milli-Q water. The hypothesis was that the organic functionalized surface would readily combine with the organic solvent dissolved CNTs. However, upon SEM of both different surfaces, it was apparent that a thin film of randomly disturbed SWNTs adhered to the surface and not of aligned CNTs, as shown in Fig. 2.21a, b. The LB method could once again be attempted with various other dipping conditions or solutions of dissolved CNTs for further investigations.

a) b)

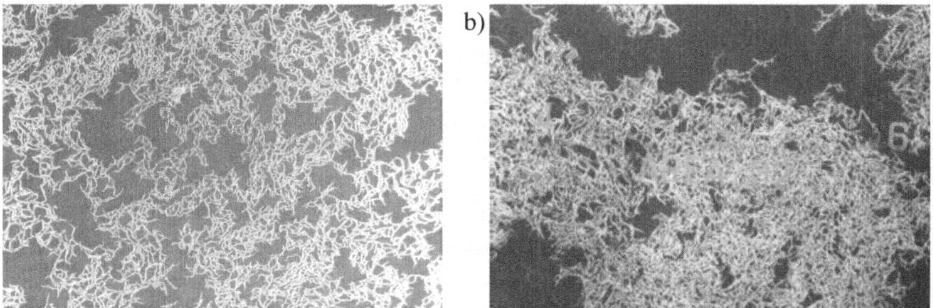

Fig. 2.21 (**a**) sample that only had a monolayer of iptycenes treated with the LB method, (**b**) sample functionalized with ART and CNTs then subsequently treated with LB method

2.2.3 Summary

Overall, although applying LBL was hypothesized to be an easy way of increasing SWNTs density and works well for nanosheets [26, 27], it is ineffective with the ART method. The number of tubes did increase on the sample surface, but they were in an increasingly random orientation. However, it was interesting to observe that when a newly prepared solution of **1**/5CB and SWNTs solution was used, the results for the 5-min sonicated data is also increased in density and SD, but the overall % of aligned SWNTs is 55.7%. The percentage is decreased from the obtained values in Sect. 2.1.1.2 but is approximately 1.5× the typical ~35% obtained when not applying sonication to the ART samples. The LB conditions could also be promising in increasing the density of CNTs if optimized in terms of solvents and dipping conditions.

2.3 Pre-carbon Nanotube Deposition: Filler Molecules

Post-deposition sonication for ART proved to be successful in terms of having a higher degree of aligned CNTs on the surface, whereas layer-by-layer dipping was not efficient. In this section, the attempted investigations of increasing the density of the SWNTs in the ART processes is investigated through a pre-SWNT deposition methodology on the silica substrate.

2.3.1 Concept and Experiment

The ART method covers an oxidized substrate with a monolayer of iptycene molecules that have π–π stacking interactions with carbon nanotubes. Without the iptycenes, it has been proven that there is no alignment of the SWNTs [1, 31]. However, the density of the deposited tubes still does not cover the entire surface. As observed through the many AFM images of many ART samples, certain areas of the substrate have an even layer of aligned SWNTS while others do not. This observation indicates that it is highly likely that the surface has gaps between the iptycenes, where SWNTs are not adhering to the surface and, therefore, not remaining there. As a way to fill in the gaps between the iptycenes and therefore increase the likelihood of π–π stacking interactions with CNTs, one idea is to deposit an extra layer of molecules that have similar binding strength and sp^2 backbone.

These gap-filling or "filler" molecules could potentially increase the number of deposited CNTs, without interfering with the original iptycene monolayer. The change is shown in Fig. 2.22a. Ideally, the addition of these molecules would help strengthen the iptycene monolayer for aromatic stacking interactions with the SWNTs. As the goal was to "fill-in" any spaces where the original molecule **1** did

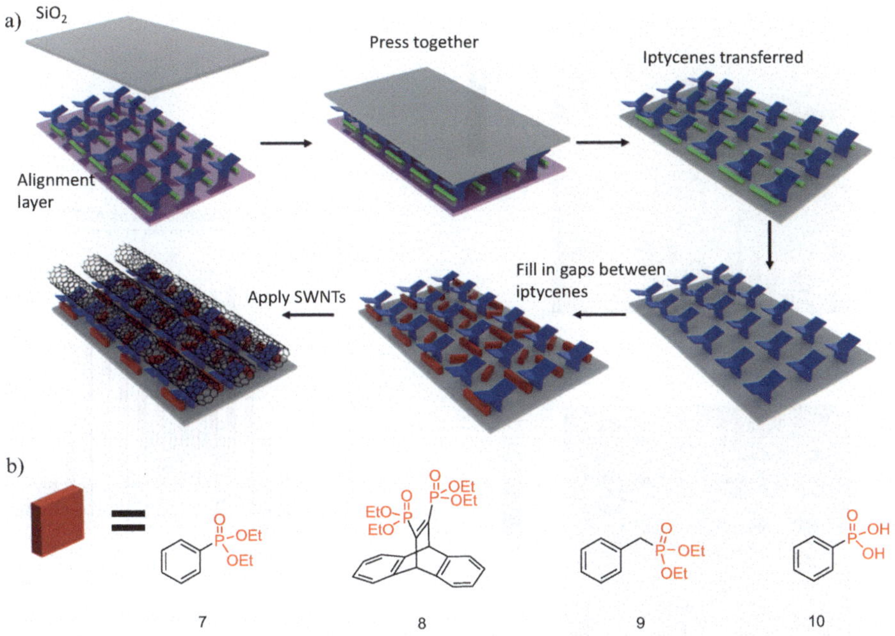

Fig. 2.22 (**a**) Graphical representation of the concept of modifying ART using intermediary molecules represented as red squares for SWNTs alignment, (**b**) attempted filler molecules 7–10 to test the hypothesis

not bind, molecules were chosen as smaller versions of the iptycene alignment molecule. Primarily, they had a similar binding group, such as a phosphonate ester or phosphonic acid, and an aromatic backbone for π–π stacking. The molecules explored are shown in Fig. 2.22b.

2.3.2 *Results and Discussion*

Diethyl phenylphosphonate (**7**) was a readily available compound and easy to test adaptation as it is a liquid at room temperature. The first attempt consisted of drop-casting **7** on an ART functionalized surface prior to the addition of the CNTs. The solution on the sample was then left to evaporate. The method was also attempted by allowing the substrate to dry in the oven, as a form of low-temperature annealing (120 °C) to be more time-efficient. The sample was rinsed with DCM and treated two more times, for a total of three drops of the filler molecule.

The degrees away from average alignment for the two different drying methods are showcased in Fig. 2.23c alongside a representative AFM image (a, b). The density with room temperature drying of the nanotubes was 0.28 SWNTs/$\mu m^{2,}$ whereas, with oven drying, the number of nanotubes was 0.52 SWNTs/μm^2. For the air-dried

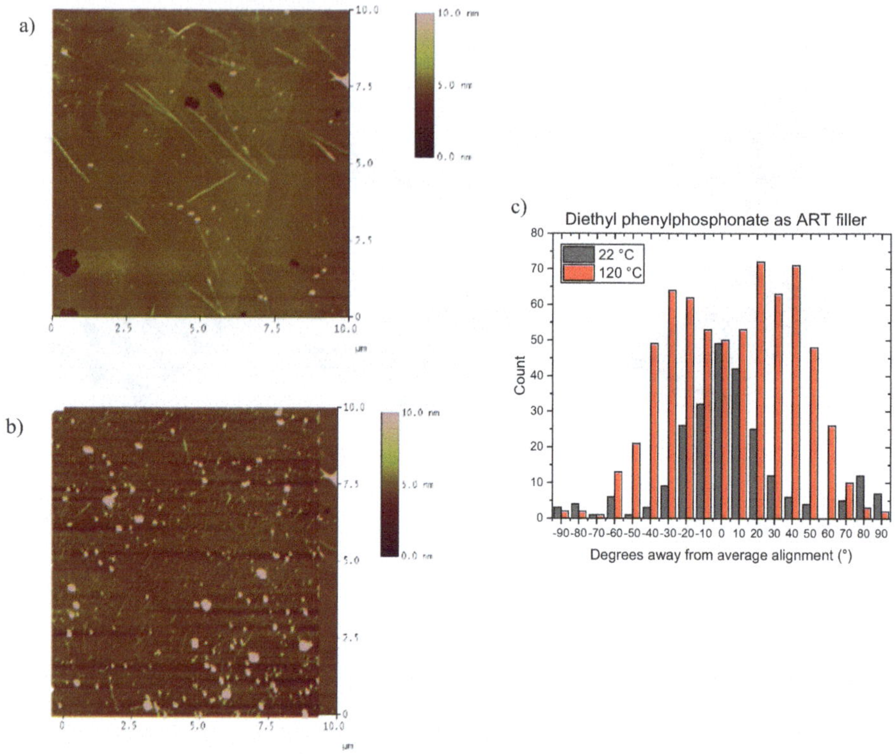

Fig. 2.23 AFM images of (**a**) room temperature drying, (**b**) oven drying, (**c**) comparison of degrees away from average orientation data when using molecule 7

sample, the SD was 14.6°, and 48.9% of the counted tubes were within the average alignment ±10°. The average length of SWNTs on the surface was 1.5 μm for the air-dried sample but significantly shorter for the oven-dried sample (0.6 μm), with SD = 46.5° and 28.3% of CNTs being in the average orientation ±10°. There were also many more particles and impurities present on the oven-dried surface. Although there were shorter SWNTs on the surface of the high temperature treated sample, the alignment was improved with the room temperature filler test.

Without filler treatment, the density of CNTs on the surface ranges from 0.1 to 0.3 SWNTs/μm^2; therefore, there is not a significant improvement when air-drying the additional molecules on the surface. However, there is an increase in density when the sample is treated at high heat (oven-dried). Although the density increases at 120 °C, the alignment is not enhanced. Therefore, it was decided that different filler molecule candidates should be explored to assess the extent of the changes.

Next, a shorter version of the original iptycene **1** molecule was attempted as a spacer molecule using the as described adaptation. Molecule **8** was isolated as a colorless oil at approximately a 50% yield from refluxing a mixture of anthracene with bis(diethoxyphosphoryl)acetylene (1 equiv.) in toluene [32]. As the product

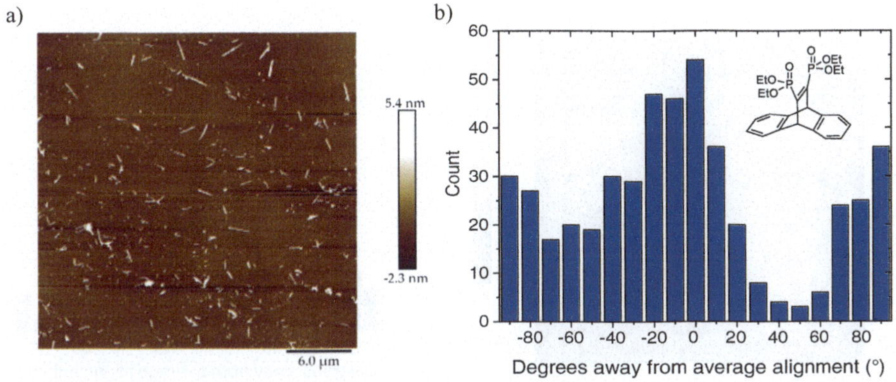

Fig. 2.24 (**a**) AFM image of using molecule 8 as filler, (**b**) alignment information obtain from using molecule 8 as filler

was an oil, it was dropcasted onto the surface, as with the diethyl phenylphosphonate (**7**). The sample was not drying at room temperature and was therefore annealed at 120 °C and rinsed with DCM. The process was once again repeated for a total of three times to ensure enough molecules could adhere and fill in any gaps on the surface. One of the obtained AFM images is shown in Fig. 2.24, along with the alignment histogram. The density was calculated to be 0.15 NTs/$\mu m^{2,}$ with an average length of 1.5 μm, the SD = 48.6°, and only 28.2% of the SWNTs were in the aligned range.

The results demonstrate that compound **8** is not a useful filler molecule, even though it appeared to have a good backbone for π–π stacking as it is very similar to the structure of the ART molecule **1**. It is possible to speculate that molecule **8** did not efficiently bind to the surface as it was much larger than molecule **7** and could not bind in between all the larger iptycene structures on the surface. Thus, compound **8** is an inefficient filler for the ART in increasing the alignment and density of SWNTs.

As smaller molecules appeared to work at increasing the density, diethyl benzylphosphonate (**9**) was synthesized and applied as a filler compound. Compound **9** was formed (95%) as a colorless oil from refluxing triethyl phosphite and benzyl bromide (1 equiv.) in 1,4-dioxane for one day [33]. The compound was observed to dry quickly when annealed at 120 °C for 10 min. Molecule **9** was chosen as it has similar properties to **7** but with the additional methylene bridge that adds flexibility in the structure. Therefore, it was selected to test if this additional flexibility in the compound would allow it to position itself in the spaces between the iptycenes to increase the SWNTs adhesion to the surface. The results obtained show a slight increase in alignment (Fig. 2.25a, b) and the density of the SWNTs (0.35 NTs/μm^2). The average length was observed to be 0.7 μm, slightly shorter than with compound **7**. The SD was calculated to be 40.5°, with 40% of SWNTs in the average alignment range ±10°.

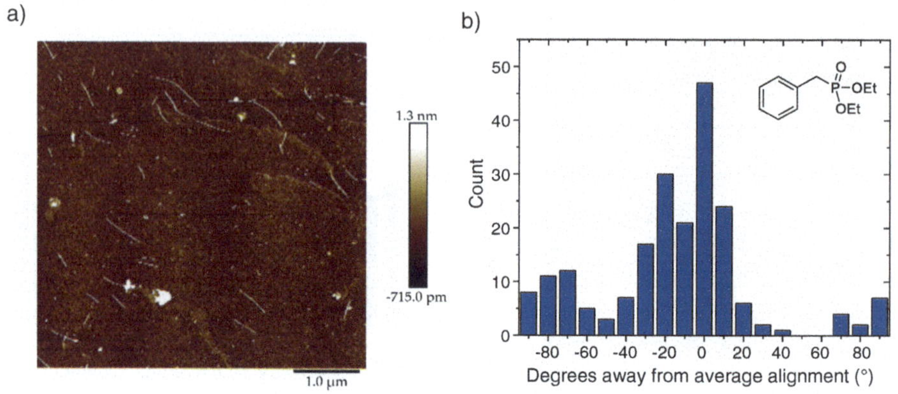

Fig. 2.25 (**a**) AFM image of using molecule 9 as filler, (**b**) alignment information obtain from using molecule 9 as filler

A phosphonic acid was also attempted, as they have been shown to bind quickly to oxidized substrates efficiently [34], and be useful in FETs [35]. Phenyl phosphonic acid (**10**) was purchased as a potential filler molecule and tested. It was first dissolved in DCM, then dropcast onto the functionalized SiO_2 surface and dried under a high vacuum at room temperature. The results are shown in Fig. 2.26a, b. The average length of the SWNTs was also calculated to be approximately 0.7 μm. From the data, the density increased nearly three-fold from the previous attempts, up to 0.95 SWNTs/μm^2. The SD was 45.3° with 42% falling in the 0° ± 10° range. The alignment of the SWNTs remains relatively the same, with only a slight increase in disorder as the distribution of Fig. 2.26b is not as evenly scattered as with the results from filler **7** at room temperature (Fig. 2.23c) or Fig. 2.12b results from the sonication. Nonetheless, the overall number of aligned tubes is a significant change. These results for compound **10** demonstrate the effectiveness of the phosphonic acid as an anchoring group for ART on the SiO_2 surface.

2.3.3 Summary of Filler Molecule Experiments on SiO_2

In summary, before applying and submerging the ART substrate in the SWNTs, an intermediate layer of molecules can be dropcast to the surface, and this additional step can significantly change the resulting deposition of the nanotubes. Commercially available phosphonic acid **10** are viable option to use for obtaining higher SWNTs density on the surface but with a less precise alignment. Additionally, the filler with a phosphonic acid anchoring component (**10**) showed a threetimes increase in density compared to conventional ART on SiO_2 when using **1/5CB** at the expense of increasing the SD by 10°. Although compounds such as **8** are also useful in increasing the number of nanotubes on the surface, they significantly decrease the alignment. Raman data for all the samples at both red and green laser frequencies were

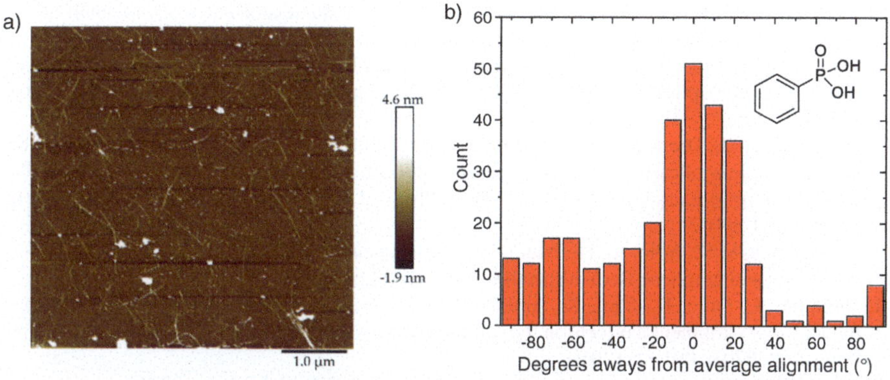

Fig. 2.26 (**a**) AFM image of using molecule **10** as filler, (**b**) alignment information obtained from using molecule **10** as filler

acquired and are shown in the Appendix Chapter 2. There were no significant differences in types compared to previous sonication results (Sect. 2.1.1.2). Each of the samples with fillers showed similar selectivity as a non-sonicated ART sample. The additional methylene link (**9**) did show a slight increase in the density of SWNTs while maintaining alignment. The phenyl phosphonic acid filler was shown to be most effective at increasing the density of the SWNTs while maintaining their aligned orientation, albeit a higher SD of 45° rather than 30° or less when using sonication.

References

1. Selmani, S., Schipper, D.: Orientation control of molecularly functionalized surfaces applied to the simultaneous alignment and sorting of carbon nanotubes. Angew. Chem. Int. Ed. **130**(9), 2423–2427 (2018)
2. Thess, A., Lee, R., Nikolaev, P., Dai, H., Petit, P., Robert, J., Xu, C., Lee, Y.H., Kim, S.G., Rinzler, A.G., Colbert, D.T., Scuseria, G.E., Tománek, D., Fischer, J.E., Smalley, R.E.: Crystalline ropes of metallic carbon nanotubes. Science. **273**(5274), 483–487 (1996)
3. Youssry, M., Al-Ruwaidhi, M., Zakeri, M., Zakeri, M.: Physical functionalization of multi-walled carbon nanotubes for enhanced dispersibility in aqueous medium. Emerg Mater. **3**, 25–32 (2020)
4. Clark, M.D., Subramanian, S., Krishnamoorti, R.: Understanding surfactant aided aqueous dispersion of multi-walled carbon nanotubes. J. Colloid Interface Sci. **354**(1), 144–151 (2011)
5. Islam, M.F., Rojas, E., Bergey, D.M., Johnson, A.T., Yodh, A.G.: High weight fraction surfactant solubilization of single-wall carbon nanotubes in water. Nano Lett. **3**(2), 269–273 (2003)
6. Fujigaya, T., Nakashima, N.: Non-covalent polymer wrapping of carbon nanotubes and the role of wrapped polymers as functional dispersants. Sci. Technol. Adv. Mater. **16**(2), 024802 (2015)
7. Sun, G., Liu, Z., Chen, G.: Dispersion of pristine multi-walled carbon nanotubes in common organic solvents. Nano. **5**(02), 103–109 (2010)
8. Saifuddin, N., Raziah, A.Z., Junizah, A.R.: Carbon nanotubes: a review on structure and their interaction with proteins. J. Chem. **2013**, 1–18 (2013)

9. Lu, X., Chen, Z.: Curved Pi-conjugation, aromaticity, and the related chemistry of small fullerenes (<C60) and Single-walled carbon nanotubes. Chem. Rev. **105**(10), 3643–3696 (2005)
10. Maultzsch, J., Telg, H., Reich, S., Thomsen, C.: Radial breathing mode of single-walled carbon nanotubes: Optical transition energies and chiral-index assignment. Phys. Rev. B. **72**(20), 205438 (2005)
11. Dresselhaus, M., Dresselhaus, G., Jorio, A., Souza Filho, A., Saito, R.: Raman spectroscopy on isolated single wall carbon nanotubes. Carbon. **40**(12), 2043–2061 (2002)
12. Saito, R., Dresselhaus, G., Dresselhaus, M.S.: Physical properties of carbon nanotubes. World Scientific (1998)
13. Dresselhaus, M.S., Jorio, A., Hofmann, M., Dresselhaus, G., Saito, R.: Perspectives on carbon nanotubes and graphene Raman spectroscopy. Nano Lett. **10**(3), 751–758 (2010)
14. Nugraha, A., Saito, R., Sato, K., Araujo, P., Jorio, A., Dresselhaus, M.: Dielectric constant model for environmental effects on the exciton energies of single wall carbon nanotubes. Appl. Phys. Lett. **97**(9), 091905 (2010)
15. Rao, C.N.R., Govindaraj, A.: Nanotubes and nanowires. RSC Publishing (2011)
16. Hennrich, F., Krupke, R., Lebedkin, S., Arnold, K., Fischer, R., Resasco, D.E., Kappes, M.M.: Raman spectroscopy of individual single-walled carbon nanotubes from various sources. J. Phys. Chem. B. **109**(21), 10567–10573 (2005)
17. Sato, K., Saito, R., Jiang, J., Dresselhaus, G., Dresselhaus, M.: Discontinuity in the family pattern of single-wall carbon nanotubes. Phys. Rev. B. **76**(19), 195446 (2007)
18. Sato, K., Saito, R., Nugraha, A.R., Maruyama, S.: Excitonic effects on radial breathing mode intensity of single wall carbon nanotubes. Chem. Phys. Lett. **497**(1–3), 94–98 (2010)
19. Ghosh, S., Bachilo, S.M., Weisman, R.B.: Advanced sorting of single-walled carbon nanotubes by nonlinear density-gradient ultracentrifugation. Nat. Nanotechnol. **5**, 443 (2010)
20. Jorio, A., Saito, R., Hafner, J., Lieber, C., Hunter, D., McClure, T., Dresselhaus, G., Dresselhaus, M.: Structural (n, m) determination of isolated single-wall carbon nanotubes by resonant Raman scattering. Phys. Rev. Lett. **86**(6), 1118 (2001)
21. Choi, J.H., Lee, J., Moon, S.M., Kim, Y.-T., Park, H., Lee, C.Y.: A low-energy electron beam does not damage single-walled carbon nanotubes and graphene. J Phys Chem Lett. **7**(22), 4739–4743 (2016)
22. Dai, H., Wong, E.W., Lieber, C.M.: Probing electrical transport in nanomaterials: conductivity of individual carbon nanotubes. Science. **272**(5261), 523 (1996)
23. Che, J., Cagin, T., Goddard III, W.A.: Thermal conductivity of carbon nanotubes. Nanotechnology. **11**(2), 65 (2000)
24. Javey, A., Guo, J., Wang, Q., Lundstrom, M., Dai, H.: Ballistic carbon nanotube field-effect transistors. Nature. **424**(6949), 654 (2003)
25. Krause, M., Blum, J., Skorov, Y.V., Trieloff, M.: Thermal conductivity measurements of porous dust aggregates: I. Technique, model and first results. Icarus. **214**(1), 286–296 (2011)
26. Hu, L., Ma, R., Ozawa, T.C., Geng, F., Iyi, N., Sasaki, T.: Oriented films of layered rare-earth hydroxide crystallites self-assembled at the hexane/water interface. Chem. Commun. **40**, 4897–4899 (2008)
27. Hu, L., Chen, M., Fang, X., Wu, L.: Oil–water interfacial self-assembly: a novel strategy for nanofilm and nanodevice fabrication. Chem. Soc. Rev. **41**(3), 1350–1362 (2012)
28. Li, X., Zhang, L., Wang, X., Shimoyama, I., Sun, X., Seo, W.-S., Dai, H.: Langmuir–blodgett assembly of densely aligned single-walled carbon nanotubes from bulk materials. J. Am. Chem. Soc. **129**(16), 4890–4891 (2007)
29. Joo, Y., Brady, G.J., Arnold, M.S., Gopalan, P.: Dose-controlled, floating evaporative self-assembly and alignment of semiconducting carbon nanotubes from organic solvents. Langmuir. **30**(12), 3460–3466 (2014)
30. Liu, L., Han, J., Xu, L., Zhou, J., Zhao, C., Ding, S., Shi, H., Xiao, M., Ding, L., Ma, Z., Jin, C., Zhang, Z., Peng, L.-M.: Aligned, high-density semiconducting carbon nanotube arrays for high-performance electronics. Science. **368**(6493), 850–856 (2020)

31. Snowdon, M., Selmani, S., Schipper, D.J.: Sonication-enhanced alignment relay technique for the orientation of single-walled carbon nanotubes. ACS Appl Nano Mater. **2**(10), 6637–6645 (2019)
32. Acheson, R.M., Ansell, P.J.: The synthesis of diethyl p-tolylsulphonylethynylphosphonate and related acetylenes, and their reactions with nucleophiles, pyridinium-1-dicyanomethylides, and dienes. J. Chem. Soc., Perkin Trans. 1, 1275–1281 (1987)
33. Saltiel, J., Redwood, C.E., Ratheesh Kumar, V.K.: The photoisomerization of cis,trans-1,2-dideuterio-1,4-diphenyl-1,3-butadiene in solution. No bicycle-pedal. Photochem. Photobiol. Sci. **18**(9), 2174–2179 (2019)
34. Queffélec, C., Petit, M., Janvier, P., Knight, D.A., Bujoli, B.: Surface modification using phosphonic acids and esters. Chem. Rev. **112**(7), 3777–3807 (2012)
35. Chang, J.-F., Hou, K.-S., Yang, Y.-W., Wang, C.-H., Chen, Y.-X., Ke, H.-D.: Enhanced mobility for increasing on-current and switching ratio of vertical organic field-effect transistors by surface modification with phosphonic acid self-assembled monolayer. Org. Electron. **81**, 105689 (2020)
36. R. Bruce, Weisman Sergei M., Bachilo (2003) Dependence of Optical Transition Energies on Structure for Single-Walled Carbon Nanotubes in Aqueous Suspension: An Empirical Kataura Plot. Nano Letters 3(9) 1235–1238 10.1021/nl034428i

Chapter 3
Different Anchoring Groups and Substrate Surfaces

Abstract Reaching this juncture, ART had only been demonstrated to function with the phosphorus group covalently binding onto silicon and indium tin oxide (ITO) samples. In this section, different substrate surfaces, as well as anchoring groups on the iptycenes, are explored. These experiments were conducted to investigate if ART had any limits, either in terms of the original design of the iptycene molecule or if the ART was solely limited toward SiO_2 and ITO Piranha-treated surfaces. Another form of iptycene-tweezer was synthesized using carboxylic acids, and this new molecule enabled better nanotube adherence to alumina surfaces than silica ones. Both alumina and gold surfaces were attempted with standard ART, and their surfaces were also treated with other molecules to fill in any gaps in order to increase the alignment and density of the SWNTs on the surface. The additional molecules were ineffective on gold but increased the SWNT density by up to 16% on alumina.

Keywords Carboxylic acids · Binding · Silica · Alumina · Oxidation · Alignment · Nanotubes · Sapphire

3.1 Alumina Coated Surface

An alumina (Al_2O_3) substrate can be used for generating short-channel organic transistors with SWNTs [1, 2]. It can also be used as an additional dielectric interface, which has been shown to work for organic FETs [2]. The ART was utilized to investigate SWNT alignment toward the Al_2O_3 surface with particles of **1**. The alumina surface used was fabricated from trimethylaluminum (TMA) using atomic layer deposition (ALD) with a plasma environment on an SiO_2 sample.

The ALD method is a subset of CVD that allows for a 10 nm layer of Al_2O_3 to be formed atop silicon. The mechanism for the reaction is illustrated in Scheme 3.1. The substrate is exposed to TMA vapors, where the molecules of TMA chemisorb to the hydroxylated surface (step i) [3, 4]. The TMA molecules do not react with one another, so the reactivity is terminated when the surface is saturated at one layer.

M. Snowdon, *Methods of the Alignment-Relay Technique for Nanosystems*,
Springer Theses, https://doi.org/10.1007/978-3-030-84413-4_3

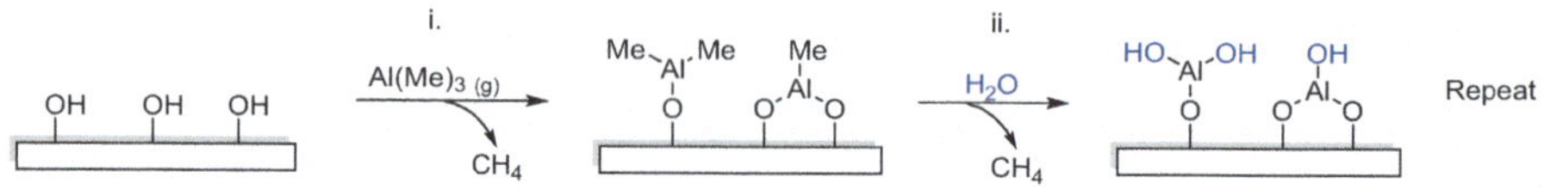

Scheme 3.1 The mechanism for the formation of Al_2O_3 from TMA using ALD with perfect stoichiometry. Adapted from Grillo et al. [3]

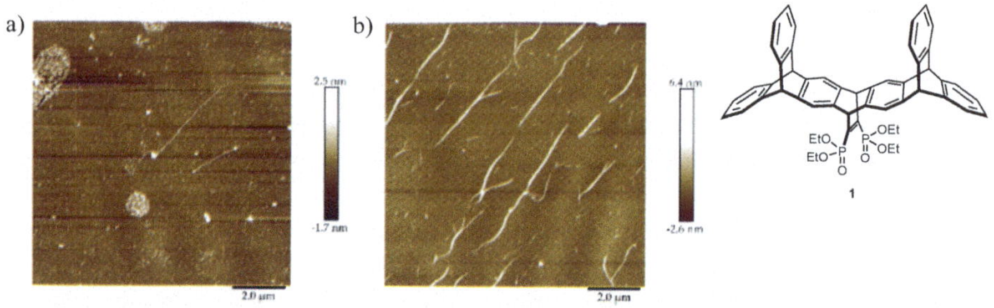

Fig. 3.1 The AFM of (**a**) ART on alumina with compound 1 and (**b**) ART on SiO_2 with compound 1

The sample is then purged with $Ar_{(g)}$ to remove excess TMA or methane by-product. Step (i) is then followed by exposure to water vapors where the water molecules react with available methyl groups, forming more Al-O bridges and a new layer of -OH groups (step ii). The substrate is once again purged. The cycle repeats until the thin film's desired thickness is reached. For the ART, the sample was briefly cleaned with Piranha solution (5 min) to ensure a freshly oxidized layer of **1** on the surface.

The ART protocol was the same as described in Chap. 1—as per previous exploits [5] and the subsequent surface underwent imaging analysis. The alumina sample from Fig. 3.1a is qualitatively compared to a deposition done on SiO_2 in Fig. 3.1b. The number of SWNTs appearing in Fig. 3.1a were significantly less than in Fig. 3.1b. The only observable difference was the ALD deposited Al_2O_3. Due to the shortage of CNTs in Fig. 3.1a, it indicated that there were very few molecules of **1** present on the wafer. This qualitative examination was likely due to a deficiency of the phosphonate ester groups effectively attaching to the Al_2O_3 wafer [6]. It has been observed that the CNTs only persist on the surface because of the fundamental iptycene monolayer and without it, the nanotubes do not align nor remain on the surface [7].

Consequently, the original configuration of compound **1** seems inefficient for the Al_2O_3 surface as insignificant results were observed as in Fig. 3.1b. In Fig. 3.1a, it was speculated that the absence of SWNTs corresponds to a deficiency of the phosphonate ester functional group to covalently bind with the hydroxyls on the wafer of freshly oxidized alumina. Although some alignment was observed, upon further tabulation of various images, there was an absence of a distinctive orientation and low density. These experiments prompted additional research towards examining ART with Al_2O_3 surfaces.

3.1.1 Substituting a Carboxylic Acid as an Anchoring Component

Different anchoring groups were subsequently investifated to overcome the decrease in covalent bonding being observed. From exploring the literature on self-assembled monolayers (SAMs) that perform comparably to ART, carboxylic acids were deemed a promising direction for the alumina surface investigations [8–10]. It has been observed that carboxylic acids and phosphonic acids mutually chemisorbed to Al_2O_3 surfaces [11, 12]. As a versatile binding component to other monolayers [8], a carboxylic acid attached iptycene (Ipt-COOH, **12**) was synthesized faster versus forming the phosphonic acid derivative. The Ipt-COOH molecule substituted compond **1** in ART, but the remainder of the procedure was unchanged, and **12** was solvated in 5CB at 2.5 wt%. Synthesis for molecule **12** is depicted in Scheme 3.2. Iptycene **6** was subjected to a Diels-alder reaction with dimethyl acetylenedicarboxylate (DMAD) to segregate **11** in a 71% yield. The methyl ester iptycene **11** underwent LiOH treatment for the formation of ART compound **12** [13].

The AFM image from using molecule **12** in the ART on Al_2O_3 and SiO_2 surfaces are given in Fig. 3.2a, b, respectively. The -COOH group proved to be successful, as both enhanced density and alignment were observed in Fig. 3.2a versus the original

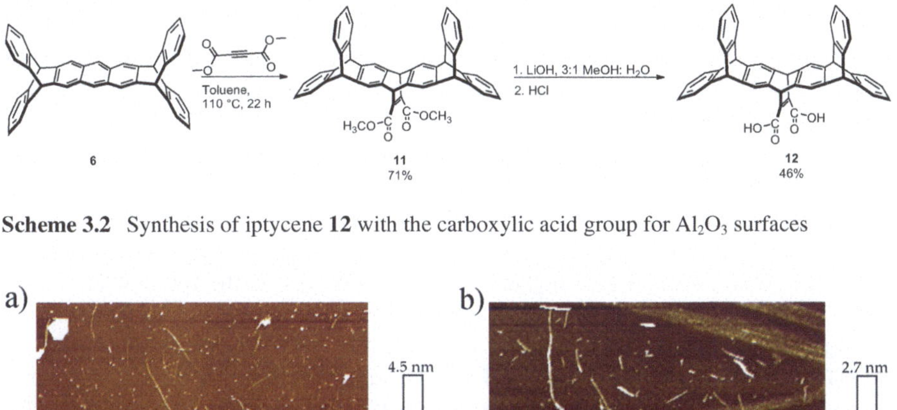

Scheme 3.2 Synthesis of iptycene **12** with the carboxylic acid group for Al_2O_3 surfaces

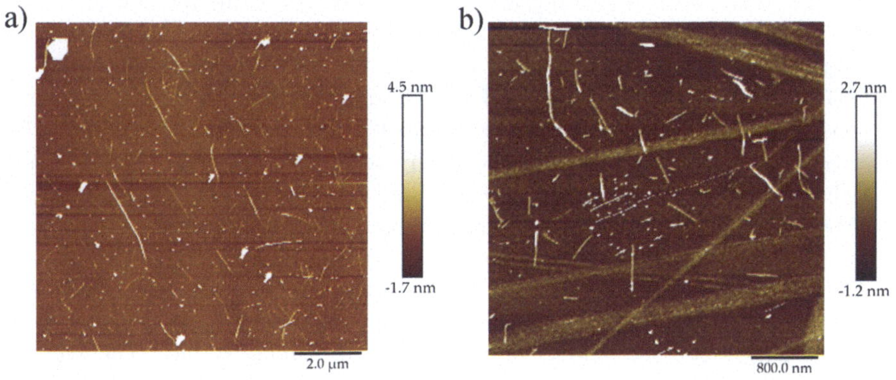

Fig. 3.2 The AFM images of (**a**) Ipt-COOH with ART on Al_2O_3 and (**b**) the Ipt-COOH with ART on SiO_2

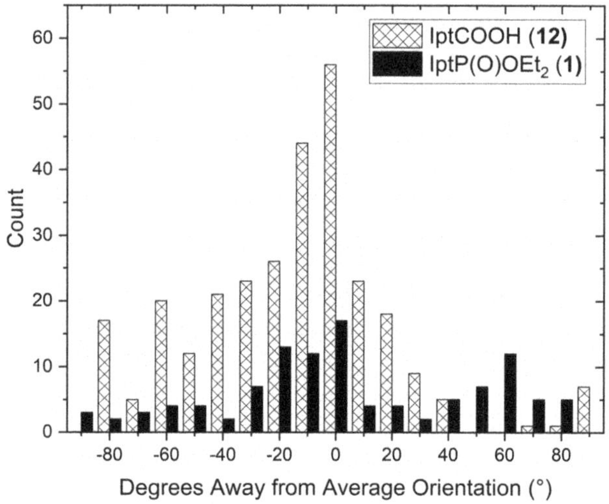

Fig. 3.3 Distribution of degrees away from average alignment of the CNTs on the surface of both the ART carboxylic acid and phosphonate ester treated surfaces

molecule **1** on alumina from Fig. 3.1a. As per Fig. 3.2b, there was a significant absence of alignment between the CNTs, contrary to what was observed for the Ipt-P(O)OEt$_2$ process on SiO$_2$ (Fig. 3.1b). This in turn proved that employing a different anchoring group was necessary for enhancing alignment on other oxidized surfaces. The deviation from the average alignment is plotted in Fig. 3.3. The target CNT alignment is similar to Gaussian distribution, where the majority of the SWNTs have minimal divergence from the average orientation. The Al$_2$O$_3$ substrate was subjected to **12,** and the nanotubes' orientation was contrasted to the alumina substrate with compound **1**.

A visual comparison of the tabulated orientation results in Fig. 3.3 further demonstrated the effectiveness of the -COOH group on its preferred chemisorption surface [6, 14]. From the histogram in Fig. 3.3, a drastic difference from the anchoring component and how it affects the ART surface, as there were far more aligned tubes present on the Al$_2$O$_3$. There was a higher number of SWNTs on the Ipt-COOH treated surface, as per the data in Fig. 3.3. The calculated densities in the samples further highlighted this observation. By using molecule **12** on the alumina, the number of SWNTs on Al$_2$O$_3$ increases to 0.25 CNTs/μm^2 from 0.10 CNTs/μm^2 when using compound **1**. In comparison, when molecule **12** for ART on the SiO$_2$ surface was used, the density of tubes decreased to 0.05 CNTs/μm^2 from 0.30 CNTs/μm^2. The standard deviation (SD) in alignment was 15.3° when using **12** on the alumina surface, whereas the SD is 4.5° when using **1** on the alumina. Although the SD was larger when **12** was used on the Al$_2$O$_3$, there were significantly more SWNTs on the surface.

Consequently, a change in the anchoring component in ART does produce a change in the number of iptycenes that are covalently bonded to the wafer surface, which in turn effects the number of SWNTs on the sample surface. The average lengths of the tubes on all substrates was under 1 μm, coherent with past results [7].

The observation in CNT lengths also suggests that the modifications in the anchoring group have no effects on the remainder of the iptycene structure. In the next section, the diameter preference is determined for the samples.

3.1.2 Spectroscopic Analysis

Raman spectroscopy (532 nm) was used to verify the tube types and their diameters when using compound **12** on Al_2O_3 and the results were then compared with those of molecule **1**. The resulting spectra are shown in Fig. 3.4 and with SWNT RBM. The dropcast SWNTs spectra are also shown, representing all the detectable types of CNTs and was sample 1 (S1). The middle spectra sample 2 (S2) shows the results for molecule **1** on the Al_2O_3 surface. The top curve represents the spectra for the sample surface of Al_2O_3 with molecule **12** and was sample 3 (S3). The peaks for the RBM region of the CNTs can be deconvoluted using Lorentzian peak fitting for finding and pinpointing the diameters and chirality (types) of the CNTs. The peak for the molecule remains at approximately 233 cm^{-1}, and the peak for the SiO_2 layer beneath the thin 5 nm Al_2O_3 is displayed at 310 cm^{-1}. Raman's typical Al_2O_3 peaks were not detected as the penetration depth does not interact with the ALD deposited alumina [15].

Thus, from Fig. 3.4, the samples include CNTs that have an approximate diameter of 1.40 nm and 1.55 nm. These diameters correspond to semiconducting SWNTs of (17,3) and (17,6). The tube types can be detected in the dropcast S1 and S3, but it was not possible to deconvolute for the (17,6) signal in S2—likely as the signal was far too weak. These findings were reasonable considering there were far fewer nanotubes on the surface of S2 versus S3. These outcomes further confirm

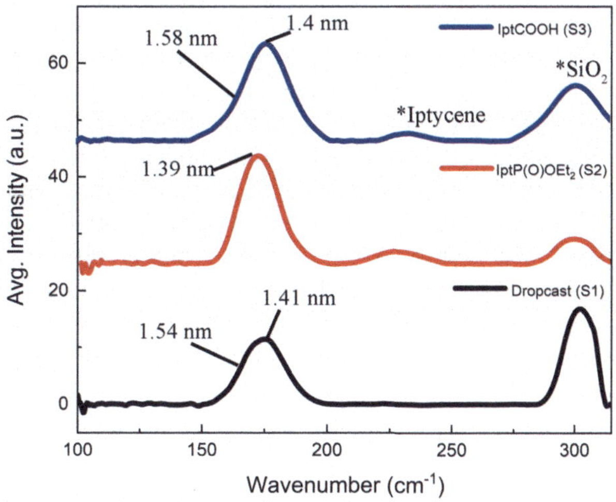

Fig. 3.4 Raman spectra with a 532 nm laser on Al_2O_3 for a solution of dropcast SWNTs (S1, black—bottom), the ART surface with molecule 1 (S2, middle—red), and the ART surface compound 12 (S3, blue—top)

that the -COOH group was beneficial for adhering SWNTs onto the Al_2O_3 substrate surface, and the SWNTs types were unchanged. Solely the number of aligned SWNTs on the surface was increased while the chirality remained the same and corroborate with previous results [7].

3.1.3 Substituting 5CB for 8CB

Furthermore, the ART was tested with 4′-Octyl-4-biphenylcarbonitrile (8CB) combined with molecule **12** using the same concentration and timeframe. This experiment was conducted because the 8CB has a longer carbon chain than 5CB and retains a smectic behavior at room temperature [16]. It has been previously attempted with ART on SiO_2 with compound **1**. The AFM image of the Al_2O_3 substrate with a solution of 8CB/**12** is shown in Fig. 3.5a. The equivalent SiO_2 surface with 8CB/**12** is given in Fig. 3.5b. These data showcase that the Al_2O_3 had an improved deposition and distribution of the CNTs. In contrast, the -COOH component was inefficient on the SiO_2 wafer with the 8CB solvent.

Additionally, the density on the alumina surface with the 8CB increased to 0.31 SWNTs/μm^2. This outcome was higher than that of the 5CB results, which had a density of 0.25 SWNTs/μm^2. Moreover, the findings with compound **12** for ART with 8CB on the SiO_2 surface showed a decreased density of about 0.07 SWNTs/μm^2, whereas using the same conditions with molecule **1** was previously shown to work very efficiently. The resulting histogram from the extrapolated results is shown in Fig. 3.5c. The plot in Fig. 3.5c represents the degrees away from the average alignment of the SWNTs on the samples on alumina and silica using 8CB/**12** as part of ART. Many more sample images with a higher number of CNTs were observed for the Al_2O_3 sample. There was a much smoother distribution of aligned nanotubes on the alumina sample with a greater number of SWNTs. The data reveals that a mixture of 8CB with compound **12** improves the ART technique for Al_2O_3 surfaces.

Furthermore, when using molecule **1** for ART on the Al_2O_3 surface in 5CB, the CNTs within a $0 \pm 10^\circ$ orientation over the total number of nanotubes was 29.7%. The number of CNTs was also not significant, as per Fig. 3.5a. However, when using molecule **12** on Al_2O_3, the number of tubes aligned increases to 38.9% (SD = 15.3°). By replacing the 5CB liquid crystal with 8CB combined with molecule **12**, the overall percentage of aligned tubes on the surface becomes 48.5% (SD = 32.2°) on the Al_2O_3 surface. A similar value of 37.2% aligned SWNTs was obtained when using 8CB/**12** on the SiO_2 substrate with a standard deviation of 3.5°. However, the number of tubes on the surface was minimal compared to that of the Al_2O_3 sample, as per the count of tubes shown in Fig. 3.5.

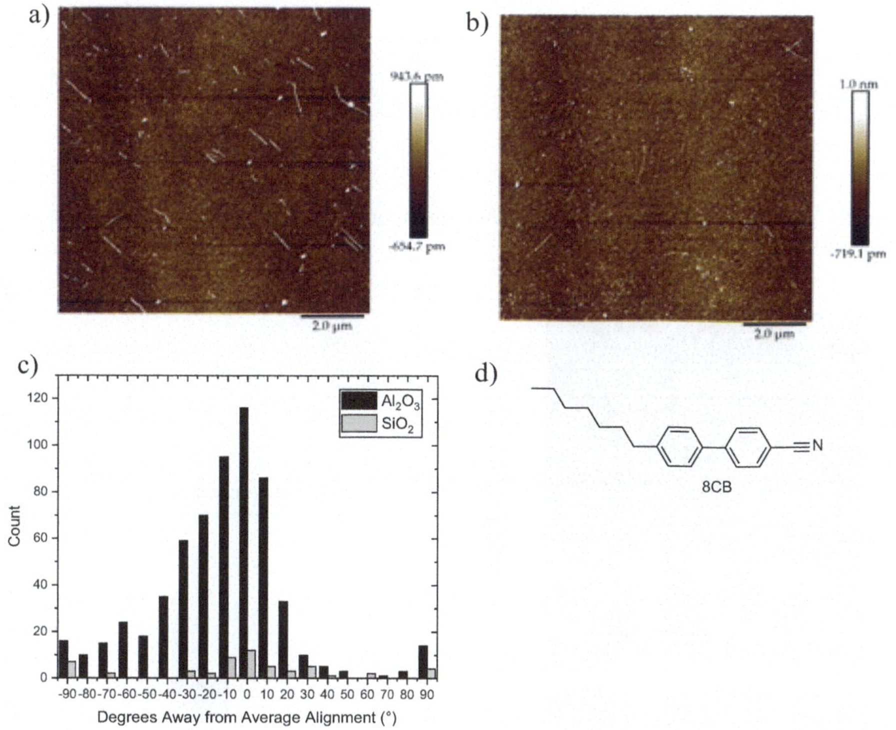

Fig. 3.5 The AFM images of (**a**) Ipt-COOH with ART on alumina using 8CB and (**b**) the Ipt-COOH with ART on silica using 8CB, (**c**) distribution of degrees away from average alignment of the CNTs on the surface of both the ART acid using 8CB instead of 5CB on the two substrates, (**d**) structure of 8CB for reference

3.1.4 Filler Exploration

The opportunity was also taken here to explore further the filler experiments discussed in Chap. 2 and carboxylic acids. The process used was the same, as shown in Fig. 2.20. However, the carboxylic acid versions of previous filler attempted molecules were tried in conjunction with compound **12/5CB** rather than molecule **1** on the Al_2O_3 surface. The filler molecules attempted are shown in Fig. 3.6. The benzoic acid (**13**) and phenylacetic acid (**14**) were expected to exhibit similar effects on the SWNTs' alignment as their phosphorus-containing counterpart from Chap. 2. These two compounds were also readily available and inexpensive to test. Compounds **15** and **17** have simple two-step synthetic processes.

However, although compound **15** was formed, based on the abysmal results regarding the equivalent structure with the phosphonate ester group (molecule **8**) (Sect. 2.3), it was decided not to pursue **15** as a filler in ART. An alternative structure, molecule **16**, was then synthesized and hypothesized to function better as there was only one spot needed for anchoring. Additionally, there was some flexibility in

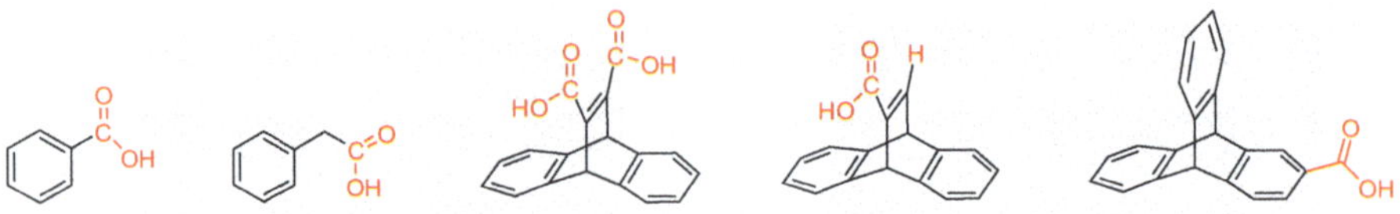

13 14 15 16 17

Fig. 3.6 Carboxylic acid group filler molecules incorporated with molecule **12/5CB** on the alumina surface using the ART

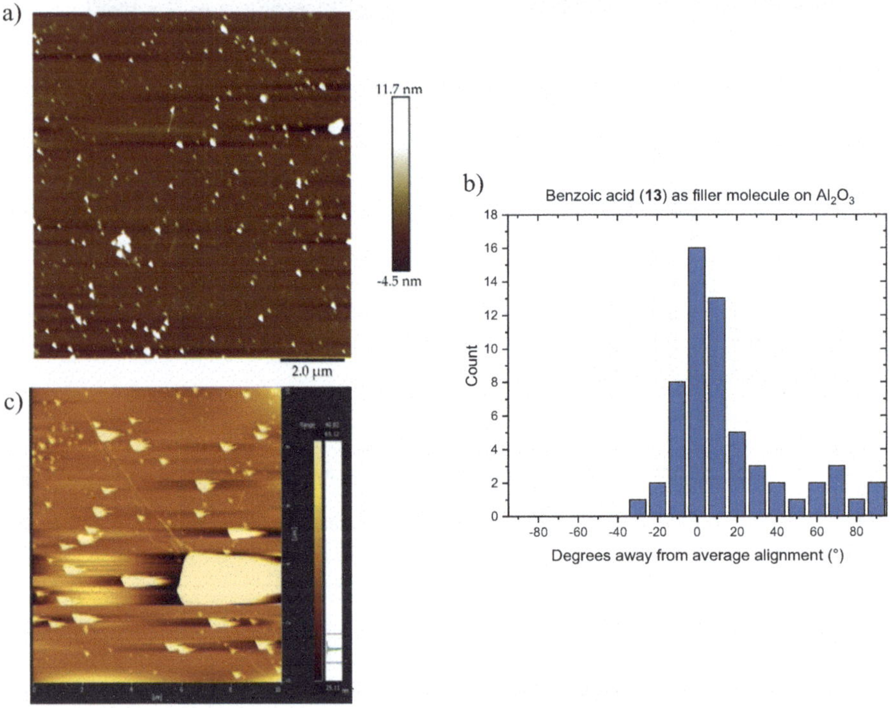

Fig. 3.7 Results from attempting molecules 13 and 14 in conjunction with 12/5CB on the Al_2O_3 surface. (**a**) representative AFM of surface treated with 13, (**b**) resulting data of aligned SWNTs on the surface using 13, (**c**) representative AFM of the surface treated with 14

the structure to enable it to fit within the iptycene monolayer gaps. Molecule **17** was thought to work similarly, as there was only one point for chemisorption onto the surface, and the remaining triptycene structure could interact to strengthen the π-π stacking interaction on the substrate.

The benzoic acid (**13**) was dropcast onto a functionalized surface of **12** using DCM and dried under ambient conditions. The alignment results are tabulated in Fig. 3.7a, b, depicting an AFM image and the histogram representing average alignment. However, the aligned tubes' density was observed to be quite low—0.05

SWNTs/μm^2—with an average length of 1 μm for the benzoic acid-treated sample. Nevertheless, the observed tubes were aligned with respect to one another (SD = 4.5°), with 62.7% of the total SWNTs on the surface being within 0 ± 10°. Although this does not show any improvements over the routine use of **12** in the regular ART with 8CB, it does indicate that when using 5CB, alignment can also be improved to the same extent with this filler adaptation.

Unfortunately, compound **14** did not yield any significant results – as seen in the poor AFM image shown in Fig. 3.7c. This result was surprising as the equivalent phosphonate ester (**9**) did exhibit some SWNT alignment when used as a filler on the SiO_2 surface. A possible reason for this observation is that there are only two available points for chemisorption on the carboxylic acid functional group (C=O, -OH), whereas the phosphonate ester has three possible points for integrations (P=O, -OEt × 2) and binds much faster as a filler on this surface. There is potential for this experiment to be optimized by increasing the time frame for increased binding on the surface; thereby increasing CNT density. However, the intention was not to significantly increase the time for the ART method, and instead the other possible fillers (**16** and **17**) were explored.

Compound **16** was synthesized from a two-step reaction starting from anthracene and ethyl propiolate. The starting materials can be either refluxed [17] or microwaved [18], followed by KOH hydrolysis to yield **16** in a 78% yield. For the synthesis of **17**, a 2-bromotriptycene precursor was formed from anthracene [19, 20], followed by a Grignard reaction with dry ice to yield the product in a 33% yield [21]. Results of SWNTs alignment from treating the sample surfaces with these molecules are shown in Fig. 3.8. For compound **16**, the AFM images acquired, such as in Fig. 3.8a, had very few tubes on the surface, which resulted in minimal alignment (SD = 1.4° with a percentage of aligned to total tubes of only 25.8%). There was an average density of 0.08 NTs/μm^2 and an average length of 1.23 μm.

On the other hand, molecule **17** showed an increase in the number of tubes, approximately 0.47 CNTs/μm^2. The tubes observed were slightly shorter, at 0.8 μm in length. Nonetheless, the average alignment was also improved to 46.7% with a SD of 26.9°, as shown in Fig. 3.8c, d. It was possible to acquire a variety of images and count the individual tubes; and as a result, the high tube count in histogram Fig. 3.8d.

From the results using **17**, an attempt was made to synthesize the equivalent phosphonate to test on the SiO_2 surface with molecule **1**. It was speculated molecule **18** could be synthesized from a Grignard reaction with 2-bromotriptycene. However, when tried, no desired product was observed, and the 2-bromotriptycene was recovered. This observation indicates there was an error in the Grignard formation, as the reaction did not proceed as indicated. Other attempts at making **18** could be further pursued and the results compared with the previous findings for the ART on the silica surface. From these space-filling studies on alumina, molecule **17** was most significant in terms of affecting SWNTs' alignment with **12**/5CB on the alumina surface.

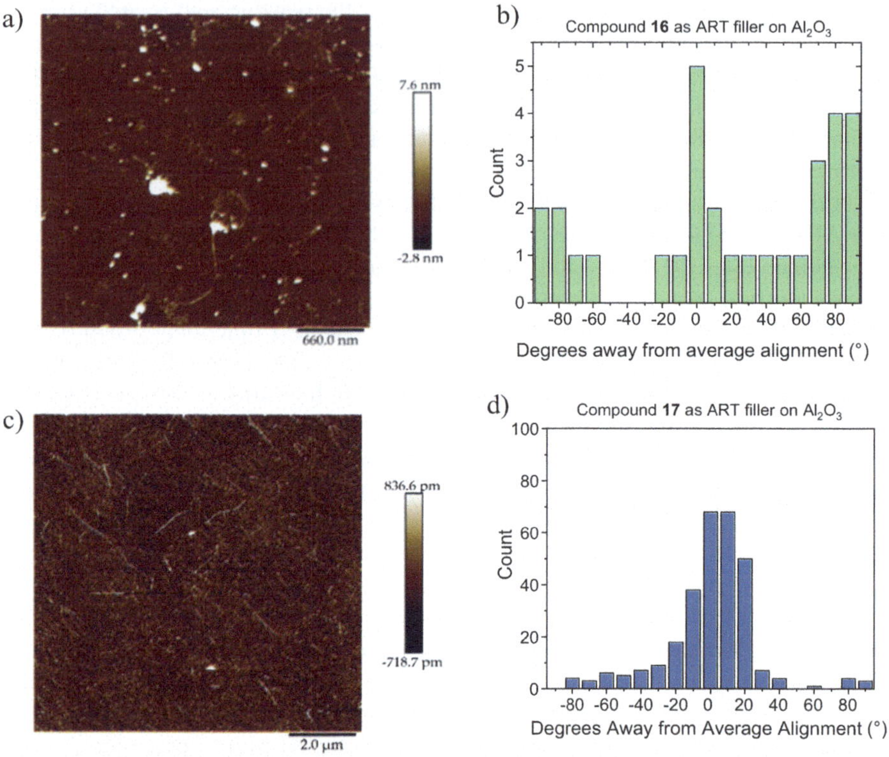

Fig. 3.8 Results from using 16 and 17 as gap-filling molecules on the Al_2O_3 surface. (**a**) AFM image using 16, (**b**) resulting histogram depicting average alignment deviation of the SWNTs using 16, (**c**) AFM image using 17, (**d**) resulting histogram depicting average alignment deviation of the SWNTs using 17

3.1.5 *α-Al_2O_3 (Sapphire)*

A sapphire substrate with a step size of 0.5° was examined because the step size could potentially assist in directing the CNTs alignment. High alignment has been obtained using α-Al_2O_3 with CVD to grow ultra-long tubes [22]. Here, it was hypothesized that the direction of the miscut angles could be an additional guide the orientation of the nanotubes. The sample's step sizes are shown in the 3D rendering of the sample surface in Fig. 3.9a. The sample was treated with UV-Ozone for 20 min to clean and oxidize the surface. The sample was imaged using AFM tapping mode before and after UV-ozone treatment to assure that the steps remained the same and were unaltered. Once this was confirmed as shown in Fig. 3.9a (and in Appendix Chapter 3 Figs. A.30 and A.31), the sample was treated with ART. The original procedure with **1**/5CB was first attempted as it was readily available, even though the first section of this chapter shows it was ineffective. The absence of orientation when using molecule **1** is an indication that the phosphonate anchoring

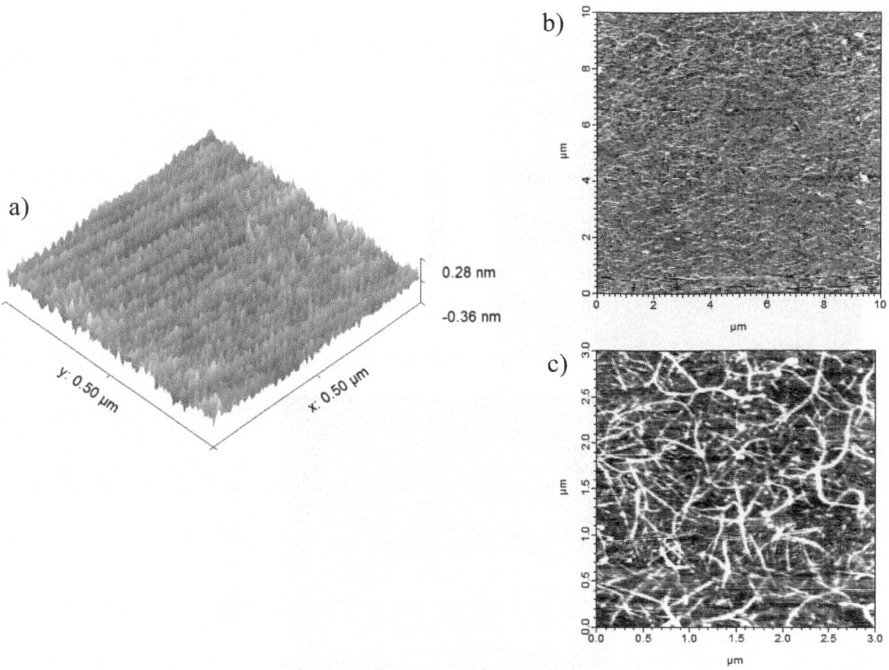

Fig. 3.9 (**a**) AFM of the sapphire surface showing the step size of 0.5 after UV-ozone treatment, (**b**) 10 × 10 µm AFM image of the sample treated with **1**/5CB that resulted in SWNTs bundles, (**c**) 3 × 3 µm AFM image of the SWNTs bundles on sapphire with ART **1**/5CB

group was ineffective on the sapphire surface with ART. Originally, no CNTs were observed to be deposited on the sample, therefore the substrate was re-exposed to the 90% sc-SWNTs solution in SDBS/SC surfactant for an additional 2 days. The subsequent treatment then resulted in many CNTs on the substrate, but none were aligned (Fig. 3.9b, c).

Compound **12** with ART was then exposed to the sapphire substrate, and the results showcase that many aligned SWNTs were deposited. Figure 3.10 shows the (a) AFM and (b) degrees away from the average alignment histogram. The SD for the alignment was 16°. The average length was 1.3 µm, and the density was 0.3 CNTs/µm^2. These results contrast those from using the original molecule on the sapphire surface. Indeed, the results from Fig. 3.10 showcase the importance of tailoring a specific anchoring group in ART to a surface. Additionally, UV-vis absorption spectroscopy was possible due to the transparency of the material and is shown in this chapters Appendix.

The SWNTs used here were also 90% sc-SWNTs from Nanointegris but were dissolved from their powdered form to create a 0.1% solution in $CHCl_3$ and were subjected to 1 minute of sonication. The sample was cleaned with UV-O_3 for 15 minutes. However, it appears that the grooves from the sapphire surface do not affect the orientation or the density of the SWNTs that are deposited. Rather, as per

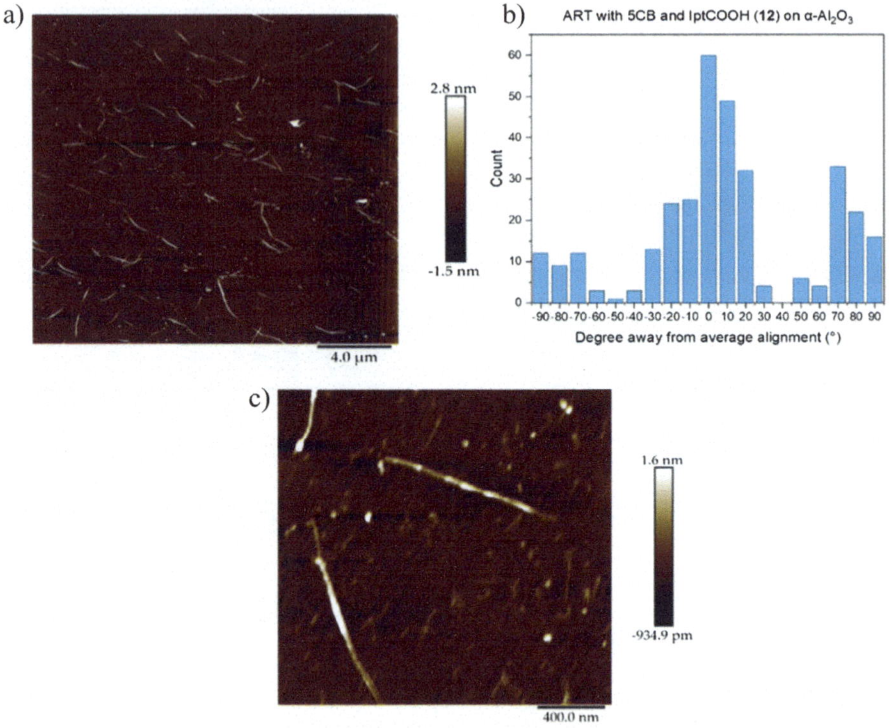

Fig. 3.10 (**a**) AFM image of aligned SWNTs on the sapphire surface using IptCOOH, (**b**) histogram representing the degrees away from average orientation for the number of counted nanotubes, (**c**) zoom in AFM image for the appearance of the grooves in the surface

Fig. 3.10c, the ridged pattern does not seem to contribute to the alignment of the SWNTs with ART. Therefore, compound **12** is responsible for the alignment on the sapphire surface, and the step size of the surface does not work in conjunction with ART.

3.2 Gold Coated Substrate

Subsequent ART experiments were then performed on a gold surface as it became freely available and expanded the library of surfaces that could be used with the ART. A silicon substrate was coated with 10 nm titanium and 50 nm gold through sputtering. The titanium (Ti) layer was necessary as otherwise, the gold (Au) does not adhere to the surface and would be readily removed as there was no stable -metal-oxide interface that formed [23]. An adhering metalized gold layer is only possible if using an adhesion layer, such as Cr or Ti. The titanium binds itself very well to surfaces, such that a sturdy Au/Ti metalized bilayer can be formed. The layer for the gold surface referred to in this section is Si-SiO_2-TiO_2-Ti-Au.

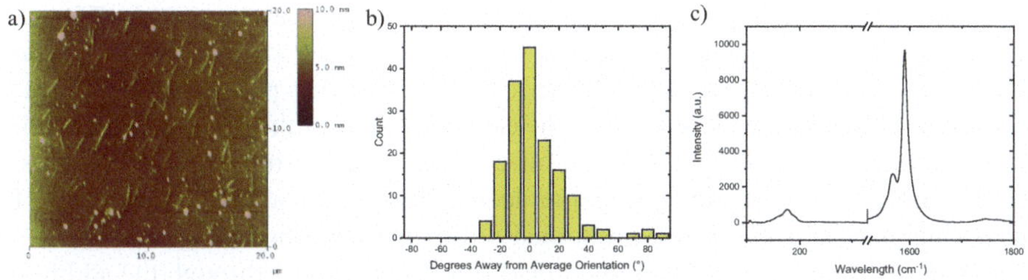

Fig. 3.11 Results of the ART using 1/5CB on a gold substrate (**a**) AFM, (**b**) average alignment histogram, (**c**) Raman spectra showing RBM and G-band using a 532 nm laser

An alteration to the ART was also necessary. The Piranha solution's surface oxidation was not possible, as Piranha etchs away the metal surface. Even with 5 min of Piranha treatment, the Au's surface could be scratched away with tweezers. Therefore, the sample was cleaned with acetone/IPA and then treated with UV-Ozone (20 min), followed immediately by exposure to the **1/5CB** solution and LC alignment layer. The UV-Ozone treatment cleans and provides a fresh surface layer of gold oxide that readily decomposes [24, 25]. Therefore, it was attempted to be used as quickly as possible. In any case, the results are shown in Fig. 3.11 and based on this data, the regular ART works very well on the Au substrate.

The average length of the SWNTs was calculated to be 1.66 μm, with an overall density of 0.23 SWNTs/μm^2. The standard deviation was calculated to be 13.5°, and 64.8% of all the SWNTs are within the average range $0 \pm 10°$. The results are very similar to the standard ART technique, resembling the data from Figs. 2.12b and A.2. Figure 3.11c shows the spectroscopic Raman information indicating similar tube type and diameter selectivity, as observed on the SiO_2. These findings were unexpected as one would not necessarily expect efficient binding of the phosphonate ester onto the gold surface, considering there are few studies currently available about the Au-P interface. The most recent work on Au-P interfaces regards black [26] or blue phosphorus layers [27, 28], interactions with gold nanoparticles [29], or dendrimers [30]. Nonetheless, Fig. 3.11 showcases an exciting observation for SWNTs alignment using the phosphorus-containing iptycene. The next direction was to observe if a better anchoring component could be created to improve these findings further.

3.2.1 Sulfur Anchoring Component for Gold

For self-assembled monolayers (SAMs) that utilize chemisorption to the gold surface, sulfur is the choice element. The Au-S interface is extensively studied [31–33], and as such, it was attempted to integrate that knowledge to ART Au surface functionalization. For instance, the replacement of the phosphonate ester (Fig. 3.12b) with a sulfur-containing group (Fig. 3.12c). Additionally, it has been shown that sulfur adsorbates, such as thiols, have excellent binding capabilities with ITO and Ag, as well as semiconducting surfaces like CdSe and InP [8]. Pursuing a sulfur substituent in this investigation could broaden the molecule's potential onto various substrates with higher preference than the current system. Although there are a variety of possible synthetic routes for sulfur-containing anchoring group derivatives, the investigations for obtaining target molecule **19** were pursued using molecule **1** as starting material for a simple preliminary investigation using ART [34].

Synthesis of molecule **19** was attempted from compound **1** through the addition of Lawesson's reagent in toluene, refluxing for 16 h. This method was previously reported to convert phosphonate esters to thiophosphonates in 36% yield [35]. Target molecule **19** was not isolated as only 15 mg of the crude mixture was recovered with multiple compound spots observable when testing through thin-layer chromatography (TLC). The crude NMR is shown in the Appendix Chapter 3, where starting material (**1**) could be present, along with other unknown materials. The un-purified residue was then dissolved in liquid crystal and used for gold functionalization with ART, following the already established protocol. It was challenging to quantify the results of alignment coming from molecule **1** (still present in the mixture) versus molecule **19**—if present in a small amount. However, no alignment was observed on the surface. Very few SWNTs were identifiable through AFM (Fig. 3.13a), and only a small area of the CNTs could be found using SEM

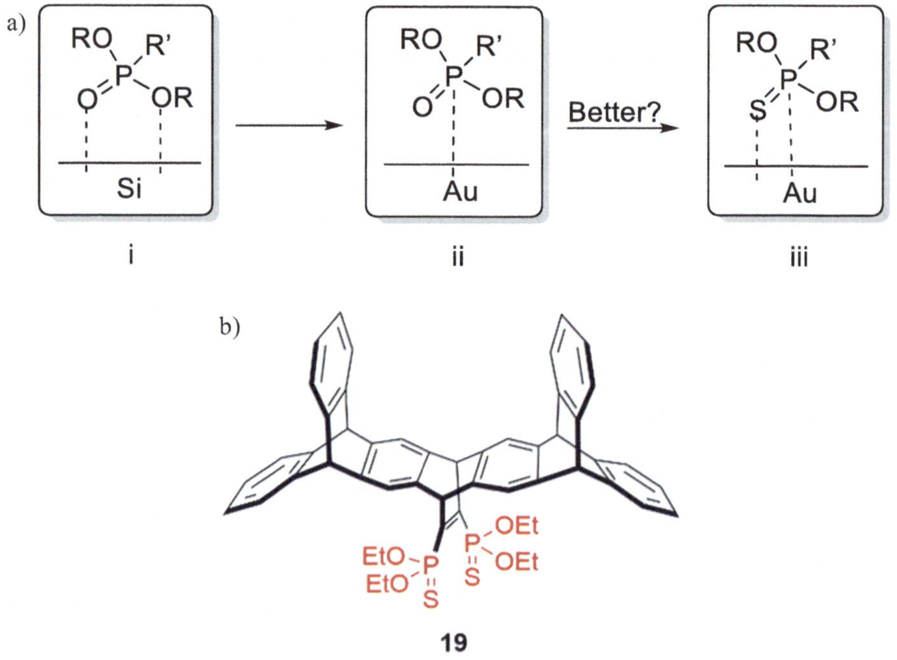

Fig. 3.12 (**a**) Hypothesis for altering the functional group for the ART, (**b**) target molecule containing sulfur for enhanced Au-S binding (**19**)

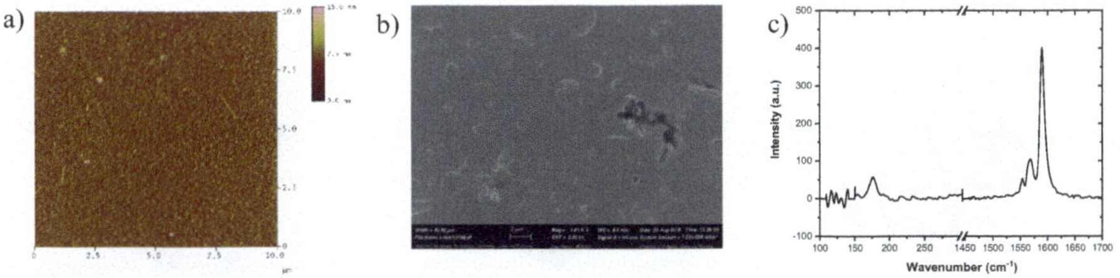

Fig. 3.13 (**a**) Tapping mode AFM showing very few SWNTs on the surface, (**b**) SEM of the sample surface depicting un-aligned tubes, (**c**) Raman spectrum of confirming presence of SWNTs on the surface for the gold substrate using a mixture of **1** and **19**

(Fig. 3.13b). A Raman spectrum of the sample was obtained using a green laser light source with a 30 × 30 μm mapped area as seen in Fig. 3.13c. The data depicts that though the alignment was not successful in this attempt, the selectivity of the SWNTs remains the same.

3.2.2 Gold with Filler Molecules Adaptation

Consequently, a different approach was taken to incorporate sulfur moieties in order to further explore the alignment of SWNTs on gold surfaces. As shown from Sect. 2.3, the ART can be adapted to incorporate different intermediate molecules (red squares) on the SiO_2 surface, and earlier in this chapter (Sect. 3.1.3), the fillers adaptation was explored on the Al_2O_3 surface. The process is re-represented on the gold surface in Fig. 3.14 [5]. The intermediates were chosen to include one representative of the phosphonate ester (**7**), one with thio-acidic properties (**20**), and one containing a thiophosphonate group (**21**). The three molecules all possess an aromatic component that is key for favorable interactions with SWNTs.

As illustrated in Fig. 3.14, subsequent functionalization occurred by dropcasting the intermediate molecules onto the already aligned surface containing molecule **1**. Liquid molecules were dropcast neat, and solid samples were dissolved in toluene. Samples were left in a ventilated area to dry and rinsed with dichloromethane to remove any molecules not covalently bound to the surface. All three molecularly functionalized surfaces were placed in a 90% semiconducting CNT solution provided by Nanointegris Inc. for 24 h.

Based on the atomic force microscopy (AFM) images (Fig. 3.15), sample 3.14a) depicts the highest degree of "alignment" and density of deposited tubes resulting from **7**, whereas Fig. 3.14b) the acid-treated (**20**), had very few SWNTs on the surface. In Fig. 3.14c), the sample treated with molecule **21** shows a preference for a few long tubes (>1 μm). However, nothing was aligned for these samples as seen in Fig. 3.15d. It appears that the phosphine sulfide has an affinity for larger diameter tubes, indicating that the three phenyl rings interact favorably for adhering to longer SWNTs. The acidic thiol surface does not appear to have any positive effects on the

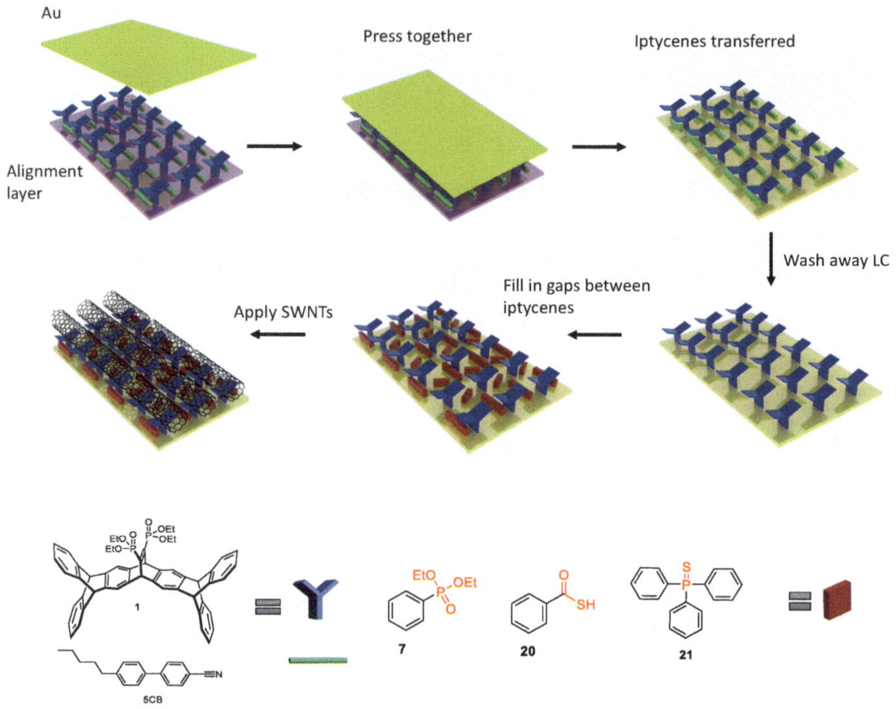

Fig. 3.14 Schematic representation of adapted ART method using intermediary molecules (red squares) for SWNTs alignment on the gold surface

ART, as it lacks deposition of any tubes, which was likely due to a bleach rinse used to remove the thiol odor. This treatment could be destroying the integrity of the surface.

To quantify and affirm the presence of NTs on the surface, Raman spectroscopy was performed on all the samples at a wavelength of 533 nm. Results of the obtained spectra are shown in Fig. 3.16, staggered along with the spectra of drop casted neat SWNTs as well as results of using no intermediate molecule in the process (i.e., Regular ART on the gold substrate). The samples were Raman mapped for a $30 \times 30\ \mu m^2$ area.

Based on the results of Fig. 3.16, the thiobenzoic acid and diethyl phenyl phosphonate treated surface contained molecule **1** intact on the surface, and the triphenylphosphine sulfide (**21**) contributed little signal of the original iptycene. Additionally, fillers **7** and **20** had a higher peak intensity. This observation was unusual as the phosphine sulfide treated surface seemed to have more nanotubes than the thiobenzoic acid treated substrate so that one would expect more intense CNT signals. This observation may be due to the CNTs present on the surface not having a strong Raman signal in the green laser wavelength. It would be interesting to observe the samples under the 632 nm red laser source to finalize the analysis—however, this was unavailable at the time. Additionally, the **21** and **7** surfaces lack a

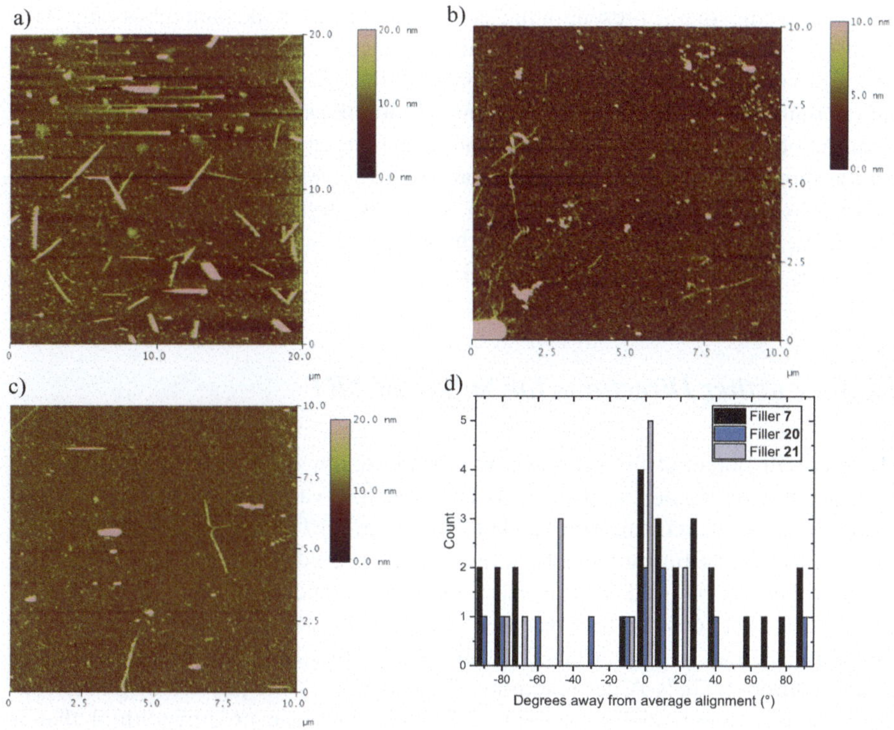

Fig. 3.15 AFM of samples treated with (**a**) diethyl phenyl phosphonate (7), (**b**) thiobenzoic acid (20), and (**c**) triphenylphosphine sulfide (21), (**d**) tabulated data for the few tubes that were found on these functionalized surfaces showing degrees away from average alignment

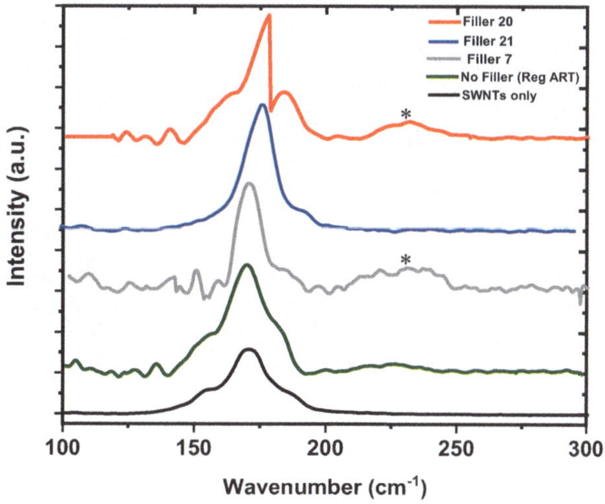

Fig. 3.16 Raman spectroscopy data at 533 nm laser wavelength for the three investigated samples, along with data of regular ART method on gold and nanotubes only for comparison purposes. *peak due to molecule 1

peak at 160 cm^{-1} that is present in the sample for **20** and in the nanotubes only. This result confirms that both the diethyl phenyl phosphonate and triphenylphosphine sulfide are more selective for tubes compared to the other intermediate sample or no intermediate at all. Based on these findings, more tubes on a gold substrate surface could be observed using **7** in comparison to the attempted sulfur compounds, and this has more selectivity for the 1.4 nm diameter (17,3) SWNT. However, no increase in SWNTs alignment occurred through these investigations on the gold surface. Currently, the best-observed ART methodology to use with the Au surface is UV-Ozone cleaning and molecule **1**/5CB.

3.2.3 Further Directions for Sulfur in ART

There is still potential for synthesis within the realm of replacing the anchoring component in molecule **1** as previously envisioned. A variety of molecules could be synthesized based on the molecular skeleton in Fig. 3.17 for a new sulfur-containing ART iptycenes. Sulfur moieties bond preferably to gold and other semiconducting surfaces that can be used for transistor applications such as CdSe, InP, and GeAs [8]. Future directions for this research entail synthesizing derivatives to achieve structures such as in Fig. 3.17. Derivatives with this kind of structure could be obtained through Diels-Alder reactions of **6** with dienophile possessing sulfur constituents like commercially available 2,5-dithia-3-hexyne or ethyndithiol that is attainable in one-step from acetylene [36]. However, the rate of the reaction is already very slow to arrive at the original ART molecule **1** with an electron-withdrawing dienophile [(diethoxyphosphoryl)acetylene, 5-day reaction] so likely this reaction would be slow to proceed, if at all.

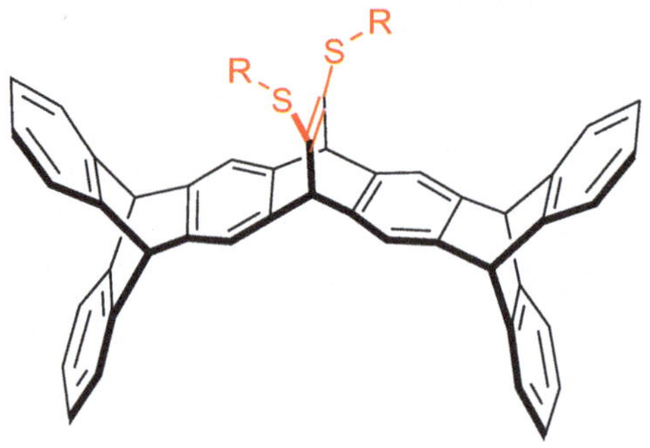

Fig. 3.17 The potential molecular backbone for further synthetic investigations in ART anchoring components

3.3 SiN_X, TiO_2 and Using UV-Ozone Cleaning

Furthermore, the suitability of using UV-Ozone as an oxidative cleaning method rather than the Piranha solution was also investigated as it worked efficiently well on the Au surface. Both cleaning techniques were first evaluated on the SiO_2 surface. The samples were treated in the same way, using **1**/5CB for functionalization for one-day and then SWNTs submersion for two-days. The results from the comparison are shown in Fig. 3.18. Although the number of SWNTs on the surface seems to have slightly decreased in this data set when comparing the UV-Ozone cleaning to the Piranha cleaning, the alignment does not seem to have altered to a great extent (SD = 7.4° using UV/Ozone versus 66° using Piranha in this case). Therefore, it is possible to use either cleaning method – whichever is readily available—for ART. Using UV-Ozone instead of Piranha could prove beneficial when performing ART on a large scale. It is straightforward to use the machine to clean numerous samples at one time, and there is no requirement for an individual to stay and observe the solution, as is the case with Piranha cleaning.

Additional surfaces of Si_3N_4 deposited through low-pressure (LP) CVD and ALD TiO_2 were tested. The UV-Ozone cleaning method (15 min) was used and evaluated on the samples with **1**/5CB. From the results in Fig. 3.19, these sample surfaces did not exhibit adequate adhesion with **1**, as indicated by the very few aligned SWNTs on the surface. This result was unusual for the TiO_2 sample (Fig. 3.19a, b), as it has excellent binding to phosphonic acids [37], but the alignment was lacking from the phosphonate ester. An attempt using IptCOOH (**12**) should be next, as TiO_2 does have similar binding to the Al_2O_3 surface [8]. Either 5CB or 8CB could work, as was shown at the beginning of this chapter. It could also be an error in using a faulty or misaligned alignment layer as the tubes on the surface have a poor orientation. The standard deviation was calculated to be 8.3°, but under 30% of all the tubes were in a similar orientation.

For the silicon nitride surface, the results are also unusual as it would be expected for this surface to have similar binding behavior to SiO_2. Although minimal, some

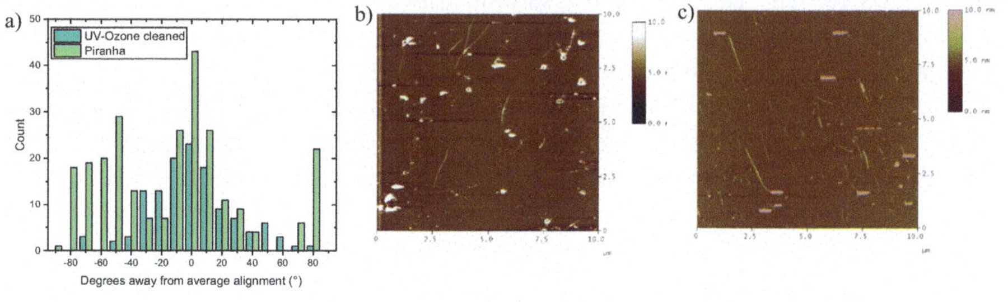

Fig. 3.18 Comparison of alignment and SWNTs numbers of the two possible cleaning methods that can be used in ART. (**a**) Histogram representing degrees away from average alignment for the samples treated with either cleaning method, (**b**) AFM image using UV-Ozone cleaning, (**c**) AFM image using Piranha cleaning

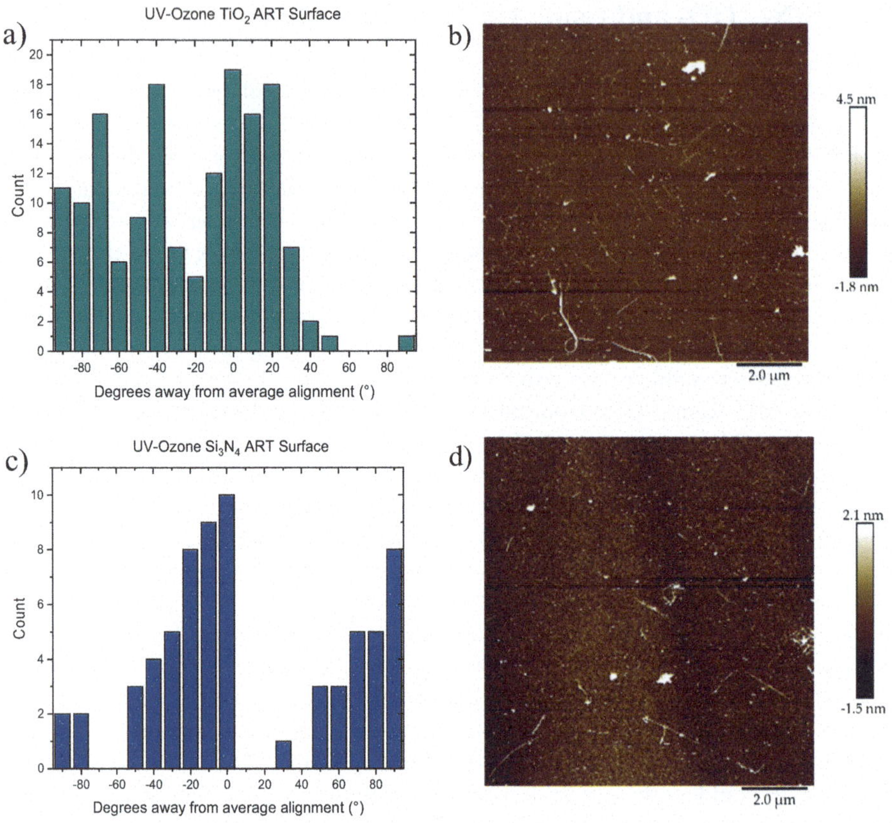

Fig. 3.19 Results from testing the Si_3N_4 and TiO_2 surfaces with 1/5CB under UV/Ozone cleaning: (**a**) alignment results for TiO_2, (**b**) AFM image for TiO_2 with ART SWNTs, (**c**) alignment results for Si_3N_4, (**d**) AFM image for Si_3N_4 with ART SWNTs

SWNTs do appear on the surface (Fig. 3.19c, d). Since Si_3N_4 is a very beneficial dielectric surface for device manufacturing [38, 39], therefore this sample should be investigated further. The SD for the sample was 5.3° calculated for the very few tubes on the surface, but again, similar to the TiO_2 sample, under 30% of all the tubes were within a specific alignment range (0 ± 10°).

As the phosphonic acid would be useful for the SiO_2 and other surfaces, an attempt to synthesize it was conducted. A trial for obtaining the phosphonic acid iptycene from hydrolysis of **1** was found to be unsuccessful with 0.1 M HCl/H_2O/EtOH. It was difficult to dissolve **1**; even at a higher temperature, the compound remained on the sides of the reaction vessel. Stronger nucleophiles should be used for future attempts, such as lithium hydroxide or sodium hydroxide, as the phosphonic acid should be easily formed [14]. It is also possible to create the phosphonic acid through reflux with 6 M HCl [40].

3.3.1 *Summary of Different Anchoring Groups and Substrate Surface Investigations*

The carboxylic acid group in **12** was shown to have improved adhesion to the TMA-ALD treated surface. This research broadens the adaptability of ART for supplementary functions in selective SWNT alignment work. The synthesis for molecule **19**, along with other sulfur-containing iptycenes, should also be revisited. The addition of dropcasting a filler molecule in some cases may be necessary, depending on the desired application. The different anchoring groups did prove to increase the density, depending on the substrate and filler chosen, but similar results were achieved when using 8CB instead of 5CB on the Al_2O_3 surface. While incorporating an intermediary sulfur-containing compound into ART did not enhance the aligned density of nanotubes on a gold surface, it was somewhat beneficial for the alumina surface.

Additionally, as shown with the Al_2O_3 substrate, the anchoring component binding is key to dictating the overall SWNTs distribution and alignment effectiveness. Two other surfaces, Si_3N_4 and TiO_2, can now be included in ART investigations and be further optimized. The silicon nitride surface can prove to be quite useful in further device applications. Another suggestion is to pursue the development of the phosphonic acid anchoring group for further improving ART.

References

1. Haeberle, T., Loghin, F., Zschieschang, U., Klauk, H., Lugli, P.: In Carbon nanotube thin-film transistors featuring transfer-printed metal electrodes and a thin, self-grown aluminum oxide gate dielectric, 2015 IEEE 15th International Conference on Nanotechnology (IEEE-NANO), 27–30 July 2015; pp 160–163 (2015)
2. Chang, J.-F., Hou, K.-S., Yang, Y.-W., Wang, C.-H., Chen, Y.-X., Ke, H.-D.: Enhanced mobility for increasing on-current and switching ratio of vertical organic field-effect transistors by surface modification with phosphonic acid self-assembled monolayer. Org. Electron. **81**, 105689 (2020)
3. Grillo, F., Moulijn, J.A., Kreutzer, M.T., Van Ommen, J.R.: Nanoparticle sintering in atomic layer deposition of supported catalysts: kinetic modeling of the size distribution. Catal. Today. **316**, 51–61 (2018)
4. Oviroh, P.O., Akbarzadeh, R., Pan, D., Coetzee, R.A.M., Jen, T.-C.: New development of atomic layer deposition: processes, methods and applications. Sci. Technol. Adv. Mater. **20**(1), 465–496 (2019)
5. Kulak, M.R., Schipper, D.J.: Carbon nanotube alignment and sorting: attempting a sulfur moiety as anchoring component. Phosphorus Sulfur Silicon Relat. Elem. **194**(7), 760–763 (2019)
6. Templeton, M.K., Weinberg, W.H.: Decomposition of phosphonate esters adsorbed on aluminum oxide. J. Am. Chem. Soc. **107**(4), 774–779 (1985)
7. Snowdon, M., Selmani, S., Schipper, D.J.: Sonication-enhanced alignment relay technique for the orientation of single-walled carbon nanotubes. ACS Applied Nano Materials. **2**(10), 6637–6645 (2019)
8. Jadhav, S.A.: Self-assembled monolayers (SAMs) of carboxylic acids: an overview. Cent. Eur. J. Chem. **9**(3), 369–378 (2011)

9. Allara, D.L., Nuzzo, R.G.: Spontaneously organized molecular assemblies. 1. Formation, dynamics, and physical properties of n-alkanoic acids adsorbed from solution on an oxidized aluminum surface. Langmuir. **1**(1), 45–52 (1985)
10. Barron, A.R.: The interaction of carboxylic acids with aluminium oxides: journeying from a basic understanding of alumina nanoparticles to water treatment for industrial and humanitarian applications. Dalton Trans. **43**(22), 8127–8143 (2014)
11. van den Brand, J., Blajiev, O., Beentjes, P.C.J., Terryn, H., de Wit, J.H.W.: Interaction of anhydride and carboxylic acid compounds with aluminum oxide surfaces studied using infrared reflection absorption spectroscopy. Langmuir. **20**(15), 6308–6317 (2004)
12. Bauer, T., Schmaltz, T., Lenz, T., Halik, M., Meyer, B., Clark, T.: Phosphonate- and carboxylate-based self-assembled monolayers for organic devices: a theoretical study of surface binding on aluminum oxide with experimental support. ACS Appl. Mater. Interfaces. **5**(13), 6073–6080 (2013)
13. Snowdon, M.R., Wang, S., Mashmoushi, N., Hopkins, S.W., Schipper, D.J.: Carboxylic acids as anchoring components on aluminum oxide for the alignment relay technique of single-walled carbon nanotubes. New J. Chem. **45**(12), 5340–5349 (2021)
14. Queffélec, C., Petit, M., Janvier, P., Knight, D.A., Bujoli, B.: Surface modification using phosphonic acids and esters. Chem. Rev. **112**(7), 3777–3807 (2012)
15. Opilik, L., Schmid, T., Zenobi, R.: Modern raman imaging: vibrational spectroscopy on the micrometer and nanometer scales. Annu. Rev. Anal. Chem. **6**(1), 379–398 (2013)
16. Patrick, D.L., Beebe, T.P.: Substrate defects and variations in interfacial ordering of monolayer molecular films on graphite. Langmuir. **10**(1), 298–302 (1994)
17. Weber, E., Csoeregh, I., Ahrendt, J., Finge, S., Czugler, M.: Design of roof-shaped clathrate hosts. Inclusion properties and x-ray crystal structures of a free host and of inclusion compounds with 1-butenol and DMF. J. Org. Chem. **53**(25), 5831–5839 (1988)
18. McNamara, Y., Bright, S., Byrne, A., Cloonan, S., McCabe, T., Williams, D., Meegan, M.: Synthesis and antiproliferative action of a novel series of maprotiline analogues. Eur. J. Med. Chem. **71**, 333–353 (2014)
19. Yang, Y., Cheng, K., Lu, Y., Ma, D., Shi, D., Sun, Y., Yang, M., Li, J., Wei, J.: A polyaromatic nano-nest for hosting fullerenes C60 and C70. Org. Lett. **20**(8), 2138–2142 (2018)
20. Chen, Z., Swager, T.M.: Synthesis and characterization of poly(2,6-triptycene). Macromolecules. **41**(19), 6880–6885 (2008)
21. Hutchins, D.O., Weidner, T., Baio, J., Polishak, B., Acton, O., Cernetic, N., Ma, H., Jen, A.K.Y.: Effects of self-assembled monolayer structural order, surface homogeneity and surface energy on pentacene morphology and thin film transistor device performance. J. Mater. Chem. C. **1**(1), 101–113 (2013)
22. Han, S., Liu, X., Zhou, C.: Template-free directional growth of single-walled carbon nanotubes on a- and r-plane sapphire. J. Am. Chem. Soc. **127**(15), 5294–5295 (2005)
23. Vogt, K.W., Kohl, P.A., Carter, W.B., Bell, R.A., Bottomley, L.A.: Characterization of thin titanium oxide adhesion layers on gold: resistivity, morphology, and composition. Surf. Sci. **301**(1–3), 203–213 (1994)
24. Ron, H., Matlis, S., Rubinstein, I.: Self-assembled monolayers on oxidized metals. 2. Gold surface oxidative pretreatment, monolayer properties, and depression formation. Langmuir. **14**(5), 1116–1121 (1998)
25. King, D.E.: Oxidation of gold by ultraviolet light and ozone at 25 °C. J. Vac. Sci. Technol. A. **13**(3), 1247–1253 (1995)
26. Zhu, S., Ni, Y., Liu, J., Yao, K.: The study of interaction and charge transfer at black phosphorus–metal interfaces. J. Phys. D. Appl. Phys. **48**(44), 445101 (2015)
27. Zhao, S., Zhang, J.L., Chen, W., Li, Z.: Structure of blue phosphorus grown on Au(111) surface revisited. J. Phys. Chem. C. **124**(3), 2024–2029 (2020)
28. Zhang, J.L., Zhao, S., Sun, S., Ding, H., Hu, J., Li, Y., Xu, Q., Yu, X., Telychko, M., Su, J., Gu, C., Zheng, Y., Lian, X., Ma, Z., Guo, R., Lu, J., Sun, Z., Zhu, J., Li, Z., Chen, W.:

Synthesis of monolayer blue phosphorus enabled by silicon intercalation. ACS Nano. **14**(3), 3687–3695 (2020)

29. Jiang, M.-S., Tao, Y.-H., Wang, Y.-W., Lu, C., Young, D.J., Lang, J.-P., Ren, Z.-G.: Reversible solid-state phase transitions between Au–P complexes accompanied by switchable fluorescence. Inorg. Chem. **59**(5), 3072–3078 (2020)
30. Majoral, J.P., Zablocka, M., Caminade, A.-M., Balczewski, P., Shi, X., Mignani, S.: Interactions gold/phosphorus dendrimers. Versatile ways to hybrid organic–metallic macromolecules. Coord. Chem. Rev. **358**, 80–91 (2018)
31. Engel, S., Fritz, E.-C., Ravoo, B.J.: New trends in the functionalization of metallic gold: from organosulfur ligands to N-heterocyclic carbenes. Chem. Soc. Rev. **46**(8), 2057–2075 (2017)
32. Pensa, E., Cortés, E., Corthey, G., Carro, P., Vericat, C., Fonticelli, M.H., Benítez, G., Rubert, A.A., Salvarezza, R.C.: The chemistry of the sulfur–gold interface: in search of a unified model. Acc. Chem. Res. **45**(8), 1183–1192 (2012)
33. Xue, Y., Li, X., Li, H., Zhang, W.: Quantifying thiol–gold interactions towards the efficient strength control. Nat. Commun. **5**(1) (2014)
34. Hennrich, F., Krupke, R., Lebedkin, S., Arnold, K., Fischer, R., Resasco, D.E., Kappes, M.M.: Raman spectroscopy of individual single-walled carbon nanotubes from various sources. J. Phys. Chem. B. **109**(21), 10567–10573 (2005)
35. Kiyokawa, K., Suzuki, I., Yasuda, M., Baba, A.: Synthesis of cyclopropane-containing phosphorus compounds by radical coupling of butenylindium with iodo phosphorus compounds. Eur. J. Org. Chem. **2011**(11), 2163–2171 (2011)
36. Trofimov, B.A., Mal'Kina, A.G., Dorofeev, I.A., Myachina, G.F., Rodionova, I.V., Vakul'Skaya, T.I., Sinegovskaya, L.M., Skotheim, T.A.: Synthesis and properties of polyenic oligosulfides derived from acetylene and elemental sulfur. Russ. J. Gen. Chem. **77**(9), 1559–1566 (2007)
37. Brodard-Severac, F.; Guerrero, G.; Maquet, J.; Florian, P.; Gervais, C.; Mutin, P. H., High-field17o MAS NMR investigation of phosphonic acid monolayers on titania. Chem. Mater. **20**(16), 5191–5196 (2008)
38. Su, F., Zhang, Z., Li, S., Li, P., Deng, T.: Long-term stability of photodetectors based on graphene field-effect transistors encapsulated with Si3N4 layers. Appl. Surf. Sci. **459**, 164–170 (2018)
39. Ha, T.-J., Chen, K., Chuang, S., Yu, K.M., Kiriya, D., Javey, A.: Highly uniform and stable n-type carbon nanotube transistors by using positively charged silicon nitride thin films. Nano Lett. **15**(1), 392–397 (2014)
40. Neff, G.A., Page, C.J., Meintjes, E., Tsuda, T., Pilgrim, W.C., Roberts, N., Warren, W.W.: Hydrolysis of surface-bound phosphonate esters for the self-assembly of multilayer films: use of solid state magic angle spinning31P NMR as a probe of reactions on surfaces. Langmuir. **12**(2), 238–242 (1996)

Chapter 4
Other Nanosystems Investigated for Alignment

Abstract To this point, ART had only been explored with a single batch of SWNTs. Other structures were probed to observe if the current ART molecular tweezer would work. Here, carboxylic acid functionalized SWNTs were attempted in order to solubilize the nanotubes in water and eliminate surfactant molecules. Nanowires (NWs) were also a material of interest for alignment. The alignment of MWNTs was additionally investigated, along with graphene nano-ribbons (GNRs), and briefly graphene and graphene quantum dots (GQDs). It was observed that the SWNTs in the diameter range of 1.4–1.8 nm were effective using the existing parameters. The MWNTs and NWs did not work as the diameters are too large to bind with the iptycene tweezer effectively. The GNRs did not bind as they have a planar structure that is a mismatch for the concave structure of the iptycene tweezers.

Keywords Graphene · Nanoribbons · Quantum dots · Surfactant · Carboxylic acid · Alignment · Multi-walled · Single-walled · Nanotube · Nanowire

4.1 Hydrophilic Carboxylic-Acid Functionalized SWNTs

The surfactant solution for the purchased carbon nanotubes was a mixture of SC and SDS in aqueous buffer. It has been determined that the hydrophobic end of these surfactant molecules adsorbs on the surface of the SWNTs. Simultaneously, the hydrophilic portion links itself with the water for suspension [1, 2]. Indeed, this additional interaction of surfactant wrapping with the SWNTs could impede the π–π stacking effect the SWNTs have with the iptycene molecules. Consequently, ART was investigated using water-soluble SWNTs. The defects inherent in the SWNTs structure from processing are advantageous for functionalization. Many chemical structures can be formed based on these defects, and CNTs can have supplementary properties depending on the functionalization [3]. These defects were advantageous for the formation of carboxylic acid SWNTs (SWNTs-COOH) [4]. The ART was attempted with SWNTs-COOH in water instead of using a surfactant buffer. Additionally, the SWNTs here were 78% SWNTs purity from Tuball Inc., unlike

M. Snowdon, *Methods of the Alignment-Relay Technique for Nanosystems*, Springer Theses, https://doi.org/10.1007/978-3-030-84413-4_4

the previous ART attempts that used 90% pure semiconducting-SWNTs. Therefore, this work also investigated the ART as applied to a less pure mixture of semiconducting and metallic SWNTs. The ratio of sc-:m-SWNTs was unknown.

4.1.1 Analysis of 78% Purity SWNTs

The as-purchased SWNTs solid was dissolved in 1% SDS/SC aqueous solution for a 0.1 mg/mL (wt/vol) concentration. This solution was the same surfactant solution used for the previous 90% sc-SWNTs from Nanointegris. The samples were applied to a SiO_2 surface that had been functionalized with molecule **1** using 5CB. The results are arranged in Fig. 4.1.

The average lengths of these untreated SWNTs were 2 μm, and the SD was 3.3°, although the data set was small as not as many AFM images were obtained. The density was 0.1 SWNTs/μm^2, and 54% of the tubes were in the 0° ± 10° range of alignment. Fewer NTs adhered to the surface with this sample. This result could be due to the mixture's impurities or that the different sizes of SWNTs do not match the iptycene **1** structure. Raman analysis was conducted on the sample in two laser frequencies ($\lambda = 632$ nm and 532 nm, Fig. 4.2a, b) to verify the diameter ranges. The full Raman spectra with G-band and the UV-visible spectrum was also collected for the raw untreated SWNTs (Appendix Chapter 4) by Dr. Heebong Yang of the Quantum Innovation (QuIN) laboratory (Kim Lab) at the University of Waterloo to compare the findings. The UV-vis data displayed diameters of SWNTs from 1.2 to 2.0 nm and chiralities from (12,5) to (25,14). Untreated SWNTs were also dropcast onto a clean Si surface and compared with the ART-treated ones' selective chiralities and diameters. This comparison was to demonstrate if ART had any effectiveness at separating diameter as was previously demonstrated with the higher purity sc-SWNTs solution [5].

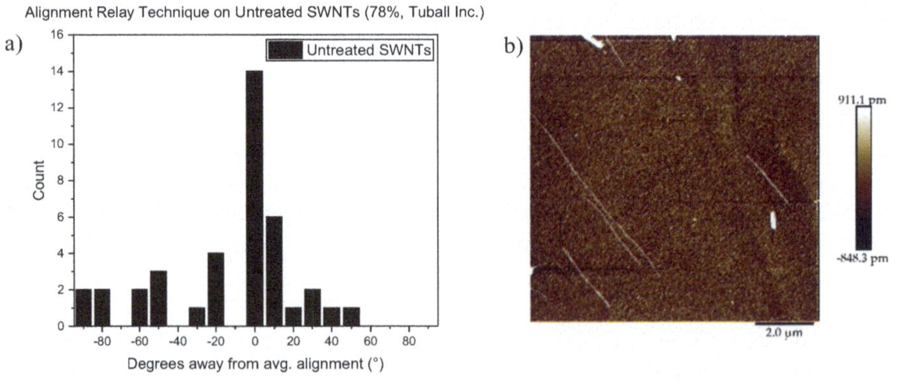

Fig. 4.1 (**a**) Degrees away from average alignment of the untreated SWNTs, (**b**) AFM of the ART processed SWNTs

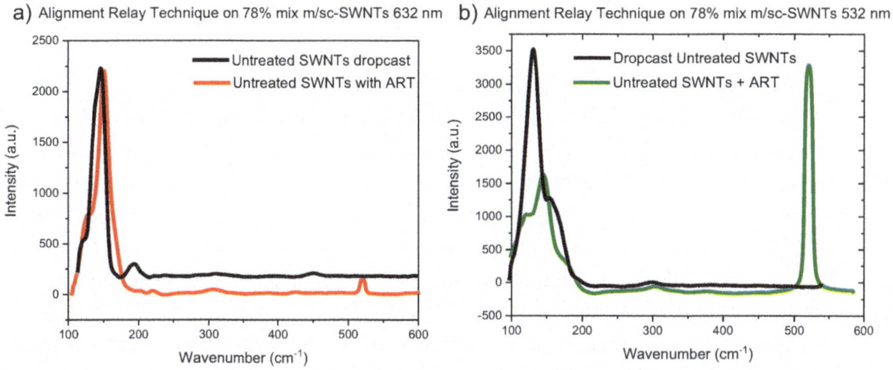

Fig. 4.2 Raman spectra comparing RBM of the dropcast versus ART functionalized 78% purity SWNTs in (**a**) red wavelength, (**b**) green wavelength. Peak at 232 cm^{-1} iptycene though very low intensity, 310 and 520 cm^{-1} from Si are for calibration; dropcast sample was thick so no reference peaks observed but can compare G-band for calibration. Mapping region was 10 × 10 μm

Table 4.1 Diameters and chiral estimates for the 78% purity mixed sc-/m-SWNTs from Tuball Inc., as is (dropcast) and ART-treated

Method	Diameter (nm)	Chiral candidates
UV-vis	1.2–2.0	(12,5)–(25,14)
Raman (632 nm)	2.02, 1.87, 1.42	(10,17)(18,9)(20,6)(14)(23,1)(22,2)(18,9)(14)(3,13)(5,14)(13,16)(19,11)(24,4)(20,9)
Raman (532 nm)	1.90, 1.56	(19,8)(5,11)(7,10)(3,14)(10,11)(3,16)(1,17)(6,16)(10,12)(19,2)(7,15)(20,0)
Raman (632 nm) (ART)	1.95, 1.64	(24,1)(21,7)(23,3)(12,17)(22,5)(18,9)(14)(3,13)(5,14)(13,16)(19,11)(11,13)(8,)(10,14)(10,15)(19,5)
Raman (532 nm) (ART)	2.06, 1.76, 1.41	(24,2)(22,7)(25,1)(24,3)(19,2)(7,15)(20,0)(10,12)(6,13)(8,12)(7,10)(3,14)

The diameters and approximate chiral candidates of the dropcast solution are disseminated in Table 4.1. It was observed that the higher wavenumber (smaller diameter) SWNTs at around 186 cm^{-1} were no longer present on the surface when analyzing with $\lambda = 632$ nm (1.96 eV). These results conform with previous findings that smaller diameter nanotubes will not have adhered to the ART surface when molecule **1** is probed in the red wavelength [5]. In the 532 nm Raman spectrum, the highest intensity observed was at 145 cm^{-1} ($d_t = 1.70$ nm) when using the ART. In contrast, the dropcast sample had a high intensity for a mix of SWNTs at 130 cm^{-1}. Accordingly, there is a shift in dominant tube types on the surface upon treatment. Additionally, metal SWNTs such as (18,9), (7,10), or (14) can still be present in the mixture. Therefore, there were many tube types on the sample. The as purchased-SWNTs were then treated under strong acidic conditions to solubilize them in water.

4.1.2 Synthesis of SWNTs-COOH

For chemical oxidation, H_2SO_4/HNO_3 in a 3:1 ratio has been reported to produce short open-ended cylinders with many carboxylic acid groups [4, 6]. The process can occur when mixing overnight [4] or within a few minutes at high pressure [7]. This treatment renders the SWNTs soluble in water and decreases their toxic effects [7]. The 78% purity Tuball, Inc. SWNTs were used and subjected to the H_2SO_4 (18 M)/HNO_3 (16 M) conditions (Scheme 4.1). The sulfonitric treatment has been described to create the most abundant carboxylic acid groups at both initial and newly created defect sites [4].

The acidic mixture produces large quantities of -COOH and -OH groups through the formation of NO_2^+. The oxidation of the SWNTs was defined as first originating from -CH_2/-CH/heptatonic rings through electrophilic addition (defect generating). Addition can also occur at hexatomic-hexatomic boundaries that create extra active sites (defect generating). The SWNTs' structure can be broken around the generated defect areas termed *defect consumption*. Defect generation occurs much faster than defect consumption, but both simultaneously occur upon acid exposure [4]. Resulting SWNTs-COOH are washed with copious amounts of water. The FTIR spectrum (Appendix Chapter 4) demonstrated the formation of the C=O stretch for carboxylic acids at ~1720 cm^{-1}. The resulting mixture was also soluble in MilliQ water between 1 and 4 wt %.

4.1.3 Results and Discussion for SWNTs-COOH

4.1.3.1 Comparing Untreated SWNTs to Acid-Treated (SWNTs-COOH)

The acid treated SWNTs were diluted in a 1% wt/vol solution with MilliQ water and dropcast onto a cleaned SiO_2 surface. Raman analysis was conducted on the samples, and the full spectra are shown in both wavelengths in Fig. 4.3. A significant increase in peaks is observed. This observation indicated the expected increase in surface defects when the SWNTs are exposed to strong oxidation conditions. Both Raman spectral frequencies reflect an increase in the CNT solution's impurities and

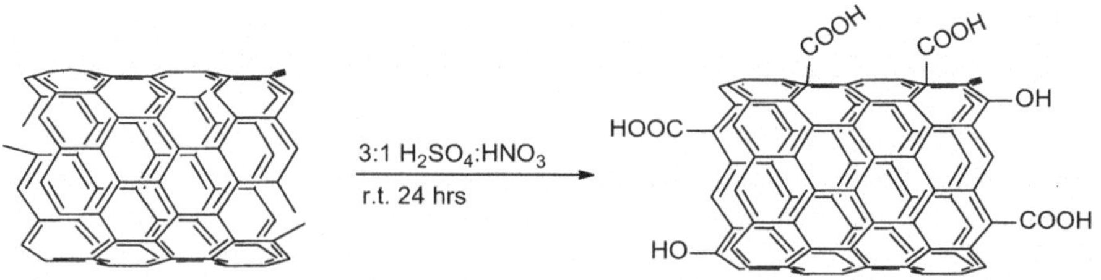

Scheme 4.1 Treatment for obtaining carboxylic acid moieties on SWNTs. The reaction begins at defect sites.

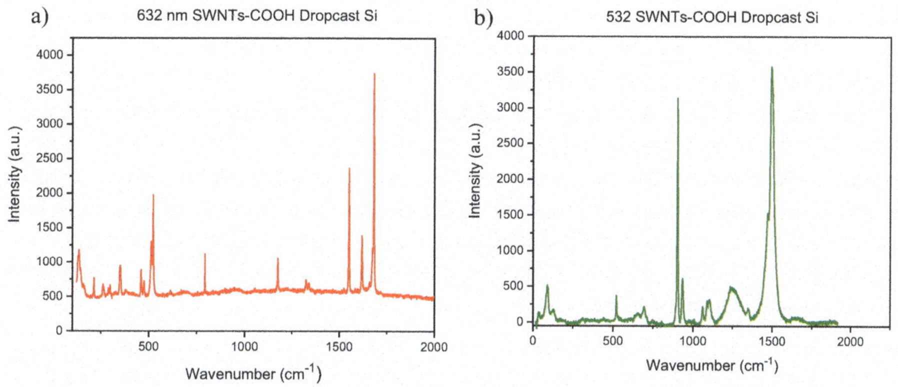

Fig. 4.3 Raman spectra of the dropcast SWNTs-COOH sample at (**a**) red wavelength, (**b**) green wavelength

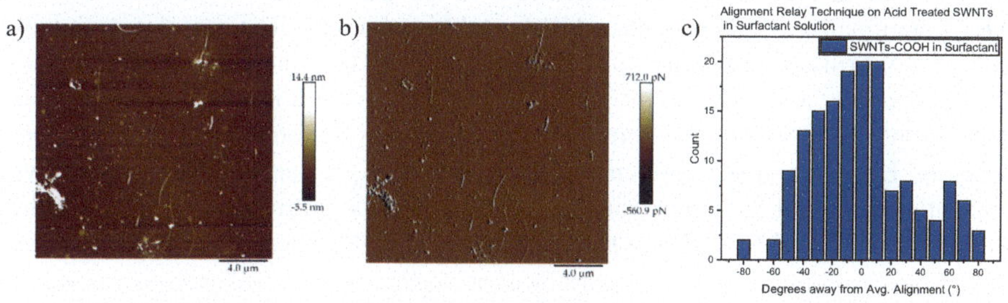

Fig. 4.4 The resulting SWNTs-COOH in surfactant solution (**a**) AFM topography image, (**b**) AFM phase image, (**c**) histogram of degrees away from average alignment over nanotube count

a disruption in C-C bonds [8]. Figure 4.3a, b showed an increase in peaks throughout the wavenumber range, and the D-band is much more prominent. The 632 nm spectra appeared to have the G-band slightly offset, but the results were the same even with multiple measurements. The same spot was taken for both the 632 nm and 532 nm frequencies. The sample is then subjected to ART in both the surfactant solution and in water to compare the differences.

4.1.3.2 Acid Treated SWNTs (SWNTs-COOH) in Surfactant

A portion of the SWNTs-COOH was dissolved in 1% SDS/sodium cholate aqueous solution for a 0.1 mg/mL (wt/vol) concentration. An ART functionalized SiO_2 was placed in the solution for 48 h. The samples were then rinsed with water and dried with $N_{2(g)}$. The findings are shown in Fig. 4.4, including an AFM phase image as some of the CNTs were not as noticeable in the corresponding topography image. Some areas depicted high alignment, whereas others had very randomly placed tubes, resulting in the histogram in Fig. 4.4c. The SD was calculated to be 38.4°, and

40% of SWNTs-COOH were in the range of 0° ± 10°. The observed length of the SWNTs-COOH is shorter compared to the untreated ART sample, at 1.3 μm, and the density on the surface was 0.2 μm^{-2}.

The surface appeared with increased impurities compared with the untreated SWNTs. This observation could be due to the treatment of the SWNTs with external reagents. Additionally, because the SWNTs are now hydrophilic, they probably interact differently with the surfactant molecules in the solution, and that could be the reason more particles on the surface are being observed. Compared to the findings in Fig. 4.1, there are more CNTs on the surface but they have a greater degree of deviation in terms of alignment.

4.1.3.3 Acid Treated SWNTs (SWNTs-COOH) in MilliQ (Water)

Samples were dissolved in water in 1, 2, and 4% wt/vol solutions. The solutions were sonicated for 10–20 min until the SWNTs-COOH appeared dissolved. An ART functionalized SiO_2 was placed in the solution for 48 h. The samples were then immediately rinsed with water and dried with $N_{2(g)}$. The concentrations at 1% and 2% yielded very few tubes on the surface. The 4% concentration displayed a greater deposition of SWNTs-COOH on the surface and was therefore chosen for further analysis. However, this treatment also resulted in a greater number of unknown impurities. Figure 4.5a depicts similar properties to the untreated SWNTs, and the surfactant dispersed SWNTs-COOH. However, Fig. 4.5b depicts the phase image

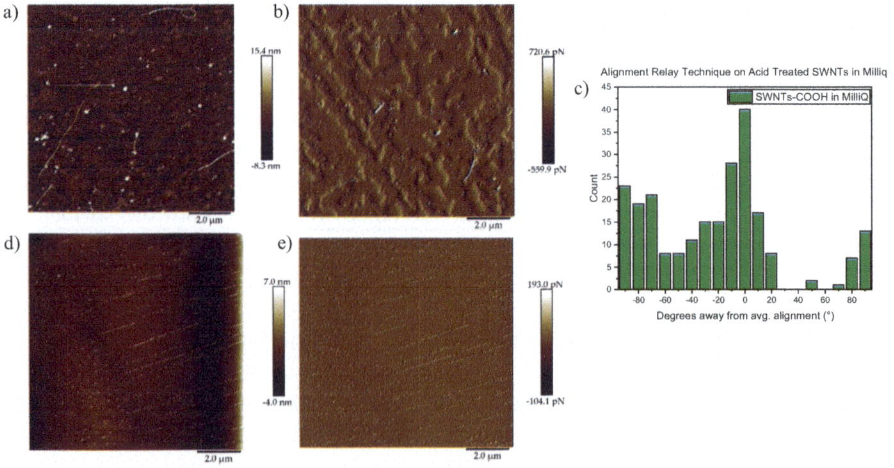

Fig. 4.5 The SWNTs-COOH dissolved in water and then deposited on the ART surface (**a**) Topography AFM, (**b**) Phase AFM showing the increased roughness and layers of impurities on the surface, (**c**) histogram of degrees away from average alignment over nanotube count, (**d**) SWNTs-COOH ART surface with alignment (topography), (**e**) SWNTs-COOH ART surface with alignment (phase)

that shows a change in surface stiffness and adhesive forces when the sample is submerged in water.

As shown in Fig. 4.4b in the surfactant solution, the phase images of the underlying surface were very smooth, but with Fig. 4.5b, that was not the case. The hydrophobic surface of the iptycenes could be poorly interacting when placed in the aqueous solution. However, some areas did show high SWNTs-COOH alignment, as shown in Fig. 4.5d with the corresponding phase image Fig. 4.5e to confirm these were real SWNTs and not AFM tip noise (height shown in the Appendix Chapter 4).

The SD was calculated to be 34.7° with 42% of the SWNTs-COOH in the 0° ± 10° range. The average length was 0.9 μm, smaller than the previous findings listed above. The density also decreased by half from the surfactant results, to 0.1 CNTs/μm^2. The results demonstrated that the water dispersion has similar results to using the surfactant solution, but with increased surface impurities. It is possible that the water rinse at the final stage of ART could be removing some of the water-soluble SWNTs-COOH from the surface, resulting in the observed low density. It was also attempted to have the water in a slightly basic environment (pH ~ 8) to increase the dispersion of the SWNTs-COOH in water. This experiment did not work and resulted in observing a single nanotube over fifteen 10 × 10 μm images.

In Fig. 4.6, the Raman spectra for the two laser wavelengths are shown for the sample from Fig. 4.5. These data displayed large width peaks, which are deconvoluted, and the chiralities are approximated and shown in Table 4.2. The dropcast untreated SWNTs are sample 1 (S1), the dropcast SWNTs-COOH are sample 2 (S2), the untreated SWNTs on ART are sample 3 (S3), the SWNTs-COOH in surfactant for ART are sample 4 (S4), and the SWNTs-COOH in water with ART are sample 5 (S5). The signals were averaged from three separate areas on the wafers rather than mapped, as the SWNTs were sparsely distributed on the surface and the AFM showed barely any tubes within the same 10 × 10 μm region, so it would not have been time efficient to conduct Raman mapping on the SWNTs-COOH ART samples.

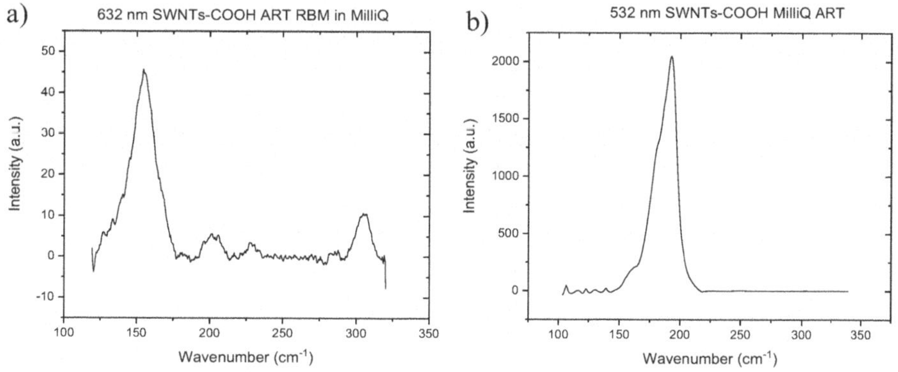

Fig. 4.6 Raman spectra in the two laser frequencies (**a**—632 nm, **b**—532 nm) of the RBM region of the SWNTs-COOH on the ART surface

Table 4.2 Compiled chiral approximations for the RBM identified peaks for samples discussed in this chapter

Sample	Wavelength (nm)/E (eV)	ω_{rbm} (cm^{-1})	Diameter (nm)	Chiral candidates (n,m)	Closest chiral approximation (n,m)
Untreated Dropcast [S2]	632/1.96	132.3	1.87	(17,10)(18,9) (20,6)(14,14) (23,1)(22,2)	(22,2)
		175.5	1.42	(18,9)(14,14) (13,3) (14,5)	(14,5)
		123.3	2.01	(16,13) (19,11) (24,4) (20,9)	(19,11)
	532/2.33	130.6	1.90	(19,8)(11,5) (10,7)(14,3)	(14,3)
		159.4	1.56	(11,10)(16,3) (17,1)(16,6) (12,10)	(12,10)
Treated Dropcast (SWNTs-COOH) [S2]	632/1.96	132.3	1.87	(17,10)(18,9) (20,6)(14,14) (23,1)(22,2)	(22,2)
		158.4	1.56	(16,6)(12,11) (20,0)(15,8)	(15,8)
	532/2.33	127.2	1.95	(16,11)(20,7) (17,9)(14,13) (22,3)	(14,13)
		172.9	1.43	(13,8)(16,5) (11,10)	(13,8)
		194.3	1.28	(15,0)(14,2)	(14,2)
Untreated with ART [S3]	632/1.96	127.0	1.95	(24,1)(21,7) (23,3)(17,12) (22,5)	(24,1)
		151.0	1.64	(13,11)(16,8) (14,10)(15,10) (19,5)	(13,11)
	532/2.33	120.3	2.06	(24,2)((22,7) (25,1)(24,3)	(24,2)
		145.9	1.70	(19,2)(15,7) (20,0)(12,10)	(19,2)
		176.0	1.41	(13,6)(12,8)	(13,6)

(continued)

Table 4.2 (continued)

Sample	Wavelength (nm)/E (eV)	ω_{rbm} (cm^{-1})	Diameter (nm)	Chiral candidates (n,m)	Closest chiral approximation (n,m)
Treated with ART in surfactant (SWNTs-COOH) [S4]	632/1.96	198.3	1.25	(13,4)(14,2)(15,0)	(13,4)
		143.5	1.73	(15,8)(20,1)(14,10)(19,3)	(14,10)
	532/2.33	140.2	1.77	(19,2)(15,7)(20,0)	(19,2)
		169.3	1.46	(13,6)(12,8)	(13,6)
		184.3	1.35	(17,1)(16,3)(12,8)	(17,1)
		191.2	1.30	(15,0)(14,2)(13,4)	(14,2)
Treated with ART in water (SWNTs-COOH) [S5]	632/1.96	156.7	1.58	(16,6)(12,11)(20,0)(15,8)	(15,8)
		149.0	1.66	(13,11)(16,8)(14,10)(15,10)(19,5)	(13,11)
		202.2	1.23	(15,3)(11,8)(14,5)	(14,3)
	532/2.33	182.3	1.36	(17,1)(16,3)(12,8)	(17,1)
		192.3	1.29	(15,0)(14,2)	(14,2)

Both S4 and S5 show similar tube types and diameters. Even smaller diameter tubes appeared to remain on the surface (Fig. 4.6a). This observation could be from the carboxylic acid functionalization that adds to the outer diameter. As such, the original smaller diameter SWNTs-COOH could be interacting with the iptycene layer. Additionally, when the SWNTs were treated with the acid (S2), the peak for the 2.0 nm diameter tube was no longer present on the surface (originally in S1), but this signal reappears in S5. It could be that the peaks were masked or not resonating due to the other nanotubes being more prominent on the surface in S2. It is difficult to confirm the tubes' metallicities as their conductivities could be disrupted by -COOH and -OH groups [9]. The (2,14) metallic tube is likely to remain on both the S4 and S5 when treating with the acid. Nevertheless, the closest chiral approximations show that the (5,14) metallic tube from S1 is no longer present on the surface of S3.

4.1.4 *Summary of SWNTs-COOH Studies*

These results demonstrated that it was possible to perform ART without the use of surfactant, but the alignment is not improved, and the density of the acid-treated SWNTs is 0.1 μm^{-2} (i.e., low). For improved results, the samples could be further purified before acid functionalization using ultracentrifugation in surfactant. Additionally, the final rinse could be substituted for an organic solvent as a means of ensuring the water-soluble SWNTs are not removed from the surface. Then, the concentration of the SWNTs-COOH in water could be adjusted without the concerning impurities.

4.2 Longer SWNTs

For transistor applications, longer SWNTs allow for easier lithographic patterning of electrodes. If longer SWNTs are aligned on the surface, although the channel length increases, it is less likely that source and drain electrodes will overlap. The less the likelihood of the source and drain overlapping, the increase in devices that can be properly fabricated. A larger gap also allows for improved resist lift-off, which leads to improved device fabrication. Here, larger SWNTs (5 μm lengths) were obtained from the Minari lab at NIMS from Dr. Qingqing Sun. These tubes were shown to be of high semiconducting purity from photoluminescence and were dissolved in toluene from the provider. Figure 4.7 shows these SWNTs dropcast onto silicon, and the Raman spectrum was recorded at 532 nm. The diameters match with what has been observed to work with the ART and molecule **1**.

These SWNTs were diluted in high purity toluene and then tip-sonicated for 30 min to de-bundle the SWNTs before use. The stock solution was concentrated SWNTs in toluene (0.38 g/L). The ART conditions were **1**/5CB on UV-Ozone

a)

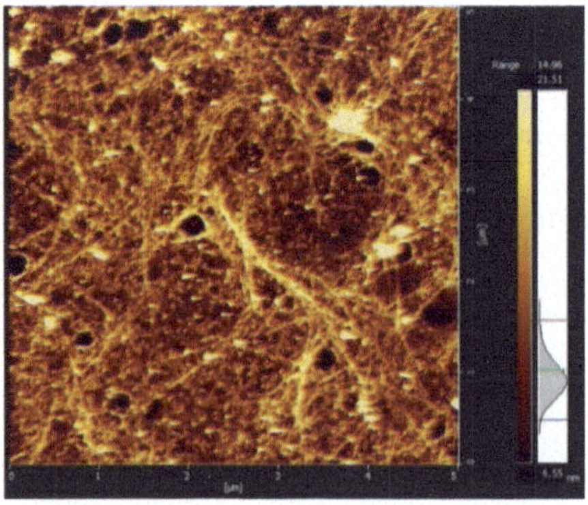

b)

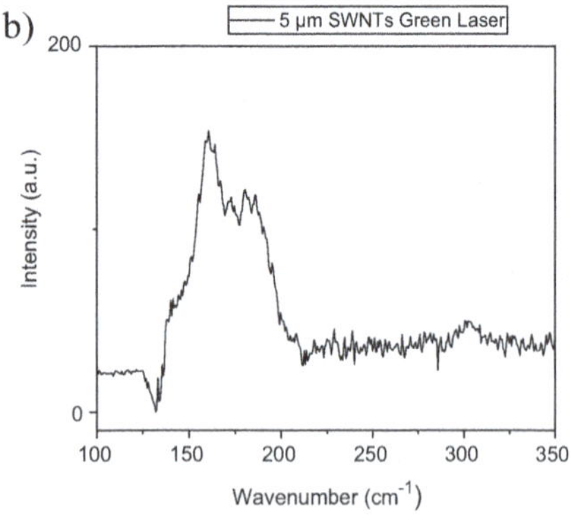

Fig. 4.7 (**a**) AFM of 5 μm SWNTs bundles dropcast on silicon, (**b**) Raman RBM of the 5 μm SWNTs at 532 nm

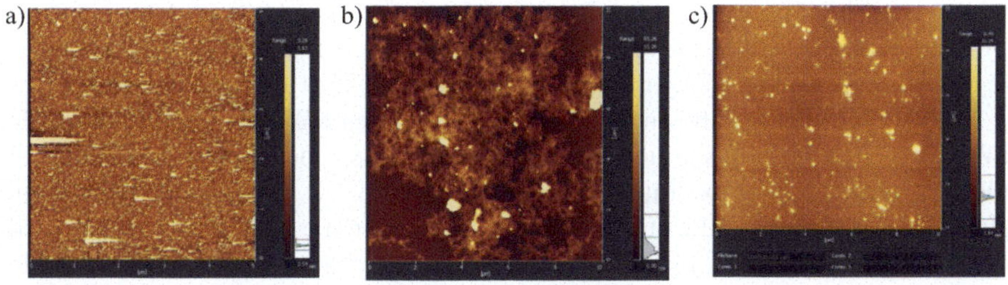

Fig. 4.8 (**a**) Toluene dissolved 5 μm SWNTs on ART surface, (**b**) hexanes dissolved 5 μm SWNTs on ART surface, (**c**) hexanes dissolved 5 μm SWNTs on ART surface at reduced concentration 0.1%

cleaned SiO_2. However, upon the treatment, the AFM results showed an un-aligned surface with few individual SWNTs (Fig. 4.8a). As from the findings in Chap. 2 when toluene was used as a solvent for the sonication experiments (Fig. 2.10, Sect. 2.1.1), similar results are being depicted herein that there are many particles and impurities on the surface with very few SWNTs. The solvent was then changed to hexanes to remove any potential interference in π–π stacking that could have been arising from the toluene. The concentration was also increased from 0.01/10 mL (v/v) toluene to 0.1/10 mL (v/v) hexanes. This increase in concentration but not an increase in tip sonication time likely the reason for the bundles of tubes being observed on the surface (Fig. 4.8b).

By decreasing the concentration to 0.01/10 mL with higher power tip-sonication in hexanes, there were fewer CNTs on the surface. However, the SWNTs in hexanes treated surface had fewer impurities and only some large particles on the surface of the SWNTs themselves. In contrast, the toluene dissolved sample possessed many smaller impurities throughout the surface. With these results, this batch of CNTs was no longer tested. A further direction for this research direction would be to remove all the toluene (rotary evaporation) from the solution and use chloroform as the solvent at a 1% concentration of SWNTs as that has shown to be efficient to disperse the CNTs individually.

4.3 Smaller Diameter SWNTs

APTE purified sc-SWNTs with smaller diameters were provided from the Kim Lab at the Institute for Quantum Computing (QuIN lab, University of Waterloo). Two solutions were obtained at unknown concentrations that were centrifuged, and the SWNTs pellet extracted and re-dissolved in aqueous surfactant (1% SDS/SC). Both samples were treated with ART (**1**/5CB). These samples showed diameter ranges in the Raman 0.85–0.94 nm (532 nm) and 0.83–0.98 nm (632 nm), provided in the Appendix Chapter 4. The three deconvoluted peak chiralities of the green solution were previously determined from UV-vis to be (4,9), (10,12), (5,8). The purple solution had SWNTs types of (3,7), (4,6), (5,6), (5,8) and (1,9) as determined by QuIN researchers. All the previous ART investigations using Nanointegris Inc. 90% sc-SWNTs, which had a larger diameter range from 1.25–1.65 nm (150–180 cm^{-1}).

The ART results are shown in Fig. 4.9 for the green SWNTs (a) and the purple SWNTs (b). The purple solution was significantly less concentrated than the green SWNTs. Both samples exhibited some alignment of SWNTs but a low distribution of CNTs. Some areas of the purple SWNTs solution on the ART substrate appeared very strange, as shown in Fig. 4.9c. What appears as aligned tubes in this figure do not equate to SWNTs based on the height parameters, and this type of pattern is present in some regions of the surface. The purple SWNTs showed very low deposition and were not analyzed further. The aligned histogram for the green SWNTs is shown in Fig. 4.9d. The SD was calculated to be 53° and a density of 0.13 NTs/ μm^{-2}. The average length was 0.6 μm. The ART appeared to promote some degree of alignment, but a higher concentration is needed for accurate results.

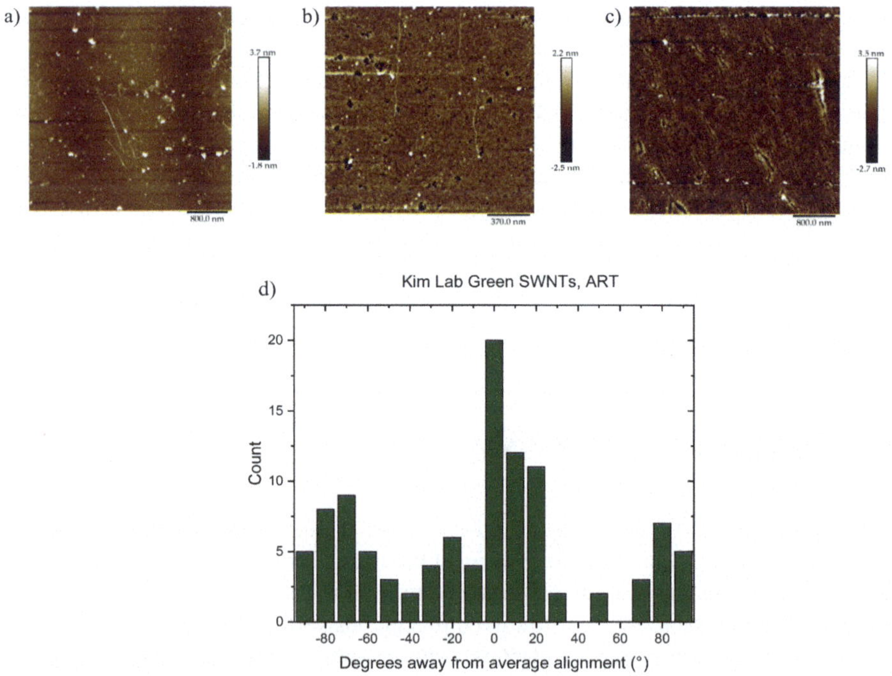

Fig. 4.9 AFM of (**a**) green SWNTs applied to ART, (**b**) purple SWNTs applied to ART at 1 × 1 μm, (**c**) purple SWNTs applied to ART at 3 × 3 μm, (**d**) histogram for degrees away from average alignment for the green SWNTs

4.4 Multi-Walled Carbon Nanotubes

Multi-walled CNTs were also attempted to observe if larger diameter structures were compatible using **1** in ART. Aligned MWNTs can perform as ballistic conductors in integrated circuits [10], therefore, research in these materials is useful for expanding ART. The MWNTs were dissolved in the same SDS/SC 1% (wt/vol) aqueous dispersion used for the SWNTs. Two diameters mixtures were tested: 8 nm (M1) and a batch with ranges of 10–20 nm (M2). The deposition of the MWNTs was first attempted at 24 h. It was observed that there were very few tubes on the surface when allowing the sample to be submerged for only one-day in both the small and large diameter samples. The number of tubes per sample image ranged from 10–20 MWNTs/30 × 30 μm^2. Longer deposition times were then tested, for 48 h and 72 h, to obtain an increase of MWNTs on the surface.

The sample with the larger diameter MWNTs slightly improved after two days to 0.3 CNTs/10 μm^2, and the smaller diameter MWNTs only increased after three-days of deposition time but only to 0.15 CNTs/10 μm^2. As shown in Fig. 4.10, some smaller diameter MWNTs were deposited on the surface (Fig. 4.10a) in very low quantities, even with increased deposition time. The larger diameter MWNTs (Fig. 4.10b) sample had significantly more tubes. The tubes were wavy and dispersed throughout the sample, with no degree of orientation. After three-days, the larger diameter MWNTs had many more impurities and unknown particles across the surface. SEM was used to image the MWNTs rather than the AFM as it was much faster, and the MWNTs were easier to recognize with this technique.

The concentration was then doubled (0.02 mg/mL) to investigate further using MWNTs to ART. All other parameters, such as deposition time and functionalization time and temperature, remained the same. By doing so, the average density of the 8 nm diameter MWNTs sample was increased to 0.1 CNTs/10 μm^2 although still not very high considering previous results for SWNTs; it was still an increase from 0.0001 μm^{-2}. On the contrary, the M2 sample did not exhibit an increase in CNT density. Therefore, a 2% solution of 8 nm diameter MWNTs and a 1% solution of the 10–20 nm diameter MWNTs mixture were subsequently used for further

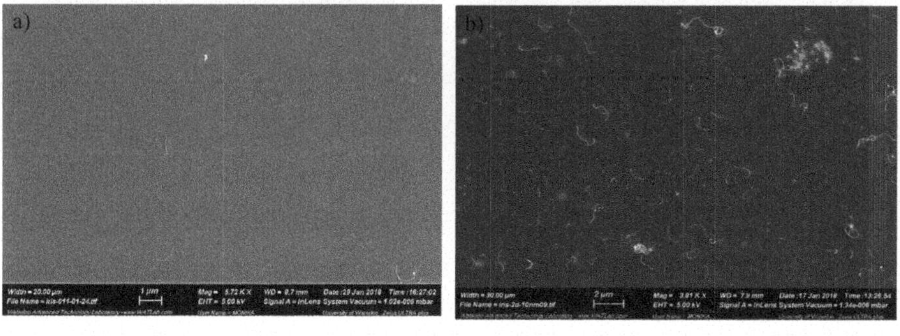

Fig. 4.10 SEM images of the (**a**) 8 nm MWNTs after 3 days depositing, (**b**) 10–20 nm MWNTs deposited for 2 days

analysis. The average lengths of the MWNTs deposited on both samples was approximately 2.5 µm.

Samples were then subjected to 5- and 10- min post-deposition sonication treatments. The results are tabulated in the histograms of Fig. 4.11a, b. The treatment was not beneficial for M1, but the few tubes accounted for seemed to be aligned when treated with 10 min of sonication. For M2, 10 min also showed a slight increase in the alignment of the MWNTs at the expense of the density. A similar trend was observed with the SWNTs in Chap. 2. For sample M1, the average lengths were shorter than without sonication, 1 µm at 5 min treatment, and 1.6 µm at 10 min treatment. For M2, the average lengths of the MWNTs remained consistently around 1 µm. The SD values for all the treatments was between 40 and 55°, and the densities were all below 0.01 μm^{-2}.

Therefore, there was little alignment, and the sonication treatment was not as effective on the larger diameter MWNTs like it was with the SWNTs. These results could be due to the larger diameter not interacting well with iptycene **1**, which leads to the MWNTs not adhering to the surface. The additional curvature observed in the

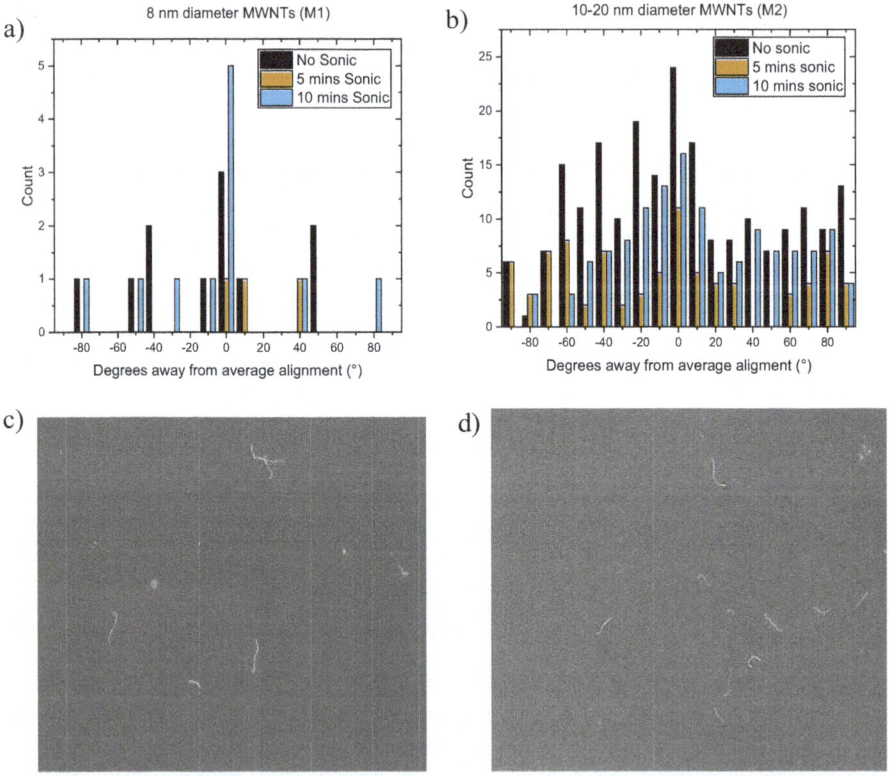

Fig. 4.11 Comparing degrees away from average alignment for the non-sonicated, 5 min and 10-min sonication treatments for (**a**) 8 nm diameter MWNTs, and (**b**) 10–20 nm diameter MWNTs, (**c**) 20 × 20 µm SEM of 5 min sonicated M2, (**d**) 20 × 20 µm SEM of 10 min sonicated M2

MWNTs in Fig. 4.11, as with Fig. 4.10, indicates an absence of interaction with the molecules. As was observed with the SWNTs, when the nanotubes aligned, they appear straight on the surface. This interaction with the iptycenes appears to be stronger than the natural tube curvature or tube bundling. Therefore, if the MWNTs had proper interactions with the iptycenes, the many curved tubes' observation would not result.

As an additional test, sonication with time for redeposition was also investigated. It was shown that allowing the SWNTs to rest in the nanotube solution after sonication did not increase the target alignment; however, as the MWNTs behave differently, it was speculated that the redeposition treatment could work. The 5 and 10 min sonication treatments were each attempted on the samples with an additional 1 h of leaving the samples in their nanotube solutions. The results for the orientation are shown in Fig. 4.12.

The average lengths were 1.5 μm for M1 in both treatments and 1 μm for M2 in both treatments. There was an increase in the number of tubes for both samples. For M1, the density increased in all the samples to slightly above 0.01 μm^{-2}, but the

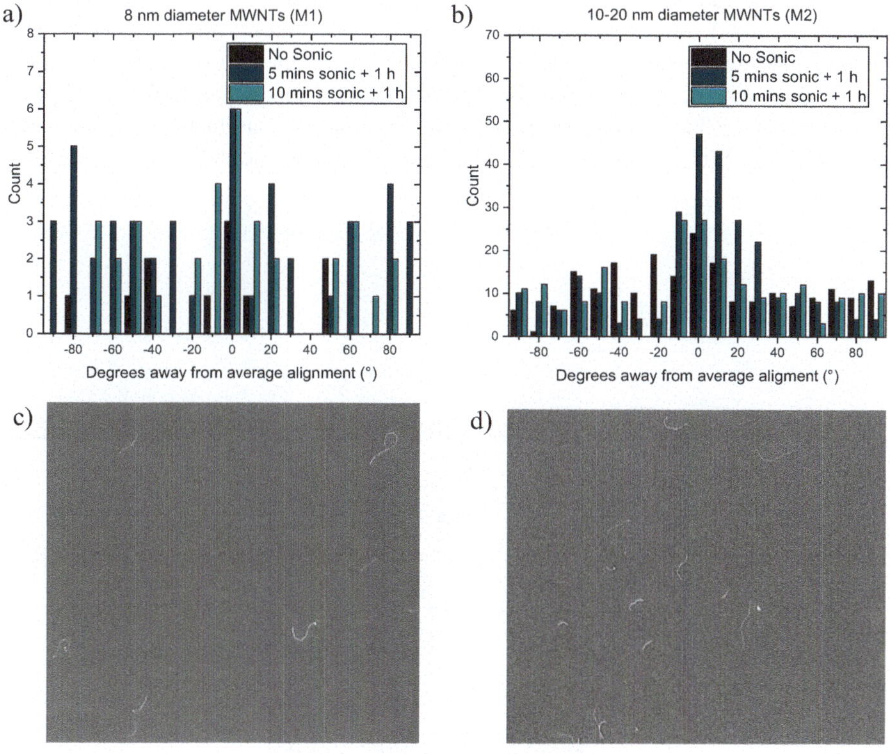

Fig. 4.12 Comparing degrees away from average alignment for the non-sonicated, 5 min with 1 h of redeposition time and 10 min sonication with 1 h redeposition treatments for (**a**) 8 nm diameter MWNTs, and (**b**) 10–20 nm diameter MWNTs, (**c**) 20 × 20 μm SEM of 5 min sonicated with redeposition of M2, (**d**) 20 × 20 μm SEM of 10 min sonicated with redeposition of M2

MWNTs were still all sparsely apart and separated, as shown by some SEM images for M2 (Fig. 4.12b). The number of CNTs remains insignificant, especially when comparing to the SWNTs results with ART. The only improvement in alignment is in M2 with 5 min of sonication with 1 h of redeposition time and the 10-min continuous sonication. These samples showed a SD decrease from 47° in the non-sonicated sample to 40° with these treatments.

Interestingly, there is a larger number of CNTs in M2 than in M1. It would be expected that neither of the MWNTs adhere to the surface as they are too large in diameter for the iptycenes. Nevertheless, the larger diameter ranged CNTs are the ones predominantly present on the surface, and the 8 nm diameter CNTs are barely observed. It is possible that some of the larger diameter MWNTs present in the M2 sample somehow have stabilizing aromatic interactions with the underlying iptycene layer that is not possible in the 8 nm diameter CNTs. Additionally, the M2 MWNTs are slightly heavier than the smaller diameter M1 MWNTs, and perhaps this slight addition in mass allows them to remain on the surface.

As a final test, the M2 MWNTs were dispersed in a 2% SDS/SC solution instead of 1% to observe if the additional surfactant would help de-bundle and distribute the MWNTs. The samples were treated with no sonication, 5 min of post-deposition sonication, and 5 min of sonication with a 1 h redistribution time. The respective SEM results are shown in Fig. 4.13a–c, and the alignment data is distributed in the histogram of Fig. 4.13d. The density of MWNTs with no sonication was 0.02 μm^{-2} with SD = 40°. With 5-min of sonication, the SD was 33°, but the density decreased to 0.01 μm^{-2}. With the additional redeposition time in the solution, the density

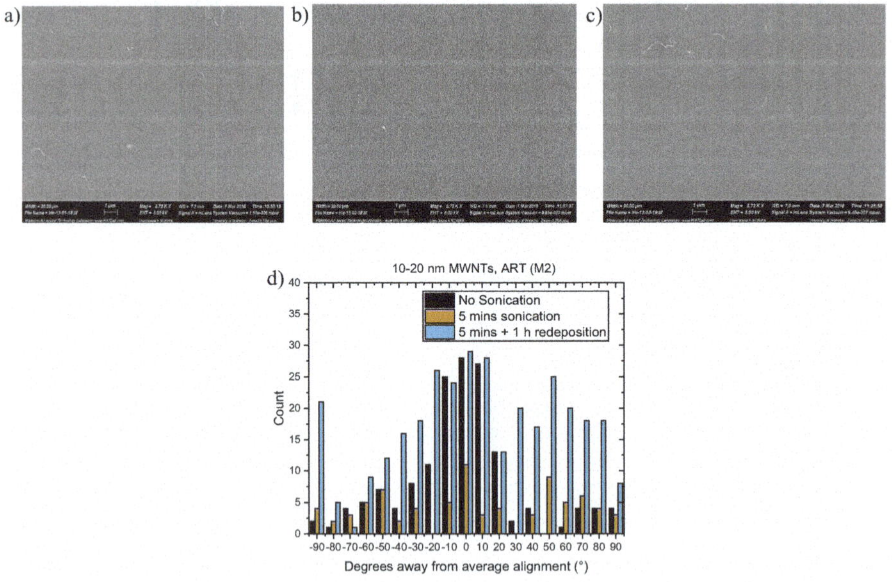

Fig. 4.13 Larger MWNTs with ART SEM images (**a**) no sonication, (**b**) 5 min sonication, (**c**) 5 min sonication +1 h redeposition, (**d**) histogram of alignment results

increased to 0.03 μm^{-2} with SD = 45°. The average length for all the samples was approximately 1.2 μm. However, if with the increase of CNTs on the surface, there is also an increase in bundles on the surface, which are not accounted for in the orientation data but are dominant on the surface, especially in the sample with sonication and redeposition.

Overall, the aligning MWNTs with ART using **1**/5CB is not effective. Changes would likely occur with the iptycene structure to increase aromatic stacking with the larger diameter tubes. Increasing MWNTs concentration and first dispersing the solution using tip sonication could also help achieve individually aligned MWNTs on the surface [11].

4.5 Graphene Nano-Ribbons

The attempted alignment of top-down and bottom-up synthesized GNRs was also investigated. The bottom-up fabricated GNRs were provided by the Chalifoux lab from the University of Nevada and had been imaged using scanning tunneling microscopy and TEM [12, 13]. A very concentrated mixture of the GNRs in toluene was dropcast on a substrate on silicon to obtain a Raman spectrum (Appendix Chapter 4). Then, a 1% wt/vol solution in toluene was created, and a functionalized substrate of **1**/5CB was left for 24 h, as shown in Fig. 4.14.

The findings from Fig. 4.14 demonstrate similar particles to cases where toluene was used as a solvent, such as a significant increase in unknown particles and impurities. The toluene was chosen as it was recommended for GNR solvation with much success. It was also speculated that the sample could not be concentrated enough, leading to very minimal GNRs being deposited. Also, the GNRs are much smaller in terms of height and length than the SWNTs, and there are only a few cases of AFM of GNRs [14]. It was then attempted to dropcast the GNRs solution

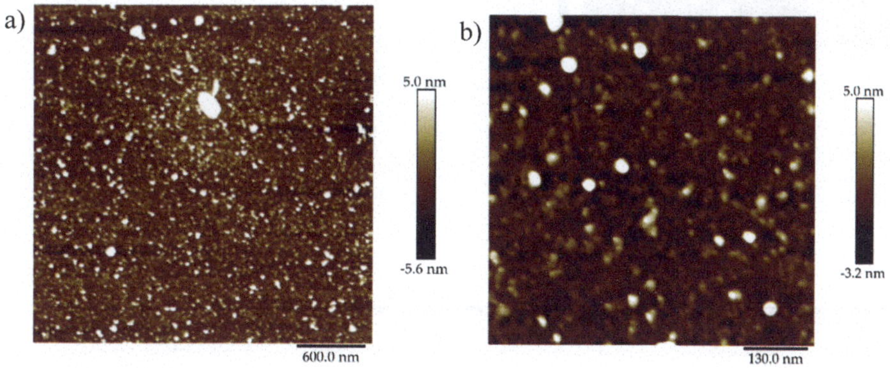

Fig. 4.14 GNRs (1% wt/vol) in toluene with an ART functionalized substrate after 24 h, (**a**) 3 × 3 μm (**b**) 500 × 500 nm scale images

onto a functionalized substrate to increase the concentration of GNRs and observe their behavior. As per the results in Fig. 4.15, the GNRs appear to adhere to the surface, but they are in a random orientation. The samples proved very difficult to image and identify the GNRs.

It was speculated that the bottom-up synthesized GNRs were much smaller and, therefore, more challenging to image using AFM than the top-down generated GNRs. Larger GNRs were obtained that had been unraveled from CheapTubes Inc. MWNTs by Dr. Saeed Habibpour from the Yu Group (Chemical Engineering) at the University of Waterloo. The sample provided was of reduced GNRs (rGNRs). The source MWNTs were a mix of 20–30 nm in diameter and 20–30 μm in length. It was speculated that these should be larger enough to image and analyze for ART effectiveness. The sample was attempted with ART and dissolved in 1% solutions with THF, rather than toluene to avoid possible π–π interferences with the iptycenes surface. Unfortunately, after many AFM imaging attempts, no GNRs were observed on the surface (one was speculated to be an rGNR and is shown in Appendix Chapter 4, but it is not confirmed). This observation was strange as the larger rGNRs arising from MWNTs unzipping are much larger and should be identifiable through AFM.

The GNRs were not pursued further; however, possible solutions for improvement are to: (1) increase the concentration of the solution and avoid toluene as the dispersing solvent for the GNRs, (2) use a freshly prepared mixture of **1**/5CB—the ART solution used here was not recently made, which could have affected the concentration of deposited iptycenes, which in turn affects the overall deposition, (3)

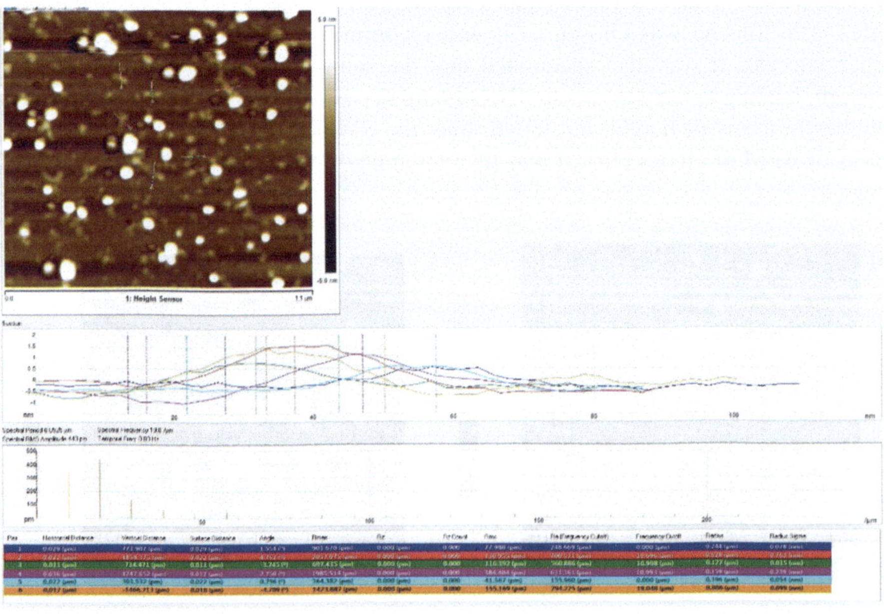

Fig. 4.15 AFM height data for dropcast GNRs on a ART functionalized surface and are approximately 0.7 nm in height (vertical distance)

change of the iptycene structure to conform with the less convex structure of GNRs. Unlike SWNTs, the GNRs are increasingly planar in their structure, and therefore, they might have better aromatic stacking with an iptycenes that better match this physical property. By considering parameters i–iii, the GNRs could potentially be characterized and optimized for ART deposition.

4.6 Nanowires

Metal oxide NWs are widely used in the semiconducting industry and optical materials due to their electronic structure and one-dimensional charge transport properties. For instance, ZnO NWs have a bandgap of 3.37 eV and an excitation energy of 60 meV for binding, making them useful for transistor systems [15, 16]. Among different nanostructures of metal oxide NWs, orderly-aligned metal oxide NW arrays have unparalleled advantages such as a high surface to volume ratio that provides a direct path for charge transport. However, it is a great challenge to align NWs on nanodevices due to NWs easily stacking together, similar to the problem faced with CNTs. The ART was also attempted in aligning ZnO and Al_2O_3 NWs. The ZnO NWs were 90 nm × 1 μm, and the Al_2O_3 NWs were a mixture of 206 nm × 200–400 nm (diameter × length). These were readily available and therefore were the first attempts at using NWs with ART. Similar to the observations with the larger MWNTs, SEM was used for characterization.

The NWs were dispersed in IPA at various concentrations using benchtop sonication for 10 min. However, upon imaging some of the samples, it was observed that the NWs were still bundled on the surface. The results in Fig. 4.16 depict a 10% wt/vol solution of the Al_2O_3 and ZnO NWs in IPA with 1 h deposition time on an ART-treated surface with **1**/5CB on silicon. The ZnO NWs were bundled on the surface, whereas the Al_2O_3 sample had single NWs randomly oriented and surrounded by many impurities. No further analysis was conducted on the systems as the research focus was shifted toward other projects.

Further studies for metal oxide NWs can be conducted by optimizing the concentration of the NWs in the solvent. Additionally, tip-sonication or longer benchtop sonication could be implemented to disperse the NWs and eliminate the bundles. Furthermore, the iptycene tweezers structure could be altered for larger convex structures, such as the NWs and the MWNTs.

4.7 Graphene and Graphene Quantum Dots

Two experiments were also conducted with graphene sheets and graphene quantum dots (GQDs) provided by Dr. Inna Novodchuk of the Yavuz lab (Mechanical Engineering) at the University of Waterloo. The samples were dispersed in 1% THF. The **1**/5CB surface remained in the sample solution for 24 h without

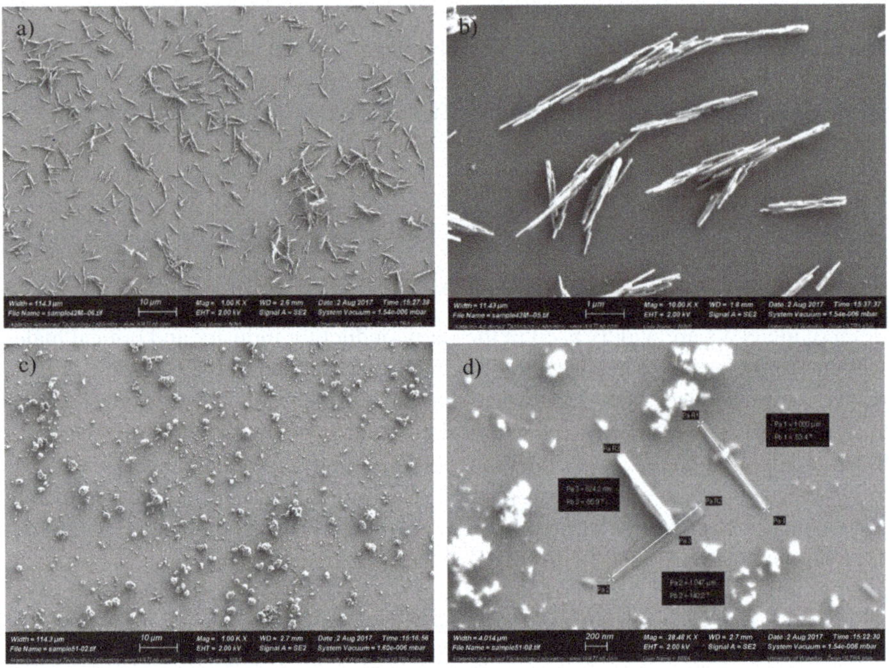

Fig. 4.16 SEM of NWs on ART surface: (a) 100 × 100 μm image of ZnO NWs, (b) 10 × 10 μm image of ZnO NWs, (c) 100 × 100 μm image of Al_2O_3 NWs, (d) 2.5 × 2.5 μm image of Al_2O_3 NWs

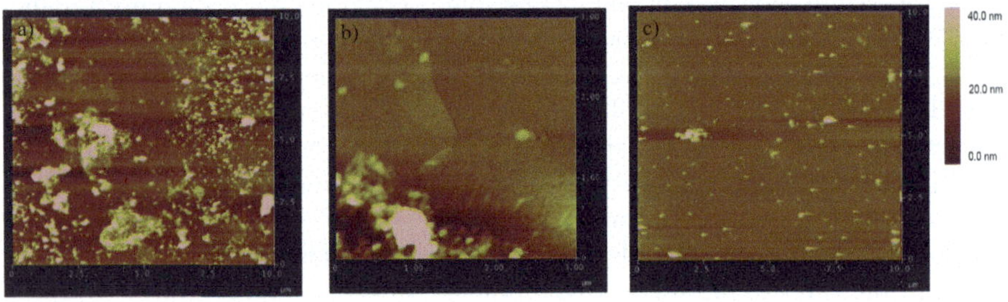

Fig. 4.17 AFM of ART treated surfaces with the following compounds: (a) graphene 10 × 10 μm, (b) graphene 1 × 1 μm possible graphene flakes, (c) GQD treated surface

sonication. Some of the AFM images are shown in Fig. 4.17. Although the graphene surface has many impurities and is very rough (Fig. 4.17a), there are possible graphene flakes deposited on the surface (Fig. 4.17b, height measurement at 4 nm shown in the Appendix Chapter 4), the results demonstrated this procedure to be inefficient.

Additionally, the images obtained for the GQDs (Fig. 4.17c) are also not beneficial. Due to the GQDs size, it is difficult to confirm their presence as the small dots appear similar to previous GQD AFM findings [17]. It would be beneficial to further analyze the sample for GQD presence through Raman, as the G-band observation would confirm that this species is present on the surface. Nevertheless, the preliminary investigations on these systems that have the same π–π stacking as SWNTs yet with a planar morphology so it is hypothesized that if the iptycene molecule in ART is adjusted to fit the planar structure of the nanosystem, then the method should work in distributing a monolayer of material across the surface.

4.8 Summary of ART for Other Nanosystems

Using SWNTs-COOH in water is possible, but the density and alignment were not as efficient as normal SWNTs in surfactant for ART. The metallicities and lengths of the SWNTs can also be altered with this treatment. The Raman results show there is no selectivity for the SWNTs-COOH diameters. However, the 78% purity SWNTs used that contained a larger diameter of 2 nm SWNTs in the dropcast were not observed on the ART-treated sample.

Any samples for ART should not be dissolved in toluene. As was the case for the 5 μm length SWNTs and the GNRs, non-aromatic containing solvent should be used as to not interfere with the π–π stacking interactions with the iptycene monolayer. This observation also conforms with the results obtained in Chap. 2. The concentration of SWNTs should not be too low in order to obtain a proper deposition of SWNTs. Smaller diameter SWNTs displayed minimal alignment and confirmed further the observation that molecule **1** predominantly adheres to SWNTs diameters in the range of 1.3–1.6 nm. Suggestions for adapting GNRs, MWNTs, NWs, graphene, and QGDs to ART include optimizing compound concentration, dispersion, and adapting the iptycene molecule structure.

References

1. Osorio, A.G., Silveira, I.C.L., Bueno, V.L., Bergmann, C.P.: H2SO4/HNO3/HCl—Functionalization and its effect on dispersion of carbon nanotubes in aqueous media. Appl. Surf. Sci. **255**(Part 1), 2485–2489 (2008)
2. Wang, H.: Dispersing carbon nanotubes using surfactants. Curr. Opin. Colloid Interface Sci. **14**(5), 364–371 (2009)
3. Peng, H., Alemany, L.B., Margrave, J.L., Khabashesku, V.N.: Sidewall carboxylic acid functionalization of single-walled carbon nanotubes. J. Am. Chem. Soc. **125**(49), 15174–15182 (2003)
4. Zhang, J., Zou, H., Qing, Q., Yang, Y., Li, Q., Liu, Z., Guo, X., Du, Z.: Effect of chemical oxidation on the structure of single-walled carbon nanotubes. J. Phys. Chem. B. **107**(16), 3712–3718 (2003)
5. Snowdon, M., Selmani, S., Schipper, D.J.: Sonication-enhanced alignment relay technique for the orientation of single-walled carbon nanotubes. ACS Applied Nano Mater. **2**(10), 6637–6645 (2019)

6. Liu, J., Rinzler, A.G., Dai, H., Hafner, J.H., Bradley, R.K., Boul, P.J., Lu, A., Iverson, T., Shelimov, K., Huffman, C.B., Rodriguez-Macias, F., Shon, Y.-S., Lee, T.R., Colbert, D.T., Smalley, R.E.: Fullerene pipes. Science. **280**(5367), 1253 (1998)
7. Saxena, R.K., Williams, W., McGee, J.K., Daniels, M.J., Boykin, E., Ian Gilmour, M.: Enhanced in vitro and in vivo toxicity of poly-dispersed acid-functionalized single-wall carbon nanotubes. Nanotoxicology. **1**(4), 291–300 (2007)
8. Zhang, K., Zhang, Q.: Raman signatures of broken C–C bonds in single-walled carbon nanotubes upon [2 + 1] cycloaddition. J. Phys. Chem. C. **119**(32), 18753–18761 (2015)
9. Skákalová, V., Kaiser, A.B., Dettlaff-Weglikowska, U., HrnčAriková, K., Roth, S.: Effect of chemical treatment on electrical conductivity, infrared absorption, and raman spectra of single-walled carbon nanotubes. J. Phys. Chem. B. **109**(15), 7174–7181 (2005)
10. Poncharal, P., Berger, C., Yi, Y., Wang, Z.L., De Heer, W.A.: Room Temperature Ballistic Conduction in Carbon Nanotubes. J. Phys. Chem. B. **106**(47), 12104–12118 (2002)
11. Yu, J., Grossiord, N., Koning, C.E., Loos, J.: Controlling the dispersion of multi-wall carbon nanotubes in aqueous surfactant solution. Carbon. **45**(3), 618–623 (2007)
12. Yang, W., Chalifoux, W.A.: Rapid π-extension of aromatics via alkyne benzannulations. Synlett. **28**(06), 625–632 (2017)
13. Yang, W., Lucotti, A., Tommasini, M., Chalifoux, W.A.: Bottom-Up synthesis of soluble and narrow graphene nanoribbons using alkyne benzannulations. J. Am. Chem. Soc. **138**(29), 9137–9144 (2016)
14. Wijeratne, S.S., Penev, E.S., Lu, W., Li, J., Duque, A.L., Yakobson, B.I., Tour, J.M., Kiang, C.-H.: Detecting the biopolymer behavior of graphene nanoribbons in aqueous solution. Sci. Rep. **6**(1), 31174 (2016)
15. Zhang, Y., Ram, M.K., Stefanakos, E.K., Goswami, D.Y.: Synthesis, characterization, and applications of ZnO nanowires. J. Nanomater. **2012**, 1–22 (2012)
16. Shen, G.-H., Tandio, A.R., Hong, F.C.-N.: Hydrothermally synthesized ultrathin zinc oxide nanowires based field-effect transistors. Thin Solid Films. **618**, 100–106 (2016)
17. Moon, J., An, J., Sim, U., Cho, S.-P., Kang, J.H., Chung, C., Seo, J.-H., Lee, J., Nam, K.T., Hong, B.H.: One-step synthesis of N-doped graphene quantum sheets from monolayer graphene by nitrogen plasma. Adv. Mater. **26**(21), 3501–3505 (2014)

Chapter 5
Applying ART Toward Single-Walled Carbon Nanotube Device Fabrication

Abstract This section explores the fabrication of CNTFETs using the ART treated surfaces. Both light-emitting diode (LED) photolithography and electron beam lithography (EBL) were used to fabricate the samples. Patterned substrates of 200 nm SiO_2 were used to ease the process for patterning and consistent testing. The nanotubes were, on average, 1 μm in length on the surface. A dropcast thin film of SWNTs was used as a means of comparison to observe for changes in electrical conductivity. Additional SWNTs were also obtained for the device fabrication. HipCo SWNTs solutions provided by the Kataura (National Institute of Advanced Industrial Science and Technology) and Ishihara (National Institute for Materials Science) groups were also used. One was a mixture of (10,3) sc-SWNTs in 0.5% SC solution (86 μg/10 mL), and the other was a mixture of many sc-SWNTs chiralities in 0.5% SC solution (0.1 mg/1 mL). Overall, top-gate and bottom-gate transistors were fabricated and tested using both electron beam lithography and photolithography. The ART nanotubes tested in a vacuum as a top-gate can achieve an I_{ON}/I_{OFF} of 10^6 with a mobility of 10 cm^2/Vs. Although greater I_{ON}/I_{OFF} and mobility values have been shown for CNTFETs, this work presents the first evidence that functioning devices are possible when using ART, and lays the ground work for the successors of this research.

Keywords Nanotubes · Carbon · Alignment · Relay · electron-beam lithography · Photolithography · Microscopy · Photoresist · LED lithography

5.1 Device Fabrication

5.1.1 Optimization of LED Lithography

In order to achieve working devices, the endpoints of the CNTs must be in contact with the source and drain electrodes. The LED patterning method required optimization to consistently obtain CNTs in the channel between the source and drain electrodes. The channel refers to the distance between the source (S) and drain (D)

M. Snowdon, *Methods of the Alignment-Relay Technique for Nanosystems*, Springer Theses, https://doi.org/10.1007/978-3-030-84413-4_5

electrodes Too large a channel resulted in the CNTs not connecting the two electrodes and, therefore, no electrical signal results. For instance, an example of a relatively successful device is depicted in Fig. 5.1; however, had the channel gap been optimized to below 1 μm, the CNTFET could have included five aligned SWNTs rather than solely one. Since the CNTs had an average length of 1 μm, a sub 1 μm channel gap was targeted.

In order to increase the success rate of the fabricated devices, optimization for achieving below 1 μm S-D gaps took place, with practice patterns conducted on bare 285 nm silicon. Once the patterns were applied to the photoresist (PR) patterned surface, the samples were kept in the dark and sputter coated with Pt. This process was done to ensure: (1) the exposure was enough such that the metals would adhere to the surface after the lift-off process, and (2) the patterned gap distance was consistently below 1 μm. However, it was observed that the parameters used for the test samples of bare silicon did not give the same results as to an ART-treated surface with CNTs [1, 2]. The parameters for untreated silicon patterns required less exposure time and a shorter channel pattern. Therefore, it was determined that test patterns had to be performed using the actual ART-functionalized surface with CNTs.

The substrate surface was 200 nm $SiO_{2,}$ as this was the surface for patterned substrates available. Additional variations were conducted by using a thicker layer of photoresist, applying a 30-s developer soak similar to chlorobenzene treatment [3], and using a bi-layer method combining polydimethylglutarimide (PMGI) for improved lift-off. However, the best parameter was to simply spin coat at 5000 rpm the OFPR-800LB PR directly, using a gap patterned distance of 2.37 μm with 7.5 μm thick electrodes, for 1 s exposure to the white light as shown in Fig. 5.2. The pattern was executed to have one massive drain and multiple sources, as shown in Fig. 5.2b, as this was the simplest to implement and exhibited the most consistent and successful lift-off.

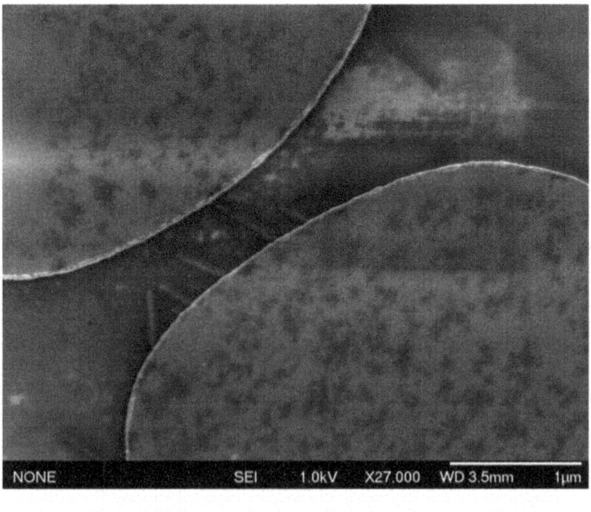

Fig. 5.1 CNTFET 1 HR-SEM image at 1 kV of one of the first few successful devices made with LED lithography

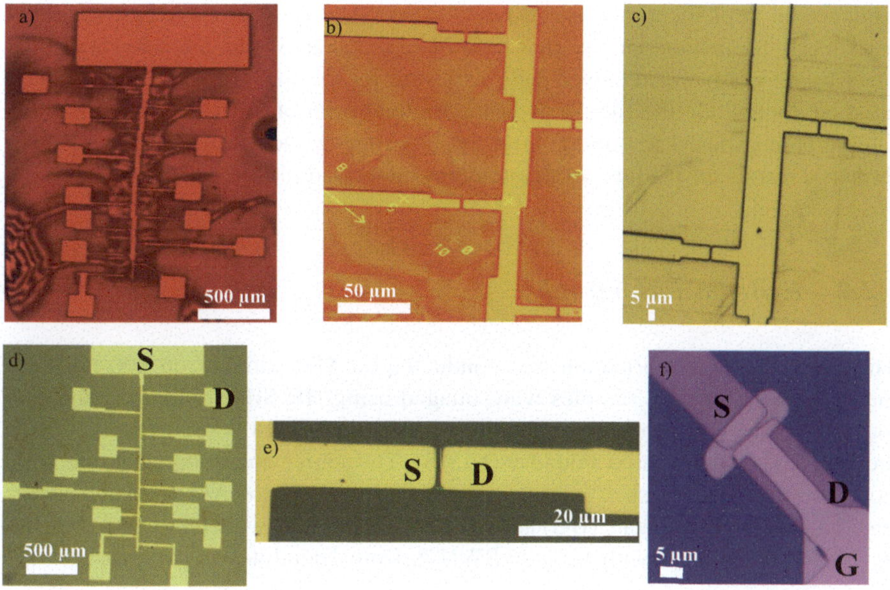

Fig. 5.2 Example of surface coated patterns (**a**) full example of the developed pattern prior to metal deposition, (**b**) close-up of the general S-D areas before metal deposition, (**c**) surface after metal deposition, (**d**) full pattern after the lift-off process, (**e**) close up of gap channel from the LED lithography process after lift-off (BG), (**f**) TG device from LED

Once the test parameters were determined, the actual samples were treated with electron beam metal deposition with either 5 nm Ti/50 nm Au or 1 nm Cr/10 nm Pd/50 nm Au. The Au and Pd allow for p-type Schottky CNTFETs and are preferred as they have a work function very close to that of CNTs [4, 5]. A direct application of Pd 10 nm/Au 50 nm was attempted following a previous protocol [6] but none of the metals adhered to the surface after the lift-off process. Within less than 1 h in acetone, the majority of the metals on the two samples were removed from the LED patterned areas. Therefore, in this case, it was not possible to avoid using the adhesion layer of either Ti or Cr for the electrodes. Ideally, the Ti or Cr would not be incorporated as they readily oxidize and can affect the conductivity between the CNT and the metal electrode [7, 8]. Nonetheless, these layers are necessary for proper adhesion of Pd and Au that do not firmly adhere to the SiO_2 surface [9]. With the optimized parameters, the success rate with standard ART to obtain working CNTFETs was about 15% using LED photolithography.

One additional observed modification was that the devices could be sonicated for 1–2 min in acetone at 30 °C to remove the excess metals/photoresist during the lift-off process. Specifically, after the solution has been in acetone and if large areas of the metal are still adhering to the surface, the sample should remain in the acetone, be sonicated briefly, then washed with IPA. Sonication for longer than 2 min damages the delicate patterns, but brief sonication is enough to remove the looser areas that have the photoresist underneath. There are some risks the devices will be damaged by implementing this additional step in the lift-off process. However, it is particularly useful for samples that have large metal areas that are adhering to the surface and not being removed with exposure to acetone. Additionally, this step must be done while the sample is in the acetone solution. It does not work if the sample is re-exposed to acetone after the IPA rinsing and nitrogen drying as then there is a residue of photoresist that remains on the samples.

5.1.2 E-Beam Lithography

The EBL CNTFET fabrication was conducted on ART samples on 200 nm SiO_2 marked substrates. The samples were imaged using HR-SEM to identify the best locations for device fabrication. The locations of the CNTs desired for the device are then sent to the Namiki Foundry (NIMS). The source and drain electrode positions for the EBL mask was then adjusted through AutoCAD software as per Fig. 5.3. Both a bottom gate (BG) and top gate (TG) device were fabricated. The BG device was fabricated with 90% sc-SWNTs from NanoIntegris, whereas the TG

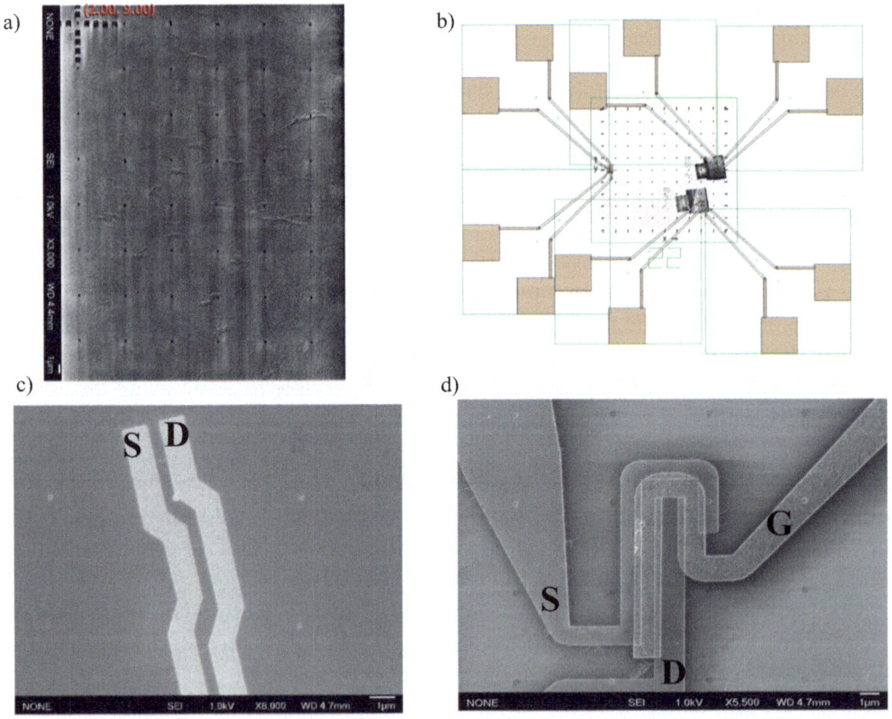

Fig. 5.3 (**a**) ART HR-SEM surface on marked substrate with (10,3) SC dispersed SWNTs, (**b**) layout for S-D electrodes for EBL using AutoCAD, (**c**) S-D electrodes for finished BG EBL CNTFET, (**d**) S-D and Gate electrodes for TG EBL CNTFET

sample was produced using a nearly single chirality batch of (10,3) sc-SWNTs. Both samples had electrodes formed with 1 nm Cr/ 10 nm Pd/ 50 nm Au.

5.2 Device Testing

Upon successful lift-off, each sample channel was imaged using HR-SEM to determine areas that had aligned CNTs between the two electrodes. Electrodes that contained the CNTs were then tested using source measurement units (SMU) in either vacuum or air.

CNTFET device 1 was fabricated using LED photolithography with Ti/Au electrodes. However, it was noted that the drain current obtained was very low (10^{-12} A) (Fig. 5.4). It was determined that tube furnace annealing in inert conditions could further improve the device [6]. In the case of CNTFET device 1, 200 °C annealing was applied in argon for 1 h. The treatment resulted in the transport curve in Fig. 5.4b. The bottom-gate (BG) device was rendered functioning in air, indicating that annealing treatment is necessary for the samples. CNTFET 1 underwent further annealing treatments at 300 °C and 400 °C. However, the tests at 400 °C were no longer better than the sample without annealing as the resistance was very high, and there was nearly no drain current.

Additionally, in order to ascertain if the iptycenes were creating any conducting signals, a device was also fabricated with only a monolayer of molecule **1** on the surface. Devices containing only the iptycenes **1** monolayer were fabricated using 5 nm Ti/50 nm Au electrodes in order to test if there was any conductivity originating from the molecules themselves. From ellipsometry, the thickness of the iptycenes has been observed to be 1.9 nm. The I-V curve demonstrates both the drain and gate current over gate voltage when $V_{SD} = 0$ V is shown in Fig. 5.5a. Additionally,

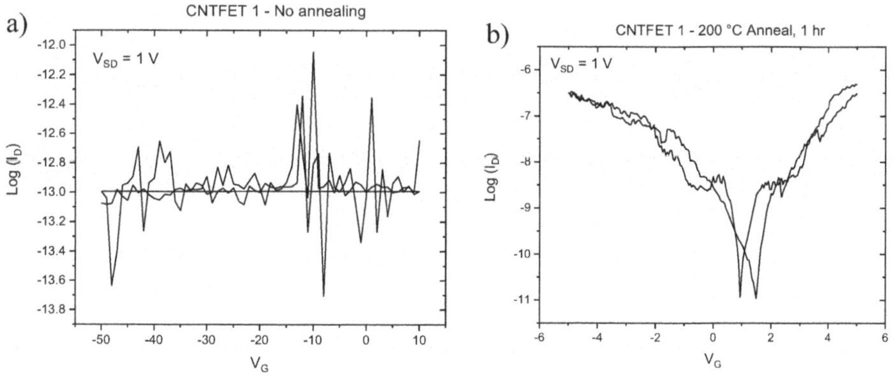

Fig. 5.4 Transport curves for CNTFET device 1 (**a**) before and (**b**) after annealing treatment, with forward and backward sweep of V_G

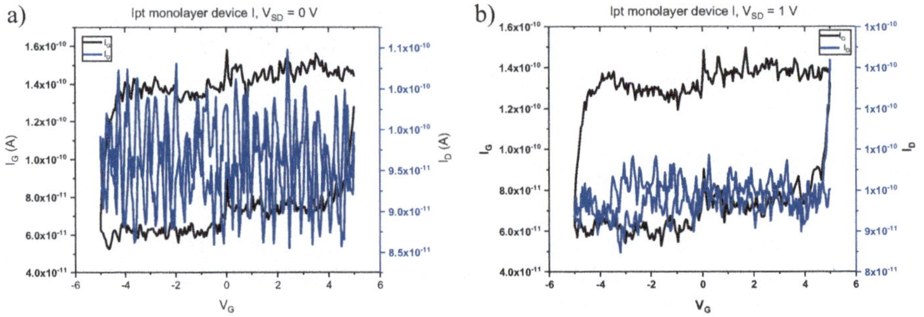

Fig. 5.5 (**a**) I-V transport data for $V_{SD} = 0$ V for iptycene only device, (**b**) I-V transport data for $V_{SD} = 1$ V for the iptycene only device

when $V_{SD} = 1$ V, as per Fig. 5.5b, the results are the same as if no voltage was applied to the electrodes. There was less noise in the signal for the drain current, but the relative current is overall very low, below the nano-ampere range. It was also apparent that both the drain and gate current are nearly identical. This result indicates that there was no conductivity coming from the iptycene molecules that can be considered disruptive toward the CNTFET.

5.2.1 Annealing Tests

Annealing tests were conducted by monitoring the resistance through I-V resistor tests. Samples were annealed under argon using a Full Tech vacuum tube furnace FT-200SP. In order to create improved contact of the CNTs with the electrodes, a decrease of the resistance is desirable after the annealing treatment.

5.2.1.1 Decreasing Resistance

Following the observation that annealing improved the contact of the CNTs with the metal electrodes, further samples were analyzed for annealing parameters. For CNTFET 2, the device was tested in air as a BG using 200 nm SiO_2 with Ti/Au electrodes (5/50 nm). All the samples were on the same wafer using 90% purity sc-SWNTs from an SDS/SC aqueous solution. Annealing was conducted for 1 h at 200 °C, 300 °C and 400 °C temperatures with a ramp rate of 5 °C/1 min under Ar. The 100 °C temperature was not included as from a previous test, there was no change in the resistance from no annealing.

Firstly, there was a slight change in the surface of the electrodes after all the annealing treatments. The surface appears to have some particles and impurities. The CNTs appear the same; however, in Fig. 5.6a the CNTs in the channel appear to intersect each other. After annealing, in Fig. 5.6b, there was only 1 CNT in the channel. It is possible that the positioning of the CNTs shifted during the heating process. From the heating, it was observed that from Fig. 5.6c, d, there was a significant decrease in the resistance of the material. It was further improved by the 300 °C treatment (Fig. 5.6e). However, the resistance increases again upon annealing at 400 °C. An additional 13 devices were tested for the same parameters, and the data is accumulated in Table 5.1. When the gate was connected with the source and drain, there was minimal resistance, and the drain current and gate current were the

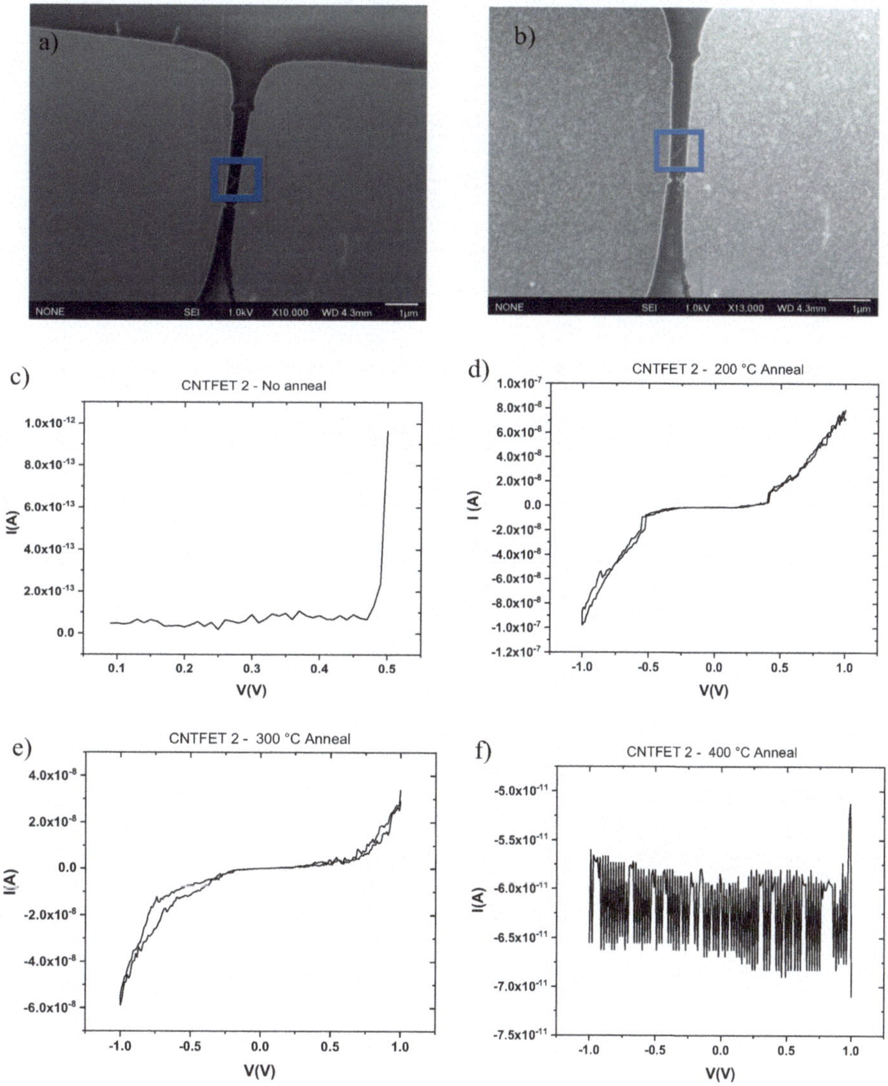

Fig. 5.6 (**a**) CNTFET 2 HR-SEM before annealing, (**b**) CNTFET 2 HR-SEM after all annealing tests, the CNTs in the channel are indicated, (**c**) I-V resistor test before annealing, (**d**) I-V resistor test with 200 °C anneal, (**e**) I-V resistor test with 300 °C anneal, (**f**) I-V resistor test with 400 °C anneal

Table 5.1 Samples tested for annealing parameters compiling the resistance at no annealing, 200 °C, 300 °C and 400 °C annealing in argon for 1 h. Resistance in Ohms (Ω). C = conductive with gate

	CNTFET number														
Annealing	2	2–2	2–3	2–4	2–5	2–6	2–7	2–8	2–9	2–10	2–11	2–12	2–13	2–14	Average
None	E+13	N/A	N/A	E+10	E+11	N/A	N/A	E+12	E+12	C	E+11	C	E+09	C	1.74E+12
200 °C	E+09	E+12	N/A	N/A	E+10	N/A	N/A	N/A	N/A	N/A	C	E+08	N/A	E+12	4.02E+11
300 °C	E+08	E+08	E+10	E+09	E+11	E+08	E+12	E+09	E+10	E+11	N/A	E+08	E+09	E+12	1.71E+11
400 °C	E+10	E+10	E+11	E+10	E+10	N/A	E+12	E+10	E+10	E+11	E+10	E+11	E+10	E+08	1.06E+11

same; hence the device is indicated as conductive. For the device to work, the CNTs cannot be connected to the gate. To disconnect the film of CNTs, the area around the S and D was gently scratched on the SiO_2 surface and annealed. Of 14 devices, 13 CNTFETs had a decrease in resistance with annealing.

From the data in Table 5.1, it was observed that 50% of devices are improved with 300 °C annealing for 1 h. By 400 °C, the resistance either increases again (worsens) or remains the same. Only in the case for CNTFET 2–14 was the performance improved at 400 °C. Some of the devices were conductive with the gate and, therefore, did not give a reliable resistance value, while others simply did not work (N/A). The lower resistance is desirable, such that the device can carry more current. Resistance in the kΩ-MΩ range is usually observed for CNTFETs [10–12]. The results indicate that it is beneficial to apply 300 °C anneal to ART formed CNTFETs.

5.2.1.2 Max Temperature

To further corroborate with annealing at 300 °C, dropcast thin films on bare silicon were tested at various annealing temperatures to determine at what instance the CNTs would become damaged or degrade. Samples were dropcast and left to air-dry onto a bare silicon wafer. Their D–/G-bands were analyzed by Raman spectroscopy at a laser wavelength of 523 nm. The starting point was 300 °C annealing for 1 h as this parameter was shown to improve the contact between the CNTs and the electrodes in a bottom gate CNTFET. The results of the changes in the 1600 cm^{-1} Raman region are shown in Fig. 5.7. The samples tested were from the usual 90% sc-SWNTs mixture from Nanointegris in a 1% aqueous mixture of SDS/SC (Fig. 5.7a). The 95% sc-SWNTs from Nanointegris were dissolved 1% in CHCl3 using tip sonication (Fig. 5.7b). SWNTs solution provided by a Kataura (AIST) - Ishihara (NIMS) group collaborative project were also used. One was a mixture of (10,3) sc-SWNTs in 0.5% SC solution (Fig. 5.7c), and the other was a mixture of many sc-SWNTs chiralities in a 0.5% SC solution (Fig. 5.7d).

From the images, on average, the samples D-band becomes significantly greater at 500 °C. This increase indicates changes in the CNT structure, such as an increase in deformities [13]. The marked wafer (5 nm Ti and 50 nm Au patterns) and a sample of the iptycene monolayer on silicon were also subjected to an annealing treatment. The marked wafer underwent 600 °C for 3 h, and the iptycenes only layer was treated with 300 °C annealing for 12 h. The ramp rate in all cases was 5 °C/min. Both samples did not exhibit any significant changes in terms of appearance for the marked substrate and in Raman signal for the iptycenes.

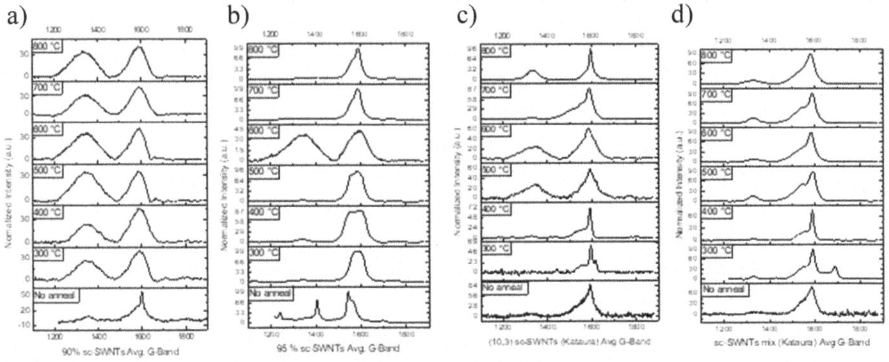

Fig. 5.7 Average G-band intensity of various CNTs treated with increasing annealing temperatures from 300 °C plotted with normalized intensity: (**a**) Nanointegris 90% SDS/SC dispersed sc-SWNTs, (**b**) NanoIntegris 95% sc-SWNTs from $CHCl_3$, (**c**) (10,3) sc-SWNTs SC dispersed sc-SWNTs, (**d**) sc-SWNTs mixture SC dispersed

5.2.1.3 Further Improvements

To further enhance and test the data, vacuum measurements were also performed. The usual vacuum conditions were 10^{-4} Pa. The combination of vacuum measurements and annealing further improved the BG devices, such as in the case of CNTFET 3 (Fig. 5.8). From testing the device as a resistor before annealing, the resistance was about 10 MΩ. With annealing, the resistance decreased to ~1 MΩ. The transport and output data are shown in Fig. 5.8a and b for testing in ambient conditions. Based on the resistance, the device should have current flowing. However, the transport results depict that the device had no electron mobility and a high R_{ON} (900 kΩ). The device was then tested in vacuum, with additional annealing (150 °C/2 h). The results are shown in Fig. 5.8 c and d. In vacuum, the device was functioning.

The mobility was calculated to be 8.76 cm²/Vs with an I_{ON} value of 0.7 μA and an I_{ON}/I_{OFF} value of 1×10^4. These data are very reasonable for CNTFETs [14, 15]. However, the R_{ON} value increased to 1762 kΩ. Regardless, it was possible to render some devices functioning using additional annealing treatments and measuring in vacuum. The CNTs possibly interact with the environment and the particles in the air, thus impeding the device performance. It is also plausible that some of the surfactant wrapped CNTs require this additional treatment as it removes additional interactions from the surfactant/environment. As the surfactant was a mixture of SDS/SC, a treatment to remove these additional molecules could be used through a water soak and additional annealing to perhaps further observe improvements in the BG system [16]. Nevertheless, by doing the measurements in an inert atmosphere and further improving the contact of the electrodes with vacuum annealing, some of the CNTFET properties can be enhanced.

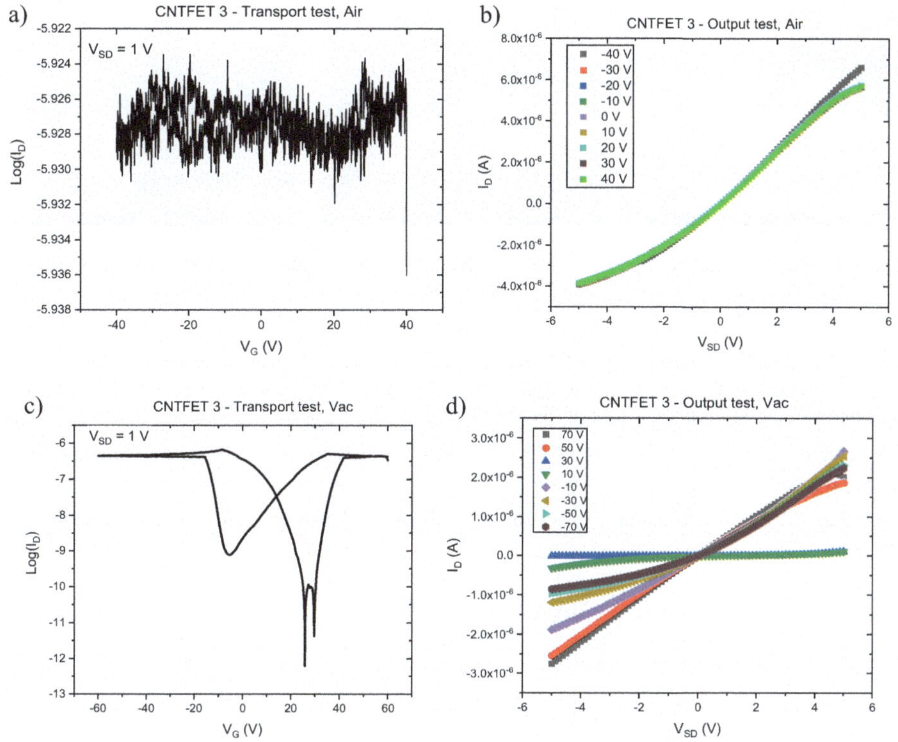

Fig. 5.8 Results from testing CNTFET 3: (**a**) Transport data in air (**b**) output data in air, (**c**) transport data in vacuum, (**d**) output data in vacuum

5.2.2 *Bottom Gate Devices*

5.2.2.1 LED Lithography Fabricated CNTFET Transport and Performance

Successful devices are shown in Fig. 5.9, using Cr/Pd/Au. These images show the sub 1 μm channels fabricated with LED photolithography, with the CNTs in the channel. Devices that have CNTs confirmed between the S and D from the HR-SEM images are then SMU tested. The average current during a resistor test before annealing was 10^{-9} A. It was reduced to 10^{-7} A upon annealing at 1 h for 300 °C for devices made either with Cr/Pd/Au or Ti/Au electrodes and with either the Kataura group SWNTs or those from NanoIntegris (both 90% and 95% sc-purity).

Under ambient conditions, the I_{ON} ranged from 1 μA to 10^{-3} μA, an I_{ON}/I_{OFF} 10 to 10^3, and the mobility ranged from 10 to 10^{-4} cm^2/Vs [Un-operational CNTFETs omitted]. On average, in air, the I_{ON} was 0.01 μA, the I_{ON}/I_{OFF} was 100, and the average mobility 0.01 cm^2/Vs. In contrast, under vacuum conditions, the LED BG CNTFETs over the range of various SWNTs types and purities resulted in an I_{ON}

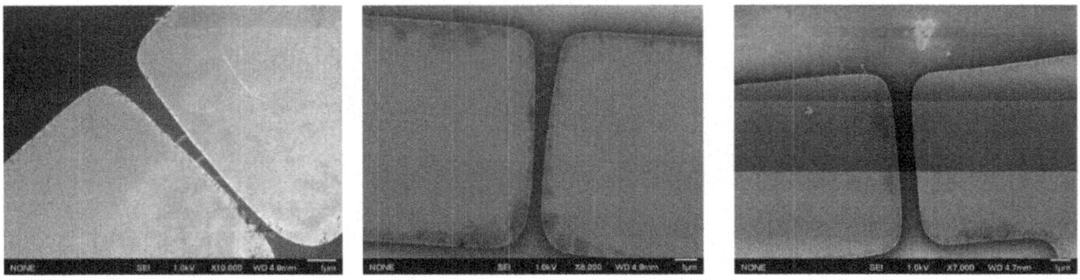

Fig. 5.9 HR-SEM of CNTFETs made with LED lithography method used for BG SMU testing

Table 5.2 Results from various LED lithography fabricated devices with various CNTs and testing conditions. Where SS is not applicable is because there was not at least a 1 log factor difference in the log transport plot of the devices

Parameter	1 CNT	3 CNTs	3 CNTs	Film	Ag gate	InGa gate
CNTFET #	4	5	5	6	7	7
Condition	Ambient	Vacuum	Ambient	Ambient	Ambient	Ambient
Mobility (cm^2/Vs)	0.2	8	1	1.3	0.4	0.4
I_{on} (μA)	0.1	0.1	5.9	5.4	6.2	4.2
I_{on}/I_{off}	100	1000	1	13	10	10
Gm (nS)	3.5	52	330	93.7	220	206
$Hysteresis_{Max}$ (V)	20	8	80	80	100	80
SS (V/dec)	N/A	1	N/A	N/A	3	16.3
R_{ON}(kΩ)	6507	4330	986	18.5	84	127

ranged from 1 μA to 10^{-4} μA, an I_{ON}/I_{OFF} 1 to 10^5 and the mobility ranged from 10 to 10^{-4} cm^2/Vs. On average in vacuum, the I_{ON} was 0.01 μA, the I_{ON}/I_{OFF} was 100 and the average mobility was 0.1 cm^2/Vs. By comparing the data, there are clear improvements in the above parameters when testing in vacuum. Yet, overall, the data under ambient conditions was still in a reasonable range for CNTFETs [17].

Additionally, some devices that do not work in air are still not functional under vacuum conditions and there was no salvaging these devices even with multiple annealing segments. Examples of some of the individual data is summarized in Table 5.2 consisting of a single CNT, multiple CNTs and a film. The hysteresis was also shown to decrease by using InGa eutectic as the back-gate instead of silver paste in ambient conditions and this was also demonstrated by the data in Table 5.2 as well as the plots in Appendix Chap. 5.

Most significantly, the mobility increases when there was more than one CNT in the channel by a factor of 10. The mobility then increases further when the device was tested in vacuum. The mobility of a CNT film was comparable to that of 3 SWNTs. However, the I_{ON}/I_{OFF} significantly increases for the individual CNTs when tested under vacuum conditions (from metallic to semiconducting behavior) even though the I_{ON} value decreases. Additionally, the resistance in the ON-state was increased in the output data of the sample that was tested in vacuum compared to its ambient tested version. The main difference when testing with the InGa gate instead of the silver one was in a decrease in the hysteresis observed in the device. In some of the cases, the subthreshold slope could not be calculated as there was not a factor of 1 change in the log scaled transport plot, and therefore, no value for the change of voltage per decade (V/factor of 1 log unit on the y-axis).

Additionally, when the same BG sample was treated and changed to a TG device, some of the devices that functioned well as BG no longer worked and vice versa. This observation will be further elaborated in Sect. 5.2.3.1. The output and transport plots are provided for each of the devices in Appendix Chap. 5.

5.2.2.2 EBL Fabricated CNTFET Transport and Performance

A sample using 90% sc-SWNTs from NanoIntegris was fabricated using the EBL method. Some of the HR-SEM images are shown in Fig. 5.10. This technique provided a 51% success rate of fabricated devices. It was much easier to control the position of the electrodes and have small precise channel gaps with this method, which lead to an increase in the number of successful devices.

In ambient conditions, the I_{ON} ranged from 60 to 10^{-6} μA, the I_{ON}/I_{OFF} varied from 1 to 10^5 and the mobility span across 70–10^{-3} cm^2/Vs. On average in air for all the EBL bottom gate fabricated devices, the I_{ON} was 0.1 μA, I_{ON}/I_{OFF} was 10^3 and average $\mu = 1$ cm^2/Vs. In contrast, the range in vacuum for the I_{ON} was 8 to 10^{-6} μA, the I_{ON}/I_{OFF} varied from 1 to 10^5 and the mobility from 48–10^{-2} cm^2/Vs. The average I_{ON} in vacuum was 1 μA, I_{ON}/I_{OFF} was 10^4 and average $\mu = 1$ cm^2/Vs. The cumulative data for all the different types of devices indicates that mobility and I_{ON} values are relatively the same for both air and vacuum. The only significant difference is observed with a slightly increased I_{ON}/I_{OFF} in vacuum conditions.

Data for single and multiple CNTFETs, as well as films, are shown in Table 5.3. The multiple CNTFET devices were also tested in nitrogen conditions to observe if there were any changes in transport and output properties. The most significant change in N_2 atmosphere was the increase in I_{ON}/I_{OFF} ratio, indicating that there was some interaction of the CNTs with the environment that can affect the OFF and ON states of the device. However, the highest overall mobility was demonstrated in vacuum conditions for the CNTFETs of individual tube. Only the film of CNTs

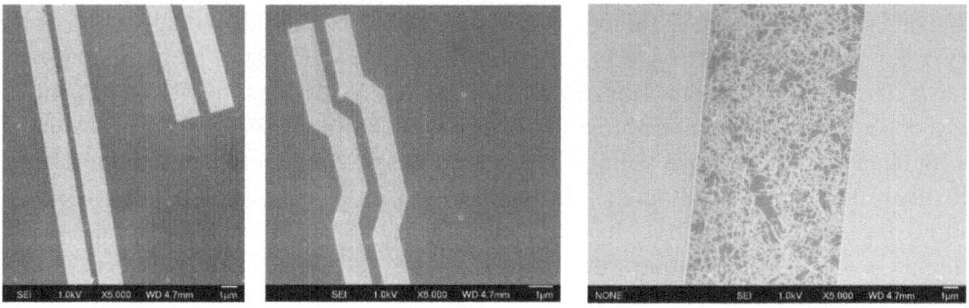

Fig. 5.10 HR-SEM at 1 kV of EBL fabricated CNTFET channels

Table 5.3 Data for various samples formed using EBL tested in vacuum, ambient and nitrogen atmosphere

Parameter	Single CNT		Multiple CNTs			Film	
CNTFET #	8		9			10	
Conditions	Air	Vac	Air	Vac	N_2	Air	Vac
Mobility (cm^2/vs)	0	0.1	0	5.7	4	1.8	0.7
I_{ON} (μA)	1.19	1.7	1.5	2.7	2.6	23	2
I_{ON}/I_{OFF}	1	10	10	450	1000	1000	100
Gm (nS)	N/A	93	N/A	210	160	786	315
$Hysteresis_{Max}$ (V)	0	5	0	92	12	44	17
SS (V/dec)	0	1.2	0	5.5	5	20	5
R_{ON}(kΩ)	132	1170	283	312	344	39	330

Table 5.4 Transport data from evaluating CNTFET #9 in vacuum by changing the sweeping and step direction

Sweep direction	−60 to + 60 V		+ 60 to − 60 V	
Step	120 mV	500 mV	120 mV	500 mV
Mobility (cm^2/vs)	5.5	6	7.5	6.9
I_{ON} (μA)	3.83	3.51	4.2	3.5
I_{ON}/I_{OFF}	1000	600	500	500
Gm (nS)	215	235	293	269
$Hysteresis_{Max}$ (V)	76	64	85	66
SS (V/dec)	4	4	5	4

performed better in air. Additionally, the film had the least resistance and a lowest mobility. Even with higher resistance in the individual CNTs, the mobility was also comparatively high indicating the benefits of having individual versus intertwined CNTs. This observation could be due to the CNTs sensitivity to a nitrogen environment [18], which has previously been reported for vertically aligned bottom gate CNTFETs [19, 20].

Further investigations in vacuum also considered the effects of sweeping direction and step speed in mV. Table 5.4 shows the difference in I_{ON}, I_{ON}/I_{OFF}, and mobility when the scan direction was either from positive to negative voltage and vice versa. The slower step gave higher mobility and I_{ON}. With this type of experiment, it was observed that the scanned log plot would be a complete full loop when scanning from negative to a positive voltage, but sometimes incomplete when scanning from positive to a negative voltage, as shown in Fig. 5.11. The R_{ON} values for sweeping V_{SD} from −5 to +5 V and + 5 to −5 V was 356 kΩ and 357 kΩ, respectively, so no change in terms of the output information.

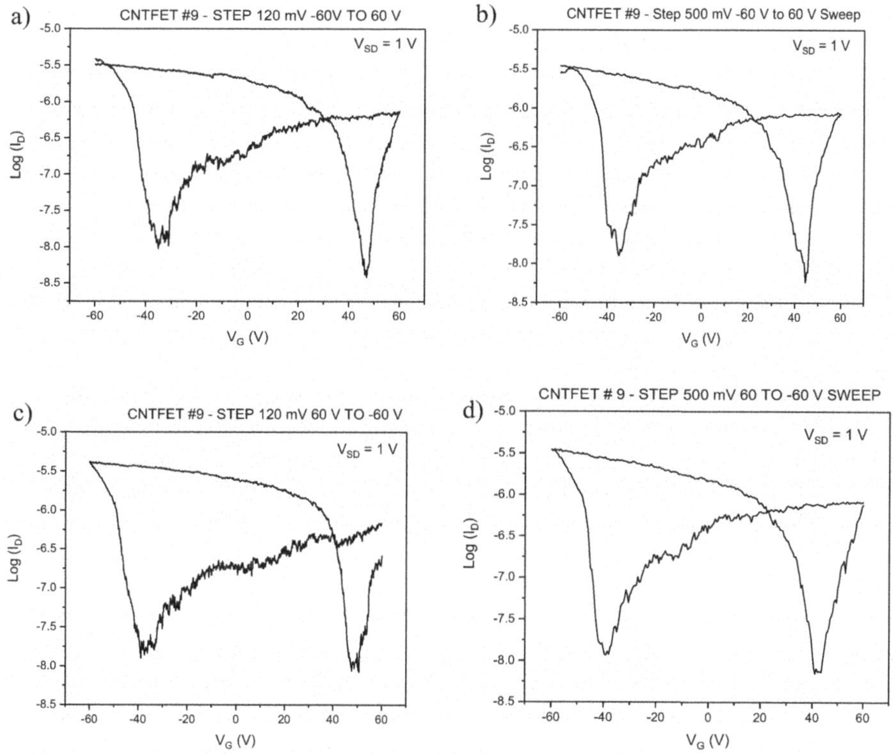

Fig. 5.11 Bottom-gate EBL sample (CNTFET#9) (**a**) sweep V_G from −60 to 60 V with 120 mV step, (**b**) sweep V_G from −60 to 60 V with 500 mV step, (**c**) sweep V_G from 60 to −60 V with 120 mV step, (**d**) sweep V_G from 60 to −60 V with 500 mV step

5.2.3 Top Gate Devices

5.2.3.1 LED Lithography Fabricated CNTFET Transport and Performance

Using an already prepared and tested device with bottom gate CNTFETs, a 20 nm alumina dielectric layer was deposited to the sample using ALD. The sample was then treated with OFPR photoresist and the standard method for photolithography used in this work. Many of the devices that were functioning as a BG in ambient conditions were no longer working in air after the TG fabrication process. The resistor testing of TG in air gave a current range of 10^{-7}–10^{-12} A, which was comparable to other BG data.

Nevertheless, the I_{ON}/I_{OFF} range was from 1–10, only with I_{ON} values from 0.1–10^{-3} μA. The sample was then tested under vacuum conditions. The I_{ON} range significantly increased to 5.7–10^{-2} μA and some of the I_{ON}/I_{OFF} ratios were increased to 10^4, which are values more in line with the BG CNTFETs. The sample was then vacuum annealed for 2 h/ 150 °C, which resulted in a further increase in the mobility of the devices upward of 10 μA, however, the I_{ON} state was still in the 0.1–10^{-2} μA range. The max I_{ON}/I_{OFF} state was 10^4, while most were in the range of 1–10.

For example, CNTFET # 4 is compared in Table 5.5 as TG and BG. The performance in air and vacuum, before and after annealing with the application of the TG can be compared. The measurements in N_2 for the bottom gate also showed an increase in $I_{ON}/I_{OFF,}$ similar to the EBL BG devices, even after the application of the dielectric layer. The device appears to behave as a metallic CNT except when in the nitrogen atmosphere with a 10^4 I_{ON}/I_{OFF}. The mobility of the BG device decreases after the fabrication of the TG. The resistance was high for all the devices. The highest mobility was observed in the TG form, after annealing in vacuum at 150 °C for 2 h.

The BG devices show a very high SS and hysteresis. In contrast, the TG CNTFETs also have a large hysteresis considering the V_G was swept from only 5 to −5 V. Another interesting observation was that the highest ON-state current is 0.1 μA and is observed for the BG device tested under ambient conditions. The I_{ON} for the TG is the highest after vacuum annealing at 0.01 μA, which is less than that of the BG by a factor of 10. Although there are some apparent differences in the data for each test, the highest mobility is in the TG device. This observation could indicate that the lower iptycene monolayer has a slightly negative effect on the mobility in the CNTs. However, the BG in air before application of the dielectric is comparable to that of the TG in air and shows a higher I_{ON} and a lower relative hysteresis. The TG devices have the most improved transport when annealed in vacuum whereas the BG devices have improved transport performance when measured in N_2. These are the first results exhibiting CNTFETs from the ART using LED photolithography, and it was interesting to observe the changes in the device and parameters.

Table 5.5 Change in transport and output properties for CNTFET # 4 in different testing conditions as both a bottom gate and top gate

Parameter	TG			BG			
		Vac			Vac		
Conditions	Air	Before vac anneal	After vac anneal	Air (before TG applied)	Before vac anneal	After vac anneal	N_2
Mobility (cm^2/vs)	2.8	5.7	10	0.2	0.09	0.1	0.9
I_{ON} (μA)	4.00E-03	9.00E-03	1.40E-02	0.1	3.80E-02	4.00E-02	1.00E-01
I_{ON}/I_{OFF}	10	30	50	100	100	100	1000
Gm (nS)	7.1	1.4	2.6	3.5	6.00E-03	0.94	5.8
$Hysteresis_{Max}$ (V)	4	3	6	20	60	47	47
SS (V/dec)	N/A	5	3	N/A	60	53	5.5
R_{ON}(kΩ)	142,857	777,202	52,631	6507	1578	25,806	15,909

5.2.3.2 EBL Fabricated CNTFET Transport and Performance

A sample was fabricated with a Ti/Au top gate (5/20) nm and 20 nm ALD Al_2O_3 layer using ART aligned (10,3) SWNTs. An example of an 11 aligned CNTFETs is shown in Fig. 5.12. Figure 5.12a shows the autoCAD depiction for masks design, whereas Fig. 5.12b demonstrates the completed device after various bouts of SMU testing and heating. It is difficult to confirm the number of CNTs in the channel as the top gate does not allow for the observation of the CNTs, and the Raman signal was also blocked.

With the TG EBL fabricated device, in ambient conditions, the I_{ON} was between 0.1–10^{-3} μA, the I_{ON}/I_{OFF} ranged from 10–10^5 and μ was between 2–0.2 cm^2/Vs. The average I_{ON} was 0.1 μA, average I_{ON}/I_{OFF} was 10^4 and the average mobility was 0.1 cm^2/Vs in air, and an I_{ON}/I_{OFF} of 10^5 was frequently observed. These data are already much higher than the average BG devices. In vacuum, the I_{ON} ranged from 5 to 10^{-3} μA, the I_{ON}/I_{OFF} was 100–10^6 and the mobility was between 30–5 x 10^{-3} cm^2/Vs. The average vacuum findings for these parameters was an I_{ON} of 0.5 μA, a 10^5 I_{ON}/I_{OFF} and an average μ of 1 cm^2/Vs. Results for a 2 SWNTs, 4 SWNTs, 11 SWNTs and a film CNTFET are tabulated in Table 5.6 under vacuum conditions after annealing. It is interesting that the 4 CNT system has a higher mobility than the 11 CNT system, and also a higher I_{ON}/I_{OFF} ratio by order of magnitude of 100. The R_{ON} is also significantly less for CNTFET 13 than all the other samples with also a higher I_{ON}.

Even the TG device may be affected by external environmental factors, which could account for the observed difference in air and vacuum values. This could also be a contributing factor when comparing the difference between CNTFET 13 and 14. It is also interesting that the CNT film also performs worse than the 4 aligned CNTFET device and has lower mobility than all the other devices in the table. This

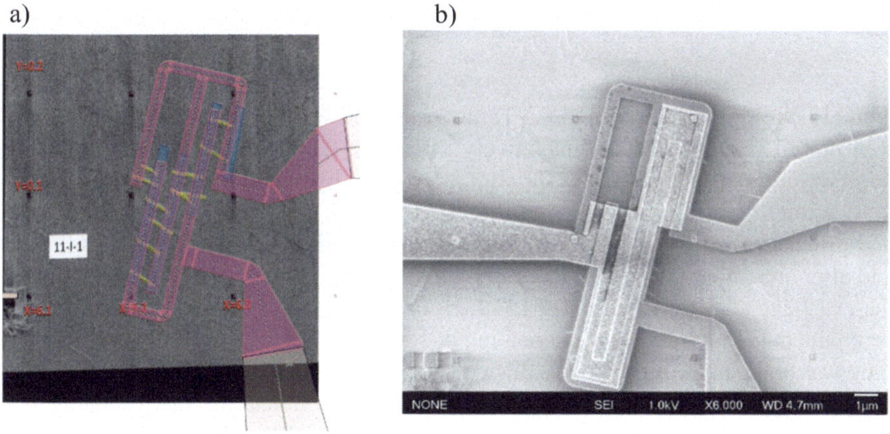

Fig. 5.12 Example of the fabrication of the top gated multi-SWNTs FET a) before and b) after patterning in HR-SEM

Table 5.6 Table demonstrating key properties for 2, 4, 11 aligned SWNT CNTFETs and a film of SWNTs based device tested as TG under vacuum conditions

Parameter	Two CNTs	Film	Four CNTs	11 CNTs
CNTFET #	11	12	13	14
Mobility (cm^2/vs)	4.4	1.20E-02	10	0.4
Ion (μA)	5.26E-03	0.02	3.8	0.3
I_{ON}/I_{OFF}	52,917	220,000	3,430,000	42,900
Gm (nS)	2.2	4.63	1260	88
$Hysteresis_{Max}$ (V)	7.3	1.02	1.2	1.5
SS (V/dec)	2.6	1.1	1.1	1.5
R_{ON}(kΩ)	214,285	29,585	210	1466

Table 5.7 Compilation of various test parameters for the device with highest mobility and I_{ON}/I_{OFF} in this study

Cntfet #13 – Four SWNTs in the channel							
	TG				BG		
		Vac				Vac	
Conditions	Air	Before anneal	After vac anneal	N_2	Air	Before anneal	After vac anneal
Mobility (cm^2/vs)	1	10	0.7	0	0.5	0.4	2.8
I_{on} (μA)	4.6	3.8	0.4	7.00E-06	0.3	4.4	0.7
I_{ON}/I_{OFF}	2,500,000	3,430,000	96,666	1	208,392	580,000	379,690
G_m (nS)	1260	1260	123	0	24	30	123
$Hysteresis_{Max}$ (v)	2.7	1.2	5	0	45	40	40
SS (v/dec)	0.6	1.1	0.6	0	0.1	0.6	3.2
R_{ON}(kΩ)	211	210	315	0	1.70E+06	210	284

is likely due to the intertwined CNTs that impeded on eachothers respective electron mobility and, therefore, negatively affect the device performance [21]. This observation further emphasizes the importance of alignment and selectivity for the fabrication of CNTFETs.

Further evaluation of the EBL devices also included testing the devices before and after annealing, as well as BG and under nitrogen atmosphere. The results from the extensive tests for CNTFET # 13–the four SWNT device, is shown in Table 5.7. From the data, before vacuum annealing at 150 °C for 2 h, the mobility and I_{ON}/I_{OFF} ratio was at the highest point. The device was annealed in argon at 300 °C/1 h before testing in air and vacuum. Therefore, extra vacuum annealing was probably not necessary. It is interesting also how in the TG, there is a 10x improvement in mobility when comparing the air and vacuum results. This observation is significant in that one would expect that because the CNT is surrounded by either SiO_2

underneath or Al_2O_3 on the sides and surface that the external environment would not have a significant impact on the device performance.

Nevertheless, the I_{ON}/I_{OFF} ratio was in the range of 10^6 for both samples, and the resistance is also the same, so the only difference is in the mobility. In contrast, testing the device in a nitrogen atmosphere proved inefficient. This N_2 I-V test was conducted after all other BG and TG testing and extra annealing, so it is likely that the device could not withstand the further testing and broke down. Another aspect is the difference of testing the same device as a TG versus a BG. The BG device has a reduced I_{ON}/I_{OFF} ratio to 10^5 but the mobility and I_{ON} is similar to that of the TG. This result could be from the thicker dielectric layer used (200 nm rather than 20 nm).

All the results were with using $V_{SD} = 1$ V unless otherwise specified. For the TG CNTFET #13, the progression in increasing the source-drain voltage is shown in Fig. 5.13 under ambient conditions. The main observation was the decrease in noise in the curve with increasing V_{SD}, as well as the change in the direction of the curve. This result is likely due to the placement of the source and drain probes on the electrode pads, and not with the actual device. It appears that the device can be switched to the ON state using 0.1 V_{SD} to achieve an I_{ON} of 0.4 μA, an I_{ON}/I_{OFF} of 10^6 and $\mu = 15$ cm²/Vs with a $G_m = 250$ nS, a max hysteresis of 2.5 V and a subthreshold slope of 0.1 V/dec. Therefore the ART CNTFETs are quite promising toward future applications of FETs [22, 23]. Unfortunately, due to the top dielectric layer, it was not possible to determine the CNT types present on the surface as the RBM signal was too weak (Fig. 5.13b).

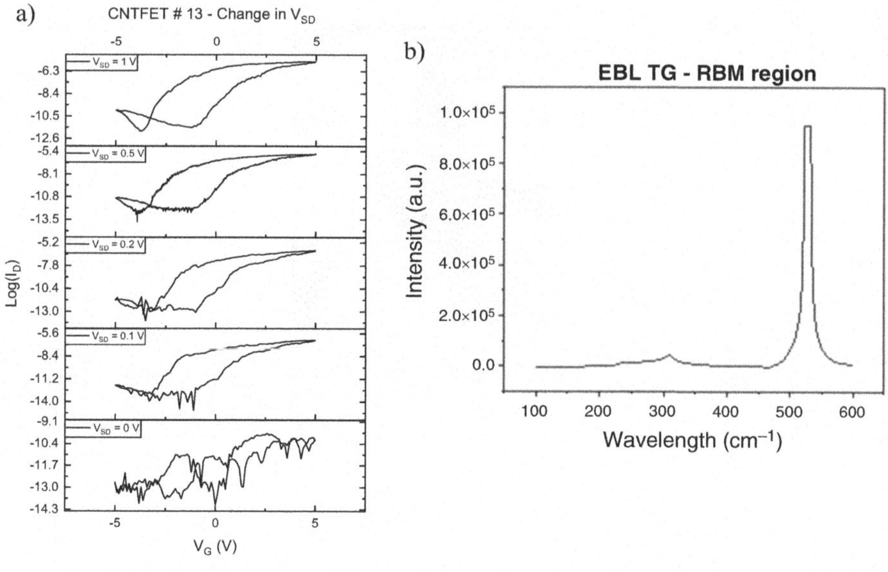

Fig. 5.13 a) Progression of increasing V_{SD} for top gate CNTFET # 13 b) RBM Raman region of the TG device but only the iptycene peak at 232 cm^{-1} and the silicon peaks at 303 and 520 cm^{-1} are visible at 100% laser intensity

Overall, it appears that from the range of devices fabricated, the highest mobility and I_{ON}/I_{OFF} were achieved using a TG device tested under vacuum conditions (10^4 Pa). On average, the TG devices provided a higher I_{ON}/I_{OFF} of 10^5 rather than 10^4 in the BG. The EBL devices had a higher success rate than the LED-based devices, but both can give reasonable results using the ART. One further aspect for improvement would be to have a consistently high density using the ART. In some cases, the deposition of aligned tubes was high, and this could be either due to the concentration of the iptycenes or the concentration of CNTs in solution. Additionally, testing the EBL methods to involve as many of the aligned CNTs in a region as possible would also be interesting for further investigation. Due to the timing, the patterns were chosen to not go over 10 μm in length that had small channel gaps below 0.5 μm to ensure an efficient lift-off that would maximize the devices fabricated for initial testing.

5.2.4 *Ionic Gel Gate*

Under ambient conditions, LED fabricated devices with 95% sc-SWNTs were tested using an ionic gel as gate instead of the Si layer. The gel was a blend of polystyrene block co-polymer and the an ionic liquid in ethyl propionate solvent [24]. The gel was applied to cover the gate electrode and the channel containing the CNTs between the S and D, as shown in Fig. 5.14. The gel was dropcast to the

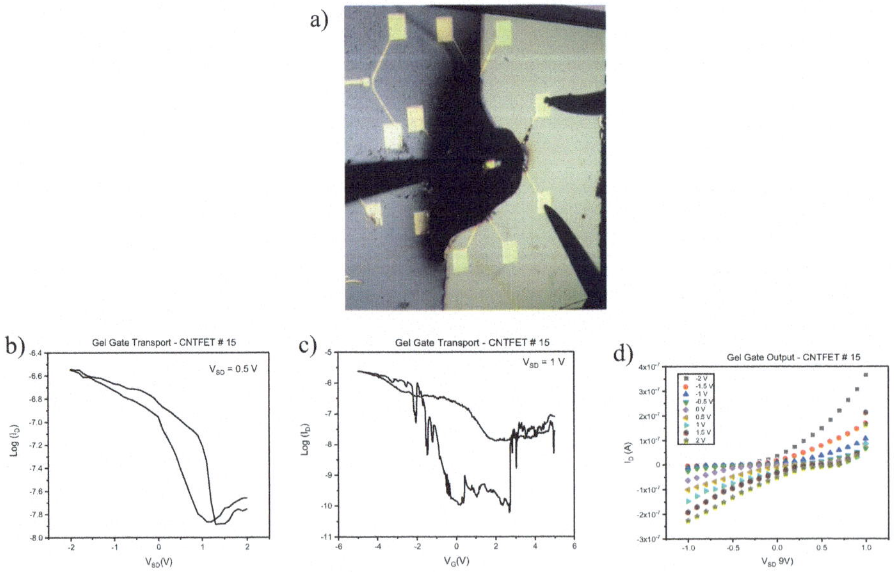

Fig. 5.14 (**a**) Image of the probes on the gel gate while testing the CNTFET, (**b**) transport data at V_{SD} = 0.5 V and (**c**) 1 V (**d**) output data

surface and allowed to air dry. The resistance test for all the devices before the application of the gel was in the range of 10^{-11} to 10^{-7} A after annealing. However, upon testing, the majority of the devices appeared to be conducting with the gate, as in, having the same drain and gate current during the transport measurement. One of the device's transport and output data is shown in Fig. 5.14d. The experiment was intended to observe any changes with the CNTFETs that would occur if the CNTs are wrapped around a charge carrying gel rather than an inert or ambient atmosphere. The I_{ON} value was 2.42 μA with an I_{ON}/I_{OFF} of 10^2, yet the R_{ON} was high, at 2857 kΩ. Although the transport plot was noisy (Fig. 5.14b) the max hysteresis was extrapolated to be 2 V, with an SS of 1 V/dec. The experiment could be revisited and improved by using spin coating and drying the sample for an extended time to ensure the gel gate does not shift during measurement.

5.2.5 Substitution of Metal Electrodes

Rather than depositing the metal electrodes through high vacuum, an attempt was made using Ag and Au inks provided by the Minari Lab (NIMS, Japan). These inks had been previously used for making thin film devices [25], and therefore it was estimated that these metal nanoparticle liquids could also function with the small patterns for the CNTFETs. Samples with ART aligned SWNTs were patterned using the LED photolithography technique. Parameters for the deposition were investigated by diluting the inks in water, dropcasting the inks, spin coating the inks, treating the substrate with UV-O_3 before the inks were applied to the surface, and different heating times to bake the inks onto the surface. However, even with all these different trials and parameters, the inks did not adhere to the surface of the ART functionalized wafer as demonstrated in Fig. 5.15a, b, c. It may be beneficial to use a printing technique similar to that found in work by Minari et al. to fully determine if the Ag/Au inks are adequate as electrodes for CNTFETs [25].

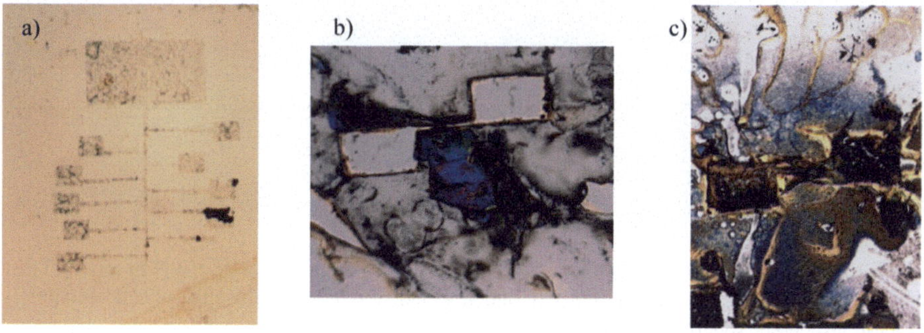

Fig. 5.15 a) Failed patterning attempt with Au ink that was spin coated onto the ART surface and lift-off in acetone after 1 h room temperature drying, b) Ag ink spin coated and baked for 9 mins at 90 °C, then removed in acetone, c) Ag ink spin coated baked at 3 mins for 60 °C after spin coating and treating the surface with UV-O_3 prior to depositing the ink

5.3 Additional Imaging

5.3.1 TEM of CNTFETs

An attempt was also made to individually place the SWNTs on a SiO_2 TEM grid using the iptycenes. This experiment was attempted in order to investigate the individual chirality and diameter of the SWNTs. Inopportunely, the thin windows in the Cu TEM grid do not allow for the standard ART with alignment as the original method involves applying pressure onto the grid which in turn destroys it. It was then attempted to dropcast the solution of Ipt **1**/8CB to allow for the molecules to deposit randomly in a 10 h time frame. The sample was gently rinsed with DCM and then a solution of SWNTs was dropcast on the surface for approximately 10 h. The sample was checked under an optical microscope first to confirm that all the windows on the grid remained intact. Next, the sample was placed under HR-SEM to identify pore areas with CNTs. Finally, the sample of both the (10,3) and mixture of sc-SWNTs was subjected to TEM analysis.

Unfortunately, the TEM results as shown in Fig. 5.16a, b, c for the samples only showed that the CNTs bundled at the edges of the grid windows and were not singly distributed across the surface of the wafer. A solution to this would perhaps be to apply only the pure iptycene molecules - which are a solid at room temperature–to a UV-O_3 cleaned surface and then heat the sample to melt the iptycenes on the surface (> 300 °C). In this way, a monolayer of the CNTs can be achieved on the surface, though not aligned, and these individual small molecules could then in theory π-π stack to randomly orient single CNTs for observation in the TEM. This experiment also circumvents the ART problem of having the liquid crystals present which can contaminate the surface of the substrate.

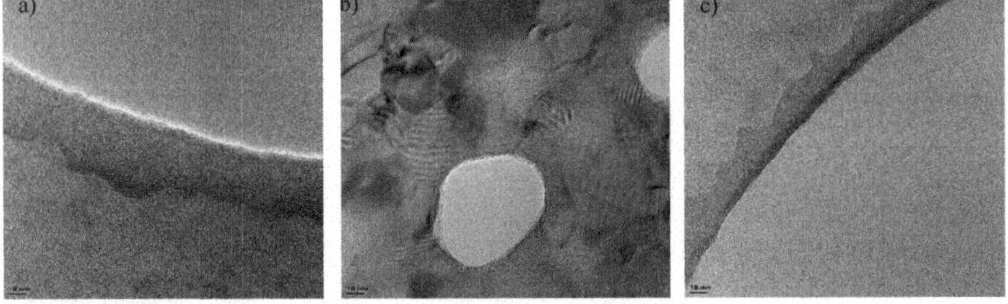

Fig. 5.16. TEM images of (**a**) (10,3) sc-SWNTs in SC, (**b**) zoom out of sc-SWNTs mix of TEM windows (**c**) zoom in of sc-SWNTs mix in one of the TEM windows. The surface had been previously treated with ART iptycenes

5.3.2 *Interesting Patterning Observation*

During the SEM imaging process, sometimes the CNTs were not effectively deposited on the surface, i.e., very few tubes on the sample. The sample was then re-exposed to a higher concentration of SWNTs solution for a longer period. It was then observed that upon a second imaging, areas that were previously exposed to the SEM electron beam were selective for the SWNTs whereas other sections were still bare as shown in Fig. 5.17a, b, c, d. Even with a high concentration of CNTs and sonication, the very distinct square-patterned lines from the electron beam remained present. This type of patterning was further investigated on surfaces that contained only iptycenes and at different SEM beam currents and voltages. However, the patterning did not appear. It only appeared on previously exposed samples to the SWNTs solutions after being exposed to the SEM beam. The contact angle for a

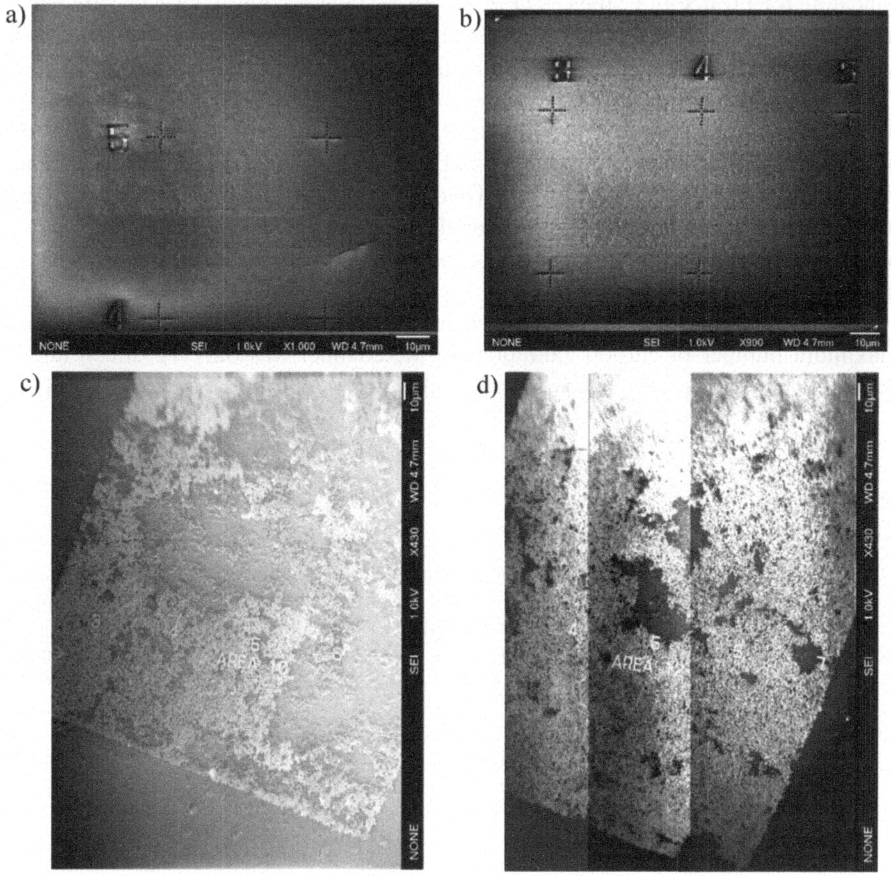

Fig. 5.17 (**a**) and (**b**) Patterned area with (10,3) sc-SWNTs in SC/aqueous solution, c and d) pattern area with a high concentration of 90% sc-SWNTs in SDS/SC aqueous solution

silicon wafer cleaned with only UV-O_3 (30 mins) was 12.13°, 36.4° with an iptycene monolayer, 31.4° for a silicon wafer exposed to a 1 kV electron beam and for a sample exposed to both the SEM beam at 1 kV and the iptycenes the contact angle then increased to 46.5°.

Though the contact angle of an iptycene functionalized surface increased, the patterning does not appear without being previously exposed to a SWNTs solution. It is speculated that the patterning was thus due to surfactants that are present in the CNTs solutions. Therefore, the next investigation should have only a silicon wafer treated with SDS aqueous solution to confirm this hypothesis.

5.4 Summary

To conclude, it was possible to create functioning CNTFETs using the ART. By annealing for 1 h at 300 °C, the CNTFETs with ART can be improved. Measurements in vacuum with additional 150 °C annealing for 2 h also improved the mobility and I_{ON}/I_{OFF} of the devices. The most significant results were observed in the case of vacuum annealed top-gate devices, using (10,3) sc-SWNTs. Either LED-lithography or EBL were efficient in device fabrication, although EBL had a higher success rate of having CNTs in contact with the S and D electrodes. Additionally, investigations with the ionic-gel gate can also work. There is room for further work regarding the deposition of the ionic gel and optimizing the parameters for V_{SD} and V_G. The device patterning in the ionic gel device can also be optimized or EBL can be implemented instead of LED lithography.

Furthermore, this was the first use of the iptycenes in a CNTFET. Although the maximum number of carbon nanotubes in the channel was only 11, much larger devices can be made next, bearing in mind the target of 125 CNTs/μm^2 [22, 26].

References

1. Snowdon, M.R., Wang, S., Mashmoushi, N., Hopkins, S.W., Schipper, D.J.: Carboxylic acids as anchoring components on aluminum oxide for the alignment relay technique of single-walled carbon nanotubes. New J. Chem. **45**(12), 5340–5349 (2021)
2. Snowdon, M., Selmani, S., Schipper, D.J.: Sonication-Enhanced Alignment Relay Technique for the Orientation of Single-Walled Carbon Nanotubes. ACS Appl. Nano Mater. **2**(10), 6637–6645 (2019)
3. Taking, S.: AlN/GaN MOS-HEMTs technology. University of Glasgow, Glasgow (2012)
4. Reddy, C.S., Maheshwaram, S., Dasgupta, S., Saxena, A., Jain, N., Manhas, S.: Schottky Barrier Carbon Nanotube FET (CNTFET) Gas Sensors. In: Extended Abstracts of the 2014 International Conference on Solid State Devices and Materials (2014)
5. Appenzeller, J., Knoch, J., Derycke, V., Martel, R., Wind, S., Avouris, P.: Field-Modulated Carrier Transport in Carbon Nanotube Transistors. Phys. Rev. Lett. **89**(12), 126801 (2002)
6. Liu, L., Han, J., Xu, L., Zhou, J., Zhao, C., Ding, S., Shi, H., Xiao, M., Ding, L., Ma, Z., Jin, C., Zhang, Z., Peng, L.-M.: Aligned, high-density semiconducting carbon nanotube arrays for high-performance electronics. Science. **368**(6493), 850–856 (2020)

7. Palacios, J.J., Tarakeshwar, P., Kim, D.M.: Metal contacts in carbon nanotube field-effect transistors: Beyond the Schottky barrier paradigm. Phys. Rev. B. **77**(11), 113403 (2008)
8. Svensson, J., Campbell, E.E.B.: Schottky barriers in carbon nanotube-metal contacts. J. Appl. Phys. **110**(11), 111101 (2011)
9. Loo, Y.-L., Willett, R.L., Baldwin, K.W., Rogers, J.A.: Additive, nanoscale patterning of metal films with a stamp and a surface chemistry mediated transfer process: Applications in plastic electronics. Appl. Phys. Lett. **81**(3), 562–564 (2002)
10. Mizutani, T., Nosho, Y., Ohno, Y.: Electrical properties of carbon nanotube FETs. J. Phys. Conf. Ser. **109**, 012002 (2008)
11. Nosho, Y., Ohno, Y., Kishimoto, S., Mizutani, T.: Evidence of Edge Conduction at Nanotube/Metal Contact in Carbon Nanotube Devices. Jpn. J. Appl. Phys. **46**(19), L474–L476 (2007)
12. Kanbara, T., Takenobu, T., Takahashi, T., Iwasa, Y., Tsukagoshi, K., Aoyagi, Y., Kataura, H.: Contact resistance modulation in carbon nanotube devices investigated by four-probe experiments. Appl. Phys. Lett. **88**(5), 053118 (2006)
13. Thomsen, C., Reich, S.: Raman Scattering in Carbon Nanotubes, pp. 115–234. Springer, Berlin (2007)
14. Penzo, E., Palma, M., Chenet, D.A., Ao, G., Zheng, M., Hone, J.C., Wind, S.J.: Directed Assembly of Single Wall Carbon Nanotube Field Effect Transistors. ACS Nano. **10**(2), 2975–2981 (2016)
15. Derenskyi, V., Gomulya, W., Talsma, W., Salazar-Rios, J.M., Fritsch, M., Nirmalraj, P., Riel, H., Allard, S., Scherf, U., Loi, M.A.: On-Chip Chemical Self-Assembly of Semiconducting Single-Walled Carbon Nanotubes (SWNTs): Toward Robust and Scale Invariant SWNTs Transistors. Adv. Mater. **29**(23), 1,606,757 (2017)
16. Peng, N., Zhang, Q., Yuan, S., Li, H., Tian, J., Chan, L.: Current instability of carbon nanotube field effect transistors. Nanotechnology. **18**(42), 424,035 (2007)
17. Bikshalu, K., Reddy, V.S.K., Reddy, P.C.S., Rao, K.V.: High-performance Carbon Nanotube Field Effect Transistors with High k Dielectric Gate Material. Mater. Today Proc. **2**(9, Part A), 4457–4462 (2015)
18. Bondavalli, P., Legagneux, P., Pribat, D.: Carbon nanotubes based transistors as gas sensors: State of the art and critical review. Sensors Actuators B Chem. **140**(1), 304–318 (2009)
19. Li, F., Wang, Y., Wang, D., Wei, F.: Characterization of single-wall carbon nanotubes by N2 adsorption. Carbon. **42**(12), 2375–2383 (2004)
20. Zhao, J., Buldum, A., Han, J., Lu, J.P.: Gas molecule adsorption in carbon nanotubes and nanotube bundles. Nanotechnology. **13**(2), 195–200 (2002)
21. Franklin, A.D., Koswatta, S.O., Farmer, D., Tulevski, G.S., Smith, J.T., Miyazoe, H., Haensch, W.: Scalable and fully self-aligned n-type carbon nanotube transistors with gate-all-around, 2012. In: International Electron Devices Meeting, 10–13 Dec. 2012, pp. 4.5.1–4.5.4 (2012)
22. Franklin, A.D.: The road to carbon nanotube transistors. Nature. **498**(7455), 443–444 (2013)
23. Peng, L.-M., Zhang, Z., Wang, S.: Carbon nanotube electronics: recent advances. Mater. Today. **17**(9), 433–442 (2014)
24. Chu, L., Schmidt, H., Pu, J., Wang, S., Özyilmaz, B., Takenobu, T., Eda, G.: Charge transport in ion-gated mono-, bi- and trilayer MoS2 field effect transistors. Sci. Rep. **4**(1), 7293 (2015)
25. Minari, T., Kanehara, Y., Liu, C., Sakamoto, K., Yasuda, T., Yaguchi, A., Tsukada, S., Kashizaki, K., Kanehara, M.: Room-Temperature Printing of Organic Thin-Film Transistors with π-Junction Gold Nanoparticles. Adv. Funct. Mater. **24**(31), 4886–4892 (2014)
26. Franklin, A.D.: Nanomaterials in transistors: From high-performance to thin-film applications. Science. **349**(6249), aab2750 (2015)

Chapter 6
Concluding Remarks & Future Directions

Abstract This work investigated different methods to adapt and optimize the alignment relay technique (ART), and presents evidence supporting various optimizations to it. It was observed that ART can be performed either before or after nanotube deposition. Through π-π stacking, different batches of various SWNTs chiralities were aligned on silica using an iptycene functionalized surface. The results presented herein show a high alignment but low density of nanotubes. However, even though there are fewer CNTs with ART, the device performance is high. CNTs are deemed a suitable replacement for Si in FET devices if their orientation, type, and density can be controlled. There are research gaps in carbon nanotube sorting and alignment, and there are still factors that are difficult to control and optimize. Therefore, further researching ART can enlighten the evolution of organic materials-based CNT devices. So far, it is one of the few methods attempting to target both nanotube alignment and sorting for transistor applications.

Keywords Carbon nanotubes · Chiralities · Alignment · Transistors · CNTFETs · Relay · Pi-stacking

Following objective I from Chap. 1 (p. 4), changes to the procedure included sonication, filler molecules, and the incorporation of other techniques with ART, such as LBL and LB (Chap. 2). Sonication treatment after the nanotubes were deposited resulted in longer SWNTs remaining on the surface. The observation indicates that the π-π stacking strength between the iptycenes and the CNTs can be broken using simple benchtop sonication [1]. The alignment of the SWNTs on the surface doubled at the expense of the nanotubes density on the surface that significantly decreased to less than 0.1 SWNTs μm^{-2}. The addition of phenyl phosphonic acid on SiO_2 before the nanotubes' deposition increased their density to 0.95 SWNTs μm^{-2}. Therefore, additional molecules can be applied to the iptycene monolayer before the exposure of the CNTs, in order to alter the surface and increase the nanotubes density. A third main finding regarding goal I refers to supplementing ART with other techniques such as LBL and LB. Although the preliminary investigations were not

M. Snowdon, *Methods of the Alignment-Relay Technique for Nanosystems*,
Springer Theses, https://doi.org/10.1007/978-3-030-84413-4_6

effective in increasing alignment or selectivity, there is potential for further research with the LB & ART combination.

Following objective II, it was observed that different anchoring groups on the iptycene molecule can be substituted to increase the reproducibility of ART on different surfaces (Chap. 3). Both ALD deposited Al_2O_3 and α-Al_2O_3 demonstrated the effective use of a carboxylic acid anchoring component rather than the phosphonate ester [2]. Using the 8CB liquid crystal in combination with IptCOOH resulted in 0.3 SWNTs μm^{-2} on the Al_2O_3 surface with a SD of ±32°, which is an improvement over 0.1 SWNTs μm^{-2} with the original molecule. Additional testing of carboxylic acid filler molecules was also conducted, along with comparing Piranha to UV-O_3 cleaning and attempts at ART with the Au, TiO_2, and Si_3N_4 surfaces. This work could be furthered by testing different surfaces, such as gold, using sulfur-derived iptycenes [3], which is an interesting future direction.

Following objective III, the results from Chap. 4 indicated that the current ART framework does not provide orientation nor alignment to MWNTs, NWs, GNRs, or GQDs. An opportunity is to synergize the tweezer structure with the target nanosystem. For instance, if using MWNTs, this would require larger molecular tweezers, whereas using graphene would require a planar structure for stacking. It appears that if the diameter of the SWNTs corresponds well with the iptycene structure, then the alignment of the CNTs can occur. The interaction could succeed if the CNTs are in the diameter range of the π-π stacking portion of the iptycene. However, it is possible to deposit SWNTs-COOH in water, eliminating the use of surfactant molecules on the surface; but, it is difficult to approximate the chirality of these functionalized CNTs. Additionally, it is necessary to avoid aromatic solvents such as toluene as it can negatively impact the ART system's π-stacking interactions.

For objective IV, FETs were successfully fabricated using ART (Chap. 5), proving that ART-oriented SWNTs and surfaces are promising for electronics. The devices can be created using not only EBL but also LED photolithography. Both Ti/Au and Cr/Pd/Au have been used for electrode fabrication. Comparing the results from SWNTs networks vs. the aligned SWNTs from the ART method, a 1000x improvement in mobility with a 10x increase in I_{ON}/I_{OFF} ratio was observed. These values are also comparable to previous findings of aligned CNTFETs data [4, 5] and demonstrate the promising scalability of ART in the nanoelectronics industry.

As emphasized throughout the thesis, CNTs are deemed a suitable replacement for Si in FET devices if their orientation, type, and density can be controlled. Research trends investigating CNTFETs have also been steadily increasing, as shown in Fig. 6.1. Even with competing compounds such as transition metal dichalcogenides, nanowires, and graphene, CNTs still manage to be a prominent research area for materials nanoarchitectronics. However, the areas of nanotube sorting and alignment for carbon nanotubes are not as highly researched, and they are still factors that are difficult to control and optimize. Therefore, the statistics in Fig. 6.1 direct towards ART having a prominent foothold in organic materials-based CNT devices. So far, it is one of the few methods attempting to target both nanotube alignment and sorting for transistor applications.

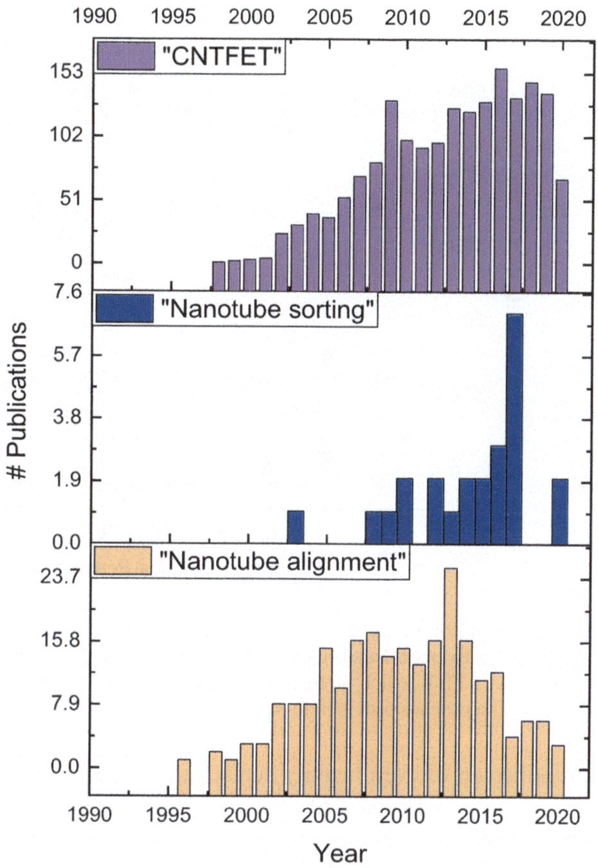

Fig. 6.1 Number of publications over time containing CNT alignment concepts, sorting, and transistors from the Web of Science database up until September 16, 2020. The search term for CNTFET includes "nanotube transistor," "CNFET," and "nanotube FET." Inspired by work from Franklin, A.D (2012, 2015)

The results of ART shown in this work can also be compared to previous findings. Figure 6.2 demonstrates the density of CNTs and I_{ON}/I_{OFF} of CNTFETs assembled by different techniques in the past decade. All the methods attempt high alignment of SWNTs, with high density for device applications. ART (red points) results have shown a high alignment but low density. However, even though there are fewer CNTs with ART, the device performance is high, as exhibited by the larger I_{ON}/I_{OFF}. This result is an indication that the iptycenes are not limiting the device performance of the material.

Additionally, this analysis showcases the room for growth and discovery with ART in that many parameters can be further optimized to enhance the density, alignment, and selectivity of the SWNTs. It is speculated that an additional direction for ART research could be to fabricate CNT sensors. The sensory applications of SWNTs should be further explored, as the interior of the CNTs could be exploited for high-sensitivity applications. The iptycenes monolayer additionally promotes

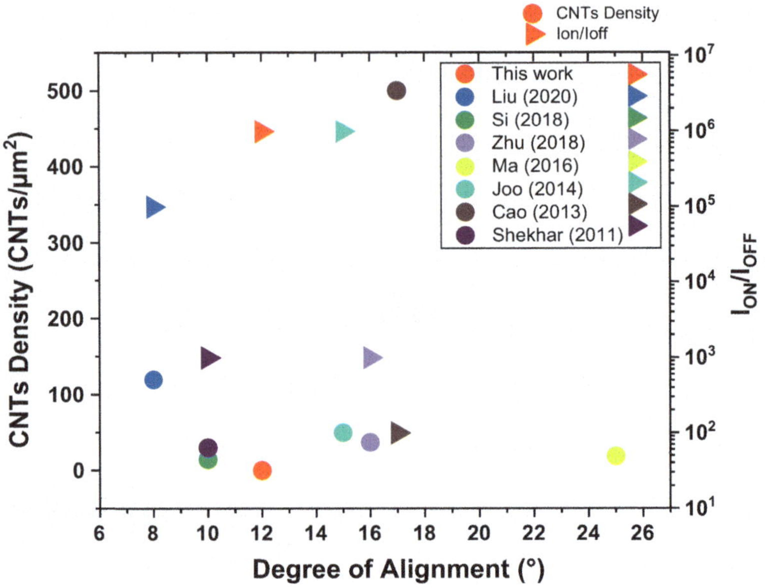

Fig. 6.2 Plot compiling the density of CNTs and I_{ON}/I_{OFF} of CNTFETs assembled by different techniques in the past decade that aim to have a high degree of alignment. Work by Ma et al. (2016) did not have a FET device, so there is no I_{ON}/IO_{FF} value. The V_{SD} ranged from 0.1 to 1 V. Additional details of the plot are provided in Appendix 6 of the back matter

the adhesion of even heavily surfactant wrapped SWNTs. This observation is particularly useful in cases requiring CNTs to adhere to the surface and overcome the removal of the nanotubes due to the presence of a surfactant. Overall, the journey of ART can follow several paths for future CNT applications.

References

1. Snowdon, M.R., Selmani, S., Schipper, D.J.: Sonication-Enhanced Alignment Relay Technique for the Orientation of Single-Walled Carbon Nanotubes. ACS Appl. Nano Mater. **2**(10), 6637–6645 (2019)
2. Snowdon, M.R., Wang, S., Mashmoushi, N., Hopkins, S.W., Schipper, D.J.: Carboxylic acids as anchoring components on aluminum oxide for the alignment relay technique of single-walled carbon nanotubes. New J. Chem. **45**(12), 5340–5349 (2021)
3. Kulak, M.R., Schipper, D.J.: Carbon nanotube alignment and sorting: Attempting a sulfur moiety as anchoring component. Phosphorus Sulfur Silicon Relat. Elem. **194**(7), 760–763 (2019)
4. Penzo, E., Palma, M., Chenet, D.A., Ao, G., Zheng, M., Hone, J.C., Wind, S.J.: Directed Assembly of Single Wall Carbon Nanotube Field Effect Transistors. ACS Nano. **10**(2), 2975–2981 (2016)
5. Derenskyi, V., Gomulya, W., Talsma, W., Salazar-Rios, J.M., Fritsch, M., Nirmalraj, P., Riel, H., Allard, S., Scherf, U., Loi, M.A.: On-Chip Chemical Self-Assembly of Semiconducting Single-Walled Carbon Nanotubes (SWNTs): Toward Robust and Scale Invariant SWNTs Transistors. Adv. Mater. **29**(23), 1,606,757 (2017)

Chapter 7
Experimental and Characterization Details

Abstract Elaborated here are all the experimental parameters and methods supporting the thesis, including all the characterization details for the alignment relay technique. Also included are the synthesis and characterization for all the molecules used during the work.

Keywords Characterization · Fabrication · Nanotubes · Carbon · Lithography · Synthesis · Alignment · Transistor · Devices · Microscopy

7.1 Sample Preparation for the Alignment Relay Technique

Silicon substrates (100) were cut to 1 cm × 1 cm dimensions and were obtained from Pure Wafer (San Jose, CA). The wafers were oxidized with a Piranha-acid bath (H_2SO_4: 30% H_2O_2, 3:1) for 30 min to form a new monolayer of atmospheric SiO_2. The iptycene **1** (2.5 wt%) in 5CB liquid crystal (Millipore-Sigma, ON) was dropped at the middle of a liquid crystal alignment wafer consisting of polyimide aligned arrays along the x-(long) direction of an ITO substrate (Instec Inc., USA). Production for iptycene **1** was completed using a previous synthesis method [1].

The refined surface of the SiO_2 was positioned on top of the iptycene/5CB solution and pushed downward to certify contact with the alignment layer to allow for the surface to functionalize. The sample remained untouched for 24–48 h. under ambient conditions to permit forming the necessary covalent bonds. The SiO_2 wafer was separated from the alignment layer template and both were washed with dichloromethane (DCM). The wash form the two substrates was then deposited into a scintillation vial, such that the DCM could be eliminated in vacuo. The solution of iptycene/5CB was stored in freezer conditions (0–2 °C). Drying of substrates was with $N_{2(g)}$ and once dry, the iptycene aligned monolayer on the wafer was positioned smooth side upwards in 10 mL of 90% pure sc-SWNTs blend (1 mg/100 mL) containing sodium cholate and sodium dodecyl sulfate (NanoIntegris Technologies Inc., QC). Additionally, 95% purity sc-SWNTs from Nanointegris Inc. were dissolved in a 1 mg/100 mL aqueous surfactant solution of 1% SDS using 30 min of

M. Snowdon, *Methods of the Alignment-Relay Technique for Nanosystems*, Springer Theses, https://doi.org/10.1007/978-3-030-84413-4_7

tip-sonication. The sc-SWNTs mixture and (3, 10) chirality specific SWNTs were dissolved in a 0.5% aqueous solution of sodium cholate.

The Nanointegris SWNTs have room temperature stability for 6 months and do not contain catalytic particle impurities. The SWNTs mixture receives 10 min of sonication prior to being in contact with the functionalized wafer prevent nanotube agglomeration and to ensure homogeneity. Alternatively, 8CB was also use instead of 5CB in the exact same manner. In cases where the original ART molecule was replaced with IptCOOH, the same parameters were also used. When applying the use of different substrate surfaces, the gold was treated with 10 min of isopropanol (IPA) sonication followed by 10 min with acetone sonication, followed by UV-O_3 for 20 min. The alumina surface was treated with 5 min of Piranha. Other substrates (TiO_2, SiN_X), including the marked wafers from Namiki Foundry (NIMS, Japan) used for transistor fabrication, were cleaned with model photosurface processor model p116–110, 110 W UV-Ozone for 30 min, rinsed with isopropanol or acetone, dry with $N_{2(g)}$. The sample remained in the CNT solution for 24–48 h at room temperature. The dropcast sample for Raman analysis was made ready by a drop from the homogeneous 90% sc-SWNTs solution onto a Piranha cleaned wafer of SiO_2 and dried in a fume hood.

Thin-films of the CNTs were prepared by dropcasting the SWNTs solution onto UV-O_3 cleaned silicon wafers, heated at 100 °C to allow for evaporation of the solution, rinsed with milliQ and dried with nitrogen. This process was repeated for a total of four times on a bare Si wafer to ensure there would be enough CNTs on the surface. Otherwise, when the CNTs were dropcast on the bare surface and rinsed, very few remained on the surface. The same treatment was used for an ART functionalized surface to obtain thin films of the CNTs with a higher success than the bare Si as the CNTs could adhere to the surface through π-π stacking with the iptycenes. Even the heavily surfactant wrapped CNTs would not be washed away. The substrates had to be rinsed after the CNTs dried on the surface for the device applications as otherwise there was too much of a thick layer of CNTs and surfactant crystals also appeared on the surface.

7.1.1 Marked Wafers

Marked wafers were fabricated in the Namiki Foundry on 200 nm SiO_2 with 10 nm Ti and 100 nm Au deposited using EBL with a Nippon Zeon 8EP-520A resist, ZED-NSO developer, ZMD-B rinse and ZDMAC remover. Outline of the patterns are shown in Fig. 7.1.

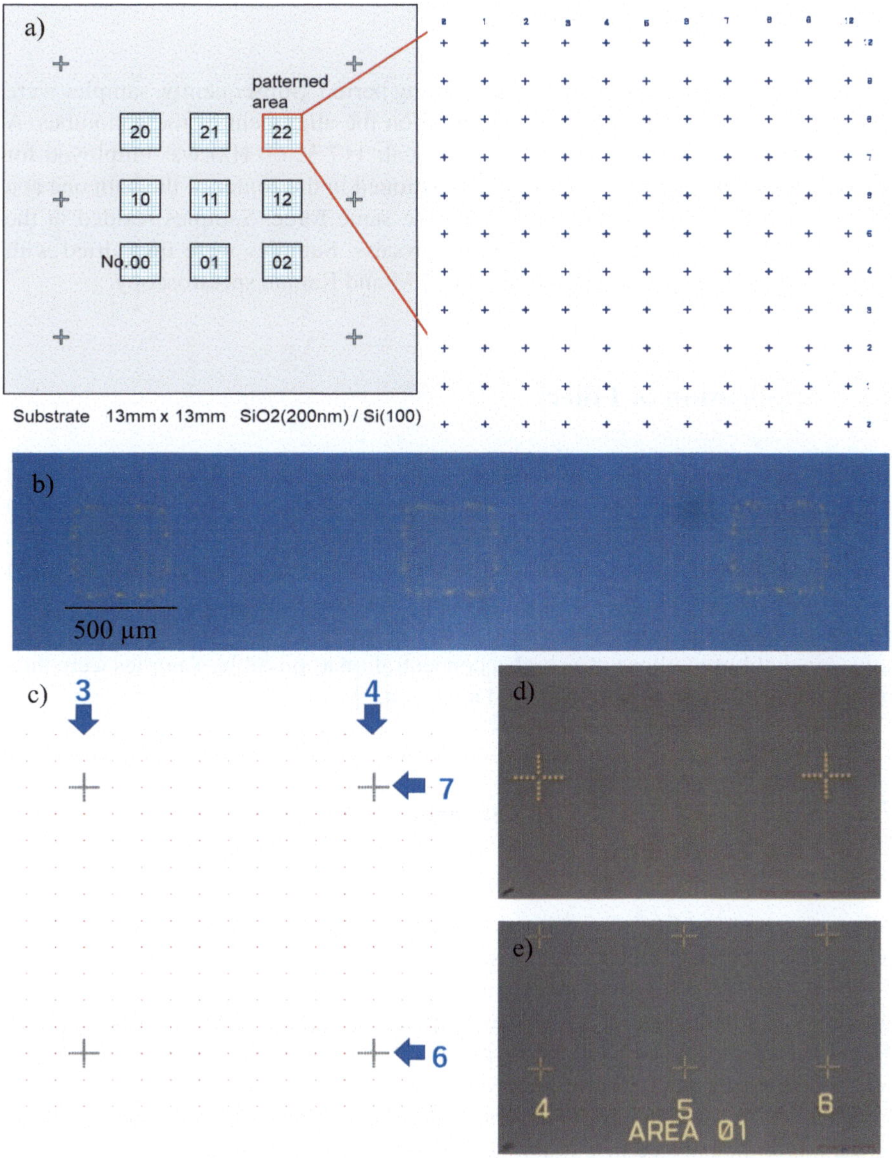

Fig. 7.1 (**a**) substrate overview for the marked wafers, (**b**) optical microscope view of the squared areas, (**c**) outline of the numbers and the markings (schematic) where the dot distance is 5 μm, (**d**) optical microscope between the numbered marks where the distance between the numbers (+ symbols) is 50 μm, (**e**) example of an area in optical microscope

7.2 Sonication Experiments

The previous process is the standard depositing period. Subsequently, samples were sonicated to observe the effects of agitation on the alignment of the nanotubes. A VWR symphony Ultrasonic Cleaner (electrical: 117 V, 60 Hz) was employed for sonication which occurred in a test tube positioned in the center of the bath one at a time to such that each sample underwent the same force. Samples resided in the same SWNTs solution for the sonication process. Samples were then dried with $N_{2(g)}$, and their surface characterized with AFM and Raman spectroscopy.

7.3 Application of Fillers

To an already treated ART wafer, one drop of chosen filler molecule was dropcast and placed to dry in an oven at 120 °C, or in the fumehood at room temperature (22 °C). If the filler compound was moisture sensitive, the sample was set in a 20 mL scintillation vial, the chosen molecule dropcast on the surface, and then left on the high-vacuum pump. The drop of compound was re-administered three times. For cases where the filler molecule was not a liquid/oil, the sample was first dissolved in dichloromethane at as high a concentration as possible. Samples were then rinsed with DCM, dried with $N_{2(g)}$ and analyzed.

7.4 SWNTs-COOH Synthesis and Surfactant Free Treatment

Adapting a protocol from Osorio et al. to a sample of 0.1 g SWNTs from Tuball, Inc. (Columbus, OH) was added a 30 mL solution of concentrated H_2SO_4: HNO_3 (3:1) in a 100 mL RBF [2]. The solution was stirred for 20 h under ambient temperature. The solution was filtered, and the acid was treated with NaOH (1 M) until neutralized. The collected nanotubes were then washed several times with MilliQ water. Water was added until a pH of 5 was achieved. The sample was then filtered outside the fumehood using a water aspirator and left to dry completely. The now SWNTs-COOH were then placed in a 20 mL scintillation vial in a desiccator for storage.

7.5 Langmuir Blodgett- ART Combination

A USI-FSD-300 computer-controlled Langmuir trough with a Teflon surface coating and affective area of 51.7 × 15 cm with a Wilhelmy balance surface pressure sensor was used. The setup was cleaned and calibrated using ultrapure water from an Elgastat UHQ-III system. The average pressure was 25 mN/m for the sample with only the ART Ipt **1** monolayer, and 15.79 mN/m for the ART surface with already aligned CNTs.

7.6 Other Nanostructures for Alignment with ART: NWs, GNRs, and MWNTs

ZnO nanowires were bought from Sigma-Aldrich, Canada, with a diameter × length of 90 nm ×1 μm. First, 5 mg of ZnO NWs were dissolved in 20 mL IPA and sonicated for 1 min before use. Al_2O_3 nanowires were acquired from Sigma-Aldrich, Canada, with a diameter × length of 2–6 nm + 200–400 nm. 5 mg Al_2O_3 NWs was dissolved in 20 mL IPA and sonicated for 1 min before using. Silica slide was placed in a solution of ZnO/Al_2O_3 NWs and left for 24 h at room temperature, covered. The same treatment occurred for a sample of graphene quantum dots (GQDs) (1% SDS/SC aqueous solution – same as the one used for SWNTs dispersion, NanoIntegris Inc). GQDs were donated by the Nano/Micro-Systems Research Laboratory (Prof Yavuz, University of Waterloo). The same dissolution occurred with the graphene nanoribbons (GNRs). Two sets of GNRs were tested for deposition: One set of pristine nanoribbons was donated from Prof Chalifoux group (University of Nevada) fabricated using their synthesis method (bottom-up approach) [3, 4]. The other reduced GNRs (1% in surfactant solution) and oxidized GNRs in water (unknown concentration) were obtained from the Professor Yu group (University of Waterloo) fabricated from CheapTubes Inc. using a top-down approach [5, 6].

The MWNTs of two different diameters, 8 nm and 10–20 nm (Sigma-Aldrich, ON) were used and compared. It was found that there was a low density and level of alignment for MWNTs with 24 h to allow for deposition, so the method was tested for 24 h, 48 h and 72 h of deposition time instead, with a MWNT concentration of 0.01 mg/mL maintained.

7.7 Characterization of ART Substrates

7.7.1 Atomic Force Microscopy

7.7.1.1 Tapping Mode

Samples were imaged with both topography and height measurements of the CNTs using a Veeco Dimension 3100 Atomic Force Microscope operating in intermittent contact (tapping) mode under at room temperature (Veeco Inc., from the WATLab at the University of Waterloo). Scans were completed with Budget Sensors TAP150G intermittent contact cantilever, with spring constant of k = 5 N/m, resonance frequency f = 146 kHz. Alternatively, a Bruker Dimension FastScan AFM (Quantum

Nanofabrication Clean Room, Quantum Nano Center, Waterloo, ON) was also used. The NanoFab AFM scans were accomplished with a ScanAsyst-Air silicon tip on nitride lever Bruker AFM probe using tapping mode, comprising of $k = 0.4$ N/m with $f = 70$ KHz. For AFM images obtained at NIMS, an AFM5200S Hitachi scanning probe microscope with an Al coated cantilever, $f = 140$ KHz, $k = 17$ N/m was used.

7.7.1.2 Conductive Mode

The instrument was a Cypher ORCA electrical AFM (Asylum Research, Inc.) that used a Ti/Ir (5/20) coated tip ($f = 73.5$ KHz). Sample was glued onto a substrate surface using 200 Silver Paint (Pella, Inc.) and placed with a magnet to ensure conductivity within the sample surface. The voltage applied ranged from 0.5 V to 2 V to create the high-resolution micrographs.

7.7.2 Scanning Electron Microscopy

Additional micrographs were imaged with a Zeiss Ultra Plus Scanning Electron Microscope (Carl Zeiss Microscopy GmbH, Germany) with In-lens secondary electron detector (5 kV, WD = 9 mm). The system vacuum was 1.5×10^{-5} mbar with Schottky filament, and stigmation was corrected for with every new sample. For the sample substrates at NIMS, a Jeol JSM-6010LA was used with 1 kV secondary-electron detector (WD = 3–4, 10 μA filament current).

7.7.3 X-Ray Photoelectron Spectroscopy

Briefly, X-ray photoelectron spectroscopy was conducted using a Thermo-VG Scientific ESCALab 250 Microprobe and a monochromatic Al Kα source (1486.6 eV) with CasaXPS software to attempt to obtain characteristic information on the binding of the iptycene with the SiO_2 surface. The only observation was that the PO peaks (P_{2p}) overlapped with those of SiO_2 and therefore, XPS was ineffective in confirming the concentration of iptycenes on the surface. Perhaps with the ITO or Al_2O_3 surfaces this characterization technique could be revisited.

7.7.4 UV-vis Spectroscopy

On an Agilent 8453 spectrometer with the assistance of Dr. Anand Lopez (Juewen Liu Lab), the sapphire functionalized wafers were attempted to be charactires through UV-vis absorption spectroscopy. The sample (blank or functionalized) was pinned to a microscopic glass slide. Prior to measuring, a clear and cleaned microscope slide was employed as a blank such that solely the sample would be a contributing factor to the absorbance. The absorbance spectrum of each sample was assessed with an Agilent 8453 spectrometer scanning from 190 nm to 1100 nm.

7.7.5 *Ellipsometry*

The Woollam M-2000DI ellipsometer (J.A. Woollam Co.) was used to calculate the film thickness for the standard ART with molecule **1**. The sample surface was mapped through seven spots on a 1 cm × 1 cm wafer. The control layer #1 was a UV-O_3 cleaned SiO_2 wafer originating from the same mother wafer as the functionalized sample and had an SiO_2 thickness of 1.73 nm. The iptycene functionalized layer (Layer #2) was approximated using a Cauchy formula with an average thickness of 1.58 nm, and mean squared error (MSE) of 2.69.

7.7.6 *Raman Spectroscopy*

Characteristic carbon nanotube stretches in the radial breathing mode location were observed with two different laser light sources (533 nm, green 25% laser filtering; 633 nm, red 50% laser filtering) with a He-Ne laser from a Horiba Jabin Yvon HR800 Raman spectrometer with Olympus BX41 Microscope (Professor X. S. Tang at University of Waterloo). The backscattering configuration was used for the measurements with a spectral resolution of 0.3 cm^{-1}. Characterization was completed in an ambient atmosphere and temperature with a laser power of 1 mW and spectra were documented with an exposure time of 5 s and totaled over 5 accumulations. The identical parameters were employed for mapping over a 30 μm × 30 μm with a step size of 1.6 μm (x and y). For the 10-min sonicated SWNTs samples, the accumulations were changed to 10 for mapping the area of 15 μm × 15 μm. The laser was centered on the substrate surface with a 20x microscope objective and a numerical aperture of 0.4. The spectral results were processed with smooth despiking tools using LabSpec 6 Spectroscopy software. The minimum power for signal measurement was restricted by the signal to noise resolution of the detector in the spectrometer. The peak positions, full-widths-at-half-maximum (FWHMs) and integral intensities were established by fitting the spectrum profiles with Lorentzian functions in OriginPro 2020. Different areas on the sample were measured to ensure repeatability. At NIMS, the Raman spectrometer was either from the Namiki Foundry Horiba Jobin-Yvon T64000 at 532 nm, 100x objective or from the Tsukagoshi lab Tokyo Instruments Raman at 532 nm, 100x objective lens and 1800 grooves/mm grating.

7.7.6.1 Diameter and Chirality Assignments

$$\omega_{RBM} = \frac{A}{d_t} + B$$

where $A = 248$, $B = 0$ for the relation of $\omega_{RBM} \propto \frac{1}{d_t}$. These values have been shown to be accurate for single SWNTs on silicon [7].

Chirality assignments were conducted following available data by Kataura [8], Weisman [9], Jorio [7], Sato [10] and Reich [11]. In principle:

1. The average diameter of tubes was identified by contacting the source manufacturers
2. Experimental Raman data was obtained at either red (632 nm), green (532 nm) or both laser excitations
3. The RBM region was peak deconvoluted
4. Diameter values were calculated
5. Vertical and horizontal lines were drawn on the experimental Kataura plot
6. The RBM frequency, diameter and excitation energy were matched along with the experimental plot and tables to identify nanotube family branches

7.7.7 Alignment Data Analysis

The SEM and/or AFM tiff images were processed for length and angle using ImageJ software. Gwyddion software was used for processing .ibw files and converting to tiff in some occasions. Measurement scales were in µm. Standard deviation was calculated from all the angles using excel STDEV.P. If the SD of the binned data is taken instead, the SD will be significantly lower, so all the individual angles were considered in the calculation in this work. The process is outlined in Fig. 7.2. The SEM was either a Zeiss Ultra (WatLab, University of Waterloo) or a Joel JSM 6700LA HR-SEM at NIMS (Mitome Lab, NIMS). The AFM was either a Hitachi AFM5000II (Sasaki Lab, NIMS), Asylum MFB-3D Origin (Tanaguchi Lab, NIMS), Bruker Dimension FastScan (QNFAB, Waterloo) or DI Nanoscope IV AFM (Watlab, Waterloo).

7.8 Synthesis

Thin Layer chromatography (TLC) analysis was completed on alumina-backed wafers pre-covered with Silica Gel 60 (250 µm). Glassware was dried in the oven (100 °C) prior to use. To visualize compounds, the TLC plates were placed under a UV lamp with a wavelength of 254 nm. Flash column chromatography was conducted with silica gel particle size 20–63 or 40–63 (two different types) (mesh

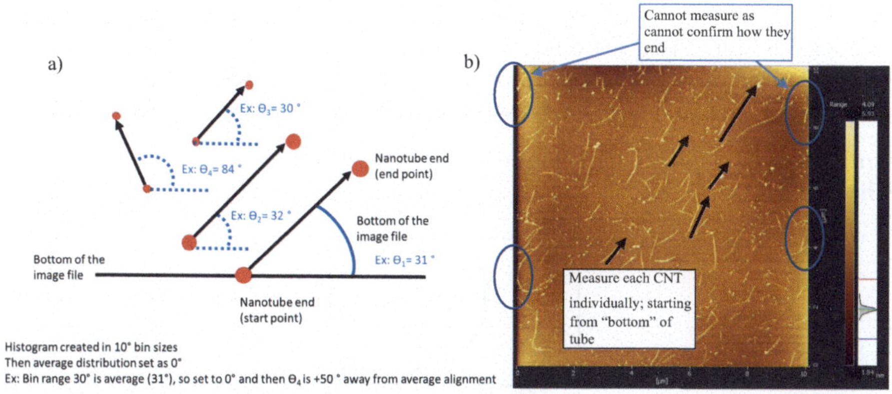

Fig. 7.2 (**a**) Diagram to illustrate how the angles and lengths were tabulated on ImageJ (**b**) Sample AFM image to represent data analysis of CNTs with indications as to what kinds of tubes were not considered in the calculation

230–400) supplied by Silicycle® using a CombiFlash Rf200i system (Teledyne Isco.) or performed manually in a glass column. Anhydrous solvents were acquired using a JC Meyer Solvent Dispensing System. Organic mixtures were dried out using anhydrous $MgSO_4$ and concentrated in a lowered pressure at 40 °C. Additionally, when removing toluene or xylenes, the temperature was set to 80 °C. Starting materials, reagents and any other commercially available chemicals were provided by Sigma-Aldrich unless otherwise noted. FT-IR analysis was conducted using a Perkin Elmer Spectrum Two provided by the University of Waterloo Organic Chemistry Teaching Lab (Science Teaching Complex) in which the sample could be placed neat on the crystal or dropcast in acetone. NMR data were obtained with a Bruker AVANCE 300 or Bruker AC300 (both 300 MHz for ^1H) NMR spectrometers.

Synthesis of cis-1,4-Dichloro-2-Butene [2] CAS 1476-11-5

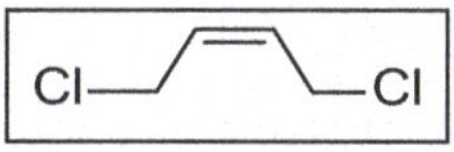

To synthesize **2**, an adapted procedure by Brandsma [12] and Hanik [13] was used. Synthesis of compound **2** was achieved by adding thionyl chloride (32.7 mL, 53.3 g, 448 mmol) dropwise to cis-1,4-butene-diol (28.2 g, 320 mmol) and pyridine (0.51 g, 6.4 mmol) in a round bottom flask over a period of 45 min. The solution became black and gas evolution occurred once the thionyl chloride was added. The solution was left to react for 25 min. The mixture was then immediately purified by fractional vacuum distillation to obtain **2** in a 70% yield (25.0 g, 224 mmol) as a clear liquid. The compound eluted at a temperature range of 80–85 °C. Literature cites that the compound can be isolated at approximately 45 °C/10 mm Hg [12]. The yield for the reaction ranged from 50–80% depending on the quality of the thionyl chloride. *Rf:* 0.77 (10% DCM/Hexane), ^{1}H NMR (300 MHz, $CDCl_3$, 7.26 ppm): δ = 5.84–5.87 (dt, 2 H, *J* = 1.6 Hz, 6.5 Hz), 4.10 (dd, 4 H, *J* = 1.6 Hz, 6.5 Hz) and ^{13}C NMR (75.5 MHz, $CDCl_3$, 77.0 ppm): δ = 128.7, 38.9. Spectral data observed matched literature values reported by Hanik [13].

Synthesis of 11,12-Bis(Chloromethyl)-9,10-Dihydro-9,10-Ethanoanthracene [3]

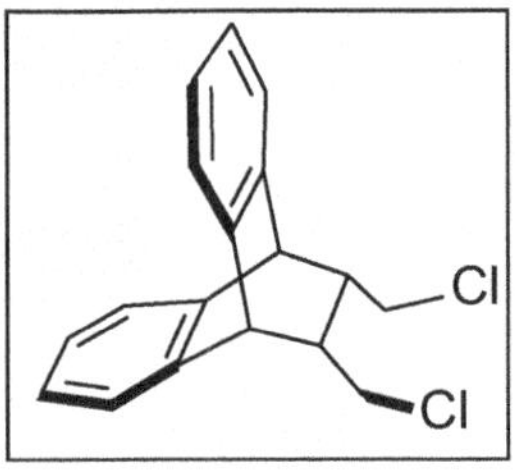

To synthesize **3**, an adapted procedure by Hart et al. was used [14]. Compound **3** was made through a Diels-Alder reaction of anthracene (1.94 g, 10.88 mmol) with cis-1,4-dichlorobutene (8.16 g, 65.29 mmol) as solvent at 190 °C for 48 h in an oven dried pressure vessel. A Celite plug was used with chloroform as the solvent to filter out unwanted particles, then purified by flash chromatography (10% DCM: Hex), R_f: 0.19. White solid of **3** (90%, 9.79 mmol, 2.96 g). ^{1}H NMR (300 MHz, $CDCl_3$, 7.26 ppm): δ = 7.32 (m, 4 H), 7.16 (m, 4 H), 4.57 (br s, 2 H), 3.49–3.46 (dd, 2 H, *J* = 4.7 Hz, 10.7 Hz), 2.83–2.89 (t, 2 H, *J* = 10.7 Hz), 1.53 (m, 2 H). ^{13}C NMR (75.5 MHz, $CDCl_3$, 77.0 ppm): δ = 142.7, 125.6, 123.8, 31.6, 22.7, 14.1. FTIR (cm^{-1}): 3100–3000 cm^{-1} (C-H, sp^2), 2999–2923 cm^{-1} (C-H, sp^3). Spectral data matched those reported by Hart et al. [14] When upscaling the reaction, often the anthracene could be recovered as white crystals and re-used. Trans-1,4-dichlorobutene was also used as purchased from Sigma-Aldrich in 95% purity and gave the trans-product. It is possible to also recover the 1,4-dichlorobutene through vacuum distillation as a clear liquid. The remaining black solid can then been purified through flash chromatography to obtain the product.

Synthesis of (9s,10s)-11,12-Dimethylene-9,10-Dihydro-9,10-Ethanoanthracene [4] CAS 36439–82-4

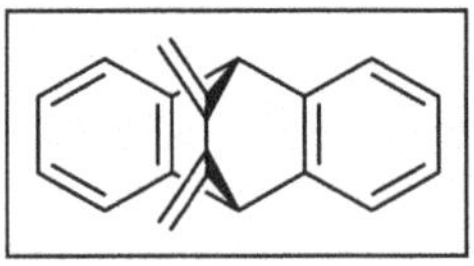

To synthesize **4**, an adapted procedure by Hart et al. was used [14]. Under an inert atmosphere, compound **3** (0.26 g, 0.87 mmol) was treated with 3 equivalents of potassium tert-butoxide (0.29 g, 2.6 mmol) in a 3:1 ratio of dimethyl sulfoxide to tetrahydrofuran anhydrous solvent mixture for 48 h. Upon addition of potassium tert-butoxide, the solution became dark purple in color. Ice water was added to quench the reaction, followed by liquid-liquid extraction done with brine extract and ethyl acetate (1:1) to purify. The organic layer was reduced on a rotary evaporator to produce an off-white powder. The product was purified further via flash chromatography (10% DCM, 90% Hex) to produce **4** (70%, 0.12 g, 0.73 mmol) as a white powder, *Rf:* 0.25 (10% DCM/Hex). ^{1}H NMR (300 MHz, $CDCl_3$, 7.26 ppm): δ = 7.31 (m, 4 H), 7.12 (m, 4 H), 5.29 (s, 2 H), 5.12 (s, 2 H), 4.86 (s, 2 H). ^{13}C NMR (75.5 MHz, $CDCl_3$, 77.0 ppm): δ = 143.9, 141.8, 126.3, 123.3, 105.3, 55.3. Spectral data equaled those described by Hart et al. [14]

Synthesis of (7aS)-5,5a,6,7a,8,9,14,15,15a,17,17a,18-Dodecahydro-5,18:9,14-Bis([1,2]Benzeno)Heptacene [5] CAS 103960–19-6

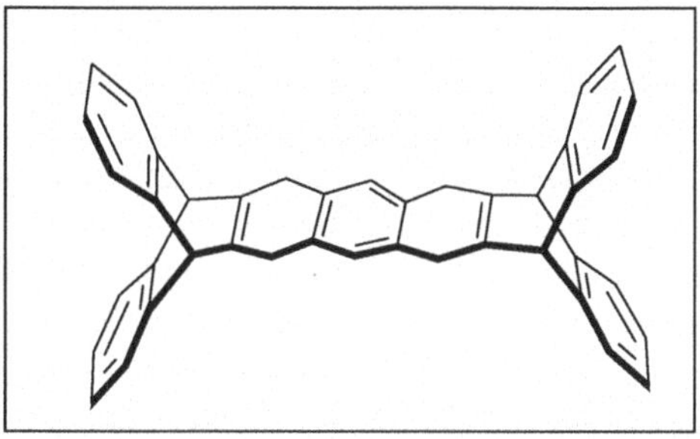

To synthesize **5**, an adapted procedure by Hart et al. was used [14]. All glassware was oven dried the day before. A solution of diene **4** (0.71 g, 3.1 mmol, 2.0 equiv.) and 1,2,4,5-tetrabromobenzene (0.30 g, 4.65 mmol, 1 equiv.) in 12 mL dry toluene under argon in an ice bath was prepared. Then, 2 equiv. n-butyllithium (2.91 mL, 0.465 mmol) in hexanes (1.6 M) was added drop wise over the course of 60 min. The reaction was then removed from the ice bath to bring then temperature to ~22 °C. The clear, yellow solution was stirred overnight, then quenched with 20 mL of water. The reaction was stirred for 30 min, then poured into 150 mL of ice-water to form precipitate. The resulting precipitate was filtered, then liquid-liquid extracted with water and ether (1:1) to give colorless **5** (0.56 g, 67% yield), *Rf:* 0.33 (40% DCM, 60% Hexanes). ^{1}H NMR (300 MHz, $CDCl_3$, 7.26 ppm): δ = 7.30 (m, 8 H), 6.90–6.94 (m, 8 H), 6.78 (s, 2 H), 4.82 (s, 4 H), 3.54 (s, 8 H). ^{13}C NMR (75.5 MHz, $CDCl_3$, 77.0 ppm): δ = 146.0, 139.8, 131.2, 125.5, 124.4, 123.8, 31.6, 22.7. Spectral data equaled those stated by Hart et al. [14]

Synthesis of Compound 5,9,14,18-Tetrahydro-5,18[1′,2′]:9,14[1″,2″]-Dibenzenoheptacene [6] CAS 87207–48-5

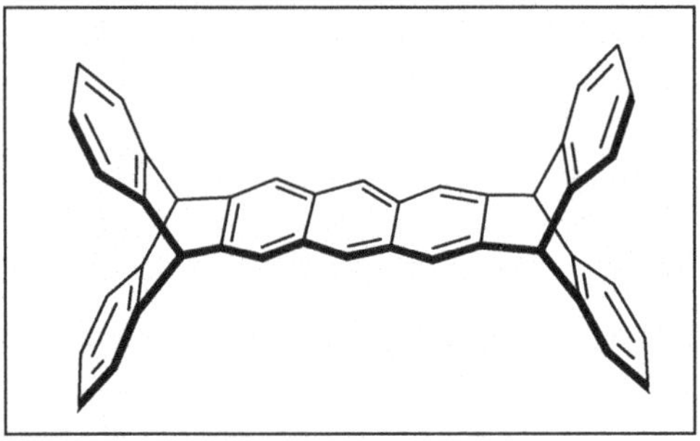

To synthesize **6**, an adapted procedure by Hart et al. was used [14]. A mixture of **5** (0.20 g, 0.37 mmol) in 30 mL xylenes containing 0.25 g 2,3-dichloro-5,6-dicyano-1,4-benzoquinone (DDQ) was warmed and refluxed at 60 °C for 48 h under argon. The hot solution was filtered and then a liquid-liquid extraction was performed with brine extract and dichloromethane (1:1) to give 60 mg of **6** as an off-white solid. The xylenes filtrate residue was extracted with brine extract and dichloromethane as well to give a light brown residue containing an additional 68 mg of **6**. The fractions were combined to yield 0.128 g of **6** (65%). ^{1}H NMR (300 MHz, $CDCl_3$, 7.26 ppm): δ = 8.05 (s, 2 H), 7.78 (s, 4 H), 7.40 (m, 8 H), 7.01 (m, 8 H), 5.45 (s, 4 H). ^{13}C NMR (75.5 MHz, $CDCl_3$, 77.0 ppm): δ = 144.3, 140.6, 130.4125.6, 125.0, 123.8, 121.3 53.6. Spectral information matched those described by Hart et al. [14]

Synthesis of Compound Tetraethyl (5,7,9,14,16,18-Hexahydro-5,18:9,14-Bis([1,2]Benzeno)-7,16-Ethenoheptacene-25,26-Diyl)Bis(Phosphonate) (IptP(O)(OEt)$_2$; 1) CAS 2243418–92-8

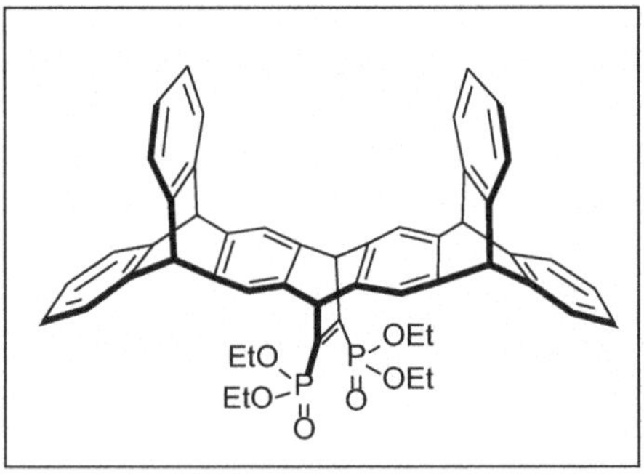

To synthesize **1**, an adapted procedure by Selmani and Schipper was used [1]. In a pressure vessel, compound **6** (0.11 g, 0.21 mmol) was refluxed in toluene (anhydrous), at 200 °C for 5 days with bis(diethoxyphosphoryl)acetylene (0.12 g,

0.41 mmol, 0.10 mL) to give molecule **1** (23%, 0.04 g, 0.048 mmol). Compound was purified using flash chromatography *Rf*: 0.1 (95% EtOAc: 5% MeOH) ^{1}H NMR (300 MHz, $CDCl_3$, 7.26 ppm): δ = 7.31 (m, 8 H), 7.20 (m, 4 H), 6.90 (m, 4 H), 6.85 (m, 4 H), 5.55 (t, 2 H, *J* = 7.2 Hz coupling to ^{31}P), 5.23 (s, 4 H), 3.90 (m, 8 H), 1.18 (t, 12 H, *J* = 7.6 Hz). *Rf:* 0.1 (5% MeOH: EtOAc). ^{13}C NMR (75.5 MHz, $CDCl_3$, 77.0 ppm): δ = 156.2 (d, *J* = 10.2), 146.7, 144.1, 142.1, 140.9, 124.8, 122.9, 118.9, 61.8 (t, *J* = 3.6 Hz), 53.5 (t, *J* = 10.2 Hz), 52.7, 15.9 (t, *J* = 3.6 Hz). ^{31}P NMR (121.5 MHz, $CDCl_3$, δ 0.00 phosphoric acid external standard): δ = 18.30. Spectral data ressebled those described by Schipper [1].

Synthesis of Tetraethyl 9, 10-Dihydro-9,10-Ethenoanthracene- 11,12- Diyl-Diphosphonate [8] CAS: 114071–16-8

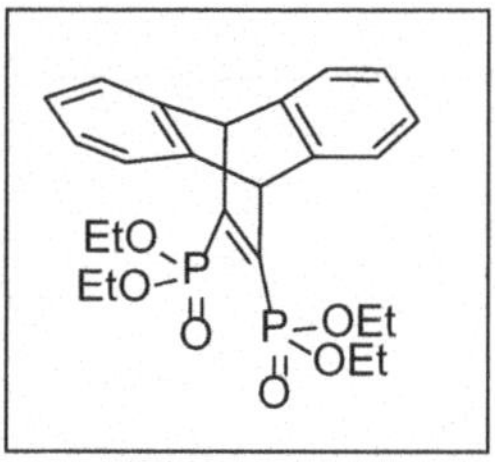

To a solution of anhydrous toluene with anthracene (3 equiv., 1.00 g) in a 50 mL pressure vessel was added bis(diethoxyphosphoryl)acetylene (1 equiv.) for 5 days in a 160 °C sand bath [15]. The reaction was allowed to cool and solvent was removed under reduced pressure. Majority of the excess anthracene was removed by trituration with methanol. The molecule was refined by flash chromatography to obtain the target molecule **8** in a 48% yield as a clear oil (100% EtOAc). *Rf*: 0.2. ^{1}H NMR (300 MHz, $CDCl_3$, 7.26 ppm): δ = 7.36 ppm (m, 4 H), 7.00 ppm (m, 4 H), 5.79 ppm (t, 2 H, *J* = 6.4 Hz coupling to ^{31}P), 4.01 ppm (m, 8H), 1.24 ppm (t, 12 H, *J* = 7.4 Hz). ^{13}C NMR (75.5 MHz, $CDCl_3$, 77.0 ppm): δ = 151.6, 144.2, 126. 8, 123.4, 61.1, 55.3, 15.9. ^{31}P NMR (121.5 MHz, $CDCl_3$, δ 0.00 phosphoric acid external standard): δ = 15.06 ppm. Spectral data matched those reported by Acheson [15].

Synthesis of Benzyl Diethyl Phosphonate [9] CAS: 1080-32-6

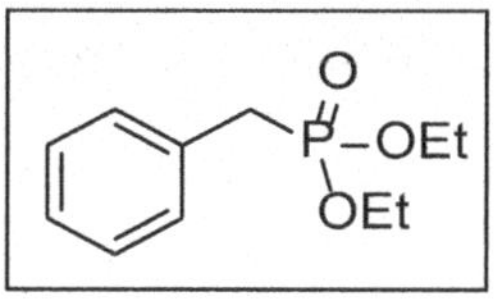

Following a previous protocol [16], a mixture of benzyl bromide (1.01 g, 5.8 mmol) and triethyl phosphite (3.5 equiv., 20.4 mmol, 3.39 g) in 1,4-dioxane (75 mL) was refluxed until the benzyl bromide was no longer visible on the TLC plate (100%

Hex). Unreacted triethyl phosphite was removed under vacuum and the product was extracted from the remaining residue using EtOAc and a 0.1 M $NaCO_3$ wash. Organic layer was dried over anhydrous $MgSO_4$ and concentrated to afford **9** (95%) as a clear oil. ^{1}H NMR (300 MHz, $CDCl_3$, 7.26 ppm): δ = 7.30–7.22 (m, 5 H), 3.96 (q, 4 H, *J* = 6.7 Hz), 3.13–3.07 (m, 2 H), 1.31 (t, 6 H, *J* = 6.7 Hz). ^{13}C NMR (75.5 MHz, $CDCl_3$, 77.0 ppm): δ = 131.8, 131.5, 129.8, 129.7, 128.9, 128.7, 126.8, 126.8, 62.1, 62.0, 34.6, 33.5, 16.3, 16.3. ^{31}P NMR (121.5 MHz, $CDCl_3$, δ 0.00 phosphoric acid external standard): δ = 27.60 ppm. Spectral data matched those reported by Saltiel [16].

Synthesis of Compound 5,7,9,14,16,18-Hexahydro-5,18:9,14-Bis([1,2]Benzeno)-7,16-Ethenoheptacene-25,26-Dicarboxylate [22] CAS: 131760–73-1

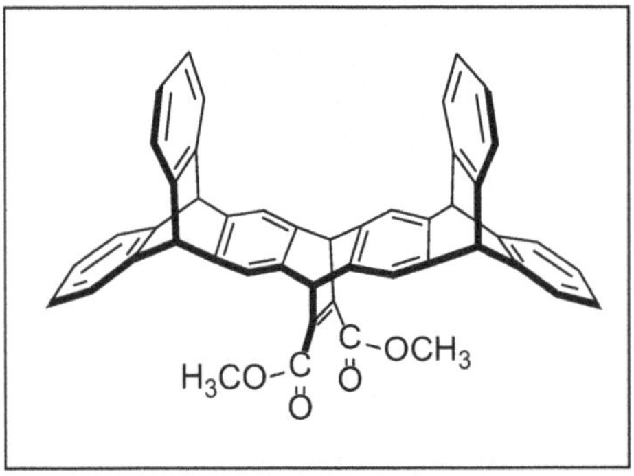

In order to obtain **22**, compound **6** (8.8 mg, 0.017 mmol) was dispersed with anhydrous toluene in an oven dried pressure vessel at 110 °C with dimethyl acetylenedicarboxylate (DMAD) (8.68 mg, 0.06105, 8 μL) for 48 h using an adapted procedure from Bouffard et al. [17] The mixture was rotary evaporated and purified by flash chromatography (40% DCM/Hexanes) yielding **22** as a white solid (71%, 8.2 mg, 0.012 mmol). *Rf*: 0.38 (40% DCM: Hex), ^{1}H NMR (300 MHz, $CDCl_3$, 7.26 ppm): δ = 7.32–7.30 (br s, 4 H), 7.28 (t, 4 H, *J* = 3.7 Hz), 7.20 (t, 4 H, *J* = 3.7 Hz), 6.94 (t, 4 H, *J* = 3.5 Hz), 6.82 (t, 4 H, *J* = 3.5 Hz), 5.24–5.22 (br s, 6 H), 3.66 (s, 6 H). ^{13}C NMR (75.5 MHz, $CDCl_3$, 77.0 ppm): δ = 165.9, 147.3, 142.8, 141.3, 125.1, 123.4, 123.4, 119.7, 76.6, 53.9, 52.2. FTIR (cm^{-1}): 3065–3014 (C-H, sp^2), 2952–2857 (C-H, sp^3), 1709 (C=O).

Synthesis of Compound (IptCOOH;12) CAS: N/A

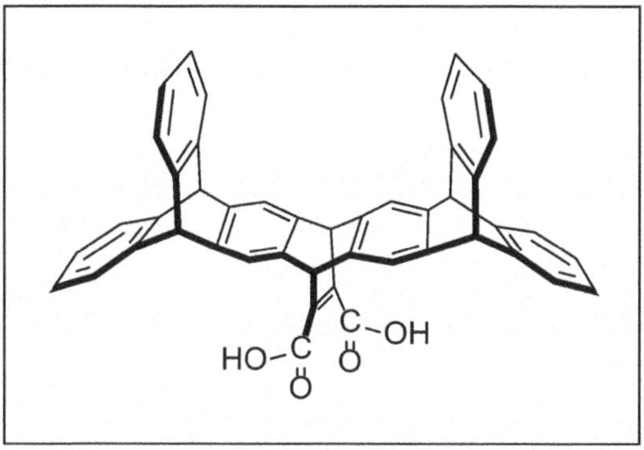

The formation of **12** was completed using an variation of published processes from Bouffard et al. [17] Molecule **12** was made by adding 10.0 equiv. of lithium hydroxide (LiOH) to **22** (8.2 mg, 0.012 mmol) and refluxing in a solution mixture of 3:1, methanol to water overnight. The resulting mixture was then chilled and stored at 10 °C for several hours to obtain **12** (46%, 3.6 mg, 0.0056 mmol) as a white precipitate. *Rf*: 0.35 (40% DCM: Hex). ^{1}H NMR (300 MHz, $CDCl_3$, 7.26 ppm): $\delta = 7.42$ (br s, 4 H), 7.39–7.35 (m, 4 H), 7.32–7.29 (m, 4 H), 6.95–6.92 (d, 4 H, $J = 4.7$ Hz), 6.86–6.84 (d, 4 H, $J = 4.7$ Hz), 5.79 (s, 2 H), 5.28 (s, 4 H). ^{13}C NMR (75.5 MHz, $CDCl_3$, 77.0 ppm): $\delta = 167.8$, 144.9, 143.3, 140.6, 125.2, 125.1, 123.5, 123.0, 53.9, 29.7. HR-MS (ESI): (+) Performed in 1:1 MeOH: H_2O + 0.1% Formic Acid. Calculated for $C_{46}H_{29}O_4$ $[M + H]^+$ 645.20604, found 645.20591; calculated for $C_{46}H_{29}O_4$ $[M + NH_4]^+$ 662.24504, found 662.23255 (likely NH_4 adduct). Recheck with negative ESI (−) Performed in 1:1 MeOH: H_2O + 0.1% NH_4OH. Calculated for $C_{46}H_{27}O_4$ $[M\text{-}H]^-$ 643.19039, found 643.19135. FTIR (cm^{-1}): 3419 (-OH), 3061–3020 (C-H, sp^2), 2922–2853 (C-H, sp^3), 1709 (C=O) [18].

Synthesis of Compound Dimethyl 9,10-Dihydro-9,10-Ethenoanthracene-11,12-Dicarboxylate [23] CAS 1625-82-7

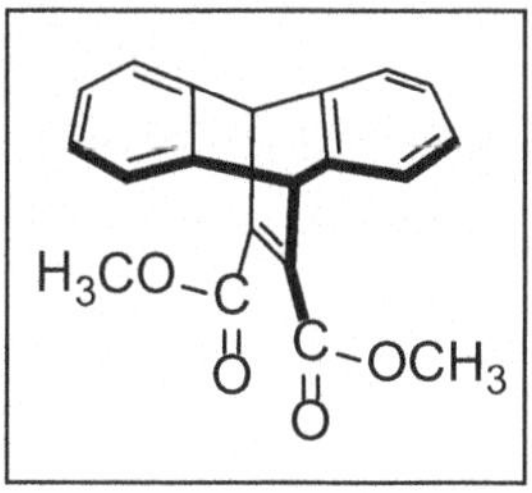

Compound **23** was made by using an adapted method of published procedures [17]. DMAD was purified before use by vacuum distillation, then added (3.24 g, 22.80 mmol, 2.79 mL) to a pressure vessel containing anthracene (3.25 g,

18.24 mmol) and dried with $N_{2(g)}$ for half an hour. This step was followed by the addition of 10 mL of toluene and the mixture was refluxed for 3 days at 180 °C. The toluene was removed using a rotary evaporator and the solid was recrystallized in methanol to purify. Upon filtration of recrystallized product, **23** was obtained as a white solid (92%, 5.36 g, 16.73 mmol). *Rf:* 0.84 (80% EtOAc, 20% Hexanes), ^{1}H NMR (300 MHz, $CDCl_3$, 7.26 ppm): δ = 7.38–7.35 (m, 4 H), 7.02–6.99 (m, 4 H), 5.46 (s, 2 H), 3.77 (s, 6 H). ^{13}C NMR (75.5 MHz, $CDCl_3$, δ 77.0 ppm): δ = 165.9, 147.0, 143.7, 125.4, 123.8, 52.4, 50.9. Spectral data matched those reported by Bouffard [17].

Synthesis of 9,10-Dihydro-9,10-Ethenoanthracene-11,12-Dicarboxylic Acid [15] CAS: 1625-81-6

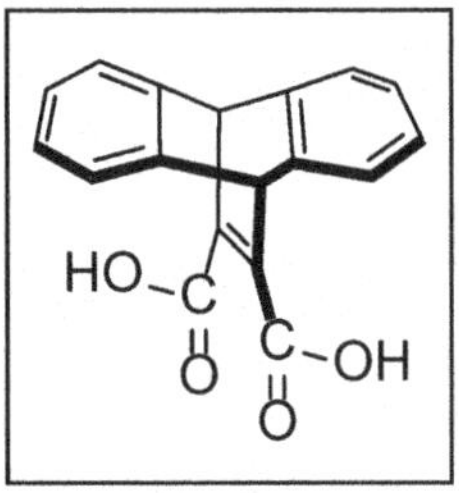

The synthesis of **15** was performed using an adaptation of Bouffard et al. [17] Compound **15** was synthesized by refluxing **23** (5.36 g, 16.73 mmol) with sodium hydroxide (6.69 g, 167.30 mmol) in a solvent mixture of 3:1 MeOH:water at 100 °C. The reaction was left overnight (14 h). The reaction was then cooled to room temperature, then placed in an 8–10 °C environment (fridge) for 6 h. A white solid appeared during the process. This precipitate was filtered and washed with small amounts (5 mL x 3) of cold methanol. The precipitate was redissolved in water and acidified with hydrochloric acid dropwise, measuring pH change along the way and stopping at pH 1. Upon acidification, white precipitate emerged and was filtered then vacuum pumped for several hours to yield **15** (94%, 4.87 g, 16.65 mmol). *Rf:* 0.15 (80% Ethyl acetate, 20% Hexanes), ^{1}H NMR (300 MHz, d_6-DMSO, 2.50 ppm): δ = 7.46–7.40 (m, 4 H), 7.03–6.90 (m, 4 H), 5.54 (s, 2 H), 4.75–4.00 (br s, 2 H). ^{13}C NMR (75.5 MHz, d_6-DMSO, 39.5 ppm): δ = 167.1, 146.8, 144.8, 125.5, 124.1, 52.0. Spectral data matched those reported by Bouffard [17].

Synthesis of 9,10-Ethenoanthracene-11-Carboxylic Acid, 9,10-Dihydro-Ethyl Ester (24) CAS: 147240–53-7

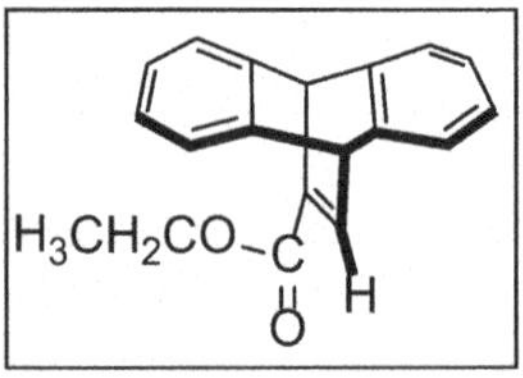

Following an adapted procedure by McNamara et al., 2014 [19]. To a 50 mL single neck round-bottom flask equipped with a magnetic stirrer and a condenser, was added anthracene (0.50 g, 2.8 mmol), and ethyl propiolate (0.28 mL, 2.8 mmol). The solution was heated to reflux for 7 days at 140 °C, while keeping conditions under argon. After cooling to room temperature, the reaction mixture solvent was removed by rotary evaporation. The crude product was purified by column chromatography (100% hex gradually increased to 33% EtOAC/Hex) to give the desired product **24**: *Rf*: 0.34 (30% EtOAc/Hex), White powder, yield 0.06 g (8%). ^{1}H NMR ($CDCl_3$, 300 MHz, 7.26 ppm): $\delta = 7.85$ (dd, 1 H, $J = 1.8$ Hz, 6.2 Hz), 7.38–7.29 (m, 4 H), 7.01–6.94 (m, 4 H), 5.67 (d, 1 H, $J = 1.8$ Hz), 5.23 (d, 1 H, $J = 6.2$ Hz), 4.17 (q, 2 H, $J = 7.1$ Hz), 1.26 (t, 3 H, $J = 7.1$ Hz). ^{13}C NMR ($CDCl_3$, 75.5 MHz, 77.0 ppm): $\delta = 164.8$, 149.2, 145.3, 144.7, 144.4, 125.0, 124.9, 123.7, 123.5, 60.7, 51.6, 50.3, 14.3. Spectral data matched those reported by McNamara [19].

Synthesis of 9,10-Dihydro-9,10-Ethenoanthracene-11-Carboxylic Acid [16] CAS: 27649–96-3

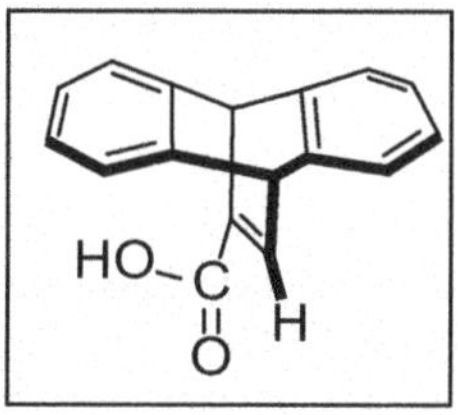

Following an adapted procedure by McNamara et al., 2014 [19]. To a 50 mL single neck RBF with a stir bar and a condenser, was added **24** (0.06 g, 0.2 mmol), KOH (0.43 g, 7.66 mmol), EtOH (4.4 mL), and MilliQ H_2O (6.6 mL). The solution was heated to reflux for 3 days at 100 °C. After decreasing the temperature to 22 °C, the reaction mixture solvent was removed by rotary evaporation. The residue was then filtered with Celite. Afterwards, the residue was dissolved in basic water (adding sodium hydroxide to MilliQ water, 1 M NaOH). Compound **16** was precipitated by acidification (HCl, 2 M). The crude product was purified by column chromatography (silica gel, 100% Hex increased to 1:9 EtOAc/Hex) to give the desired product: *Rf*: 0.46 (10% EtOAc/Hex), White powder, yield 0.047 g (78%). ^{1}H NMR ($CDCl_3$, 300 MHz, 7.26 ppm): $\delta = 7.97$ (dd, 1 H, $J = 1.6$ Hz, 6.3 Hz), 7.32 (m, 4 H), 6.97 (m, 4 H), 5.61 (d, 1 H, $J = 1.6$ Hz), 5.25 (d, 1 H, $J = 6.3$ Hz). ^{13}C NMR ($CDCl_3$, 75.5 MHz, 77.0 ppm): $\delta = 169.6$, 152.4, 145.1, 144.0, 143.9, 125.1, 124.9, 123.8, 123.7, 51.8, 50.1. Spectral data matched those reported by McNamara [19].

Synthesis of 2-Bromotriptycene (25) CAS: 20712–00-9

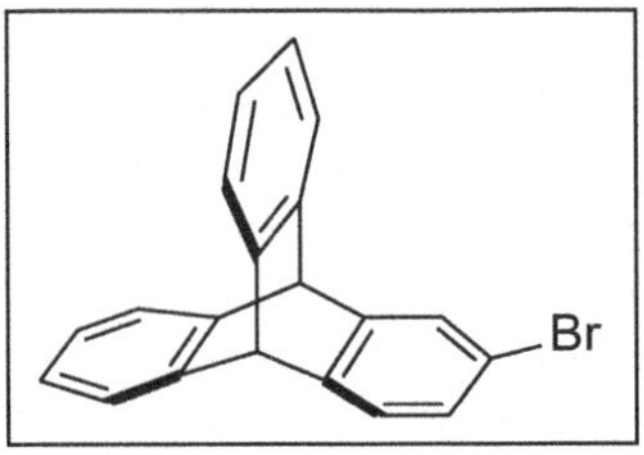

Following an adapted procedure by Yang et al. [20] To a 250 mL three neck round-bottom flask containing a magnetic stir bar and a condenser, was placed anthracene (1.00 g, 5.6 mmol) and dry 1,2-dichloroethane (15 mL), the solution was heated to reflux. A mixture of 2-amino-4-bromobenzoic acid (2.42 g, 11 mmol) in anhydrous THF (20 mL) and a solution of freshly distilled isoamyl nitrite (1.9 mL, 14 mmol) in anhydrous 1,2-dichloroethane (20 mL) simultaneously was added over 1 h, while keeping the content of isoamyl nitrite more than 2-amino-4-bromobenzoic acid in the reaction mixture. The solution was refluxed for 22 h. The solution was then cooled to room temperature, and then poured into water and extracted with DCM. The combined extracts were washed with water and brine, dried over anhydrous $MgSO_4$, and concentrated in vacuo. The crude product was purified using flash chromatography with a 5:95 DCM/Hex solution to give the desired product **25**. White powder, 0.61 g (33%).^{1}H NMR ($CDCl_3$, 300 MHz, 7.26 ppm): δ = 7.54 (d, 1 H, *J* = 1.5 Hz), 7.39 (m, 4 H), 7.23 (d, 1 H, *J* = 7.8 Hz), 7.13 (dd, 1 H, *J* = 1.5 Hz, 7.8 Hz), 7.02 (m, 4 H), 5.40 (d, 2 H, *J* = 4.6 Hz). ^{13}C NMR ($CDCl_3$, 75.5 MHz, 77.0 ppm): δ = 147.8, 145.0, 144.7, 144.7, 128.1, 127.1, 125.6, 125.6, 125.6, 125.2, 123.9, 123.8, 53.9, 53.8.

Synthesis of 2-Triptycenylcarboxylic Acid [17] CAS: 30086–25-0

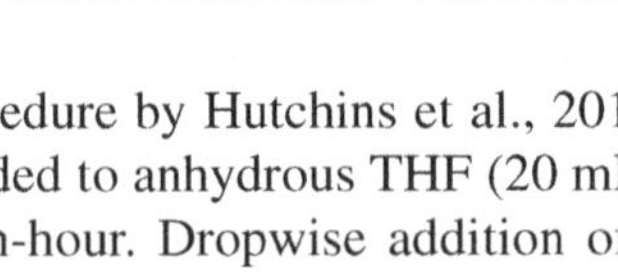

Following an adapted procedure by Hutchins et al., 2013 [21]. 2-Bromotriptycene (0.61 g, 1.8 mmol) was added to anhydrous THF (20 mL) and chilled to −78 °C in dry ice/acetone for half-an-hour. Dropwise addition of n-butyllithium (4.58 mL, 1.6 M in dry THF, 7.3 mmol) followed, and the reaction was kept at −78 °C for 30 min. The solution was warmed to 0 °C for 20 min and then re-cooled to −78 °C. After 30 min, dry ice was placed in a closed flask. The $CO_{2(g)}$ was transported with a needle and bubbled into the solution at −78 °C over a 30 min timeframe. The combination was then made acidic with the addition of 2 M HCl. The solution was washed with water, extracted with EtOAc, and dried with Na_2SO_4. The sample was rotary evaporated and the residue was purified using flash chromatography with Hex:EtOAc (10:1 to 6:1) to form **17** (0.18 g, 33%). ^{1}H NMR ($CDCl_3$, 300 MHz, 7.26 ppm): δ = 8.09 (s, 1 H), 7.80 (d, 1 H, *J* = 7.7 Hz), 7.47 (d, 1 H, *J* = 7.7 Hz), 7.40 (m, 4 H), 7.02 (m, 4 H), 5.51 (s, 1 H), 5.50 (s, 1 H). ^{13}C NMR ($CDCl_3$, 75.5 MHz, 77.0 ppm): δ = 171.6, 151.4, 145.8, 145.7, 144.2, 128.1, 126.2, 125.5, 125.4, 124.9, 123.8, 123.8, 60.3, 54.1, 53.6.

7.8.1 Attempted Yet Unsuccessful Molecular Synthesis

The following two procedures were also attempted, yet failed to recover/create the desired materials:

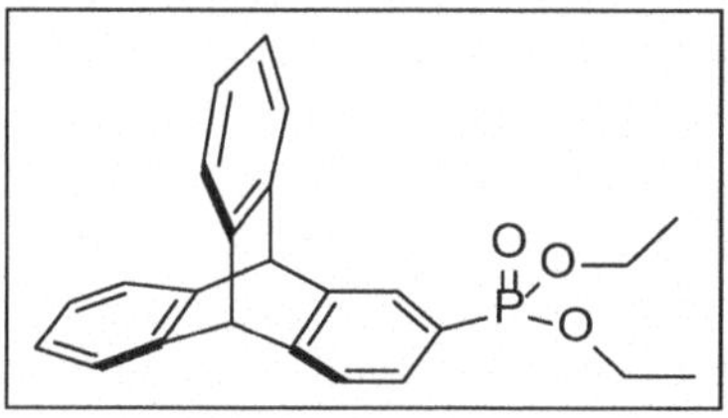

From 2-bromotryptycene (**18**), an attempted synthesis for the 2-diethylphosphonatetryptycene was conducted using Grignard chemistry following a protocol by Volle et al. using triethylphosphite [22]. However, the starting material was recovered, indicating the reaction did not proceed. No further investigations followed.

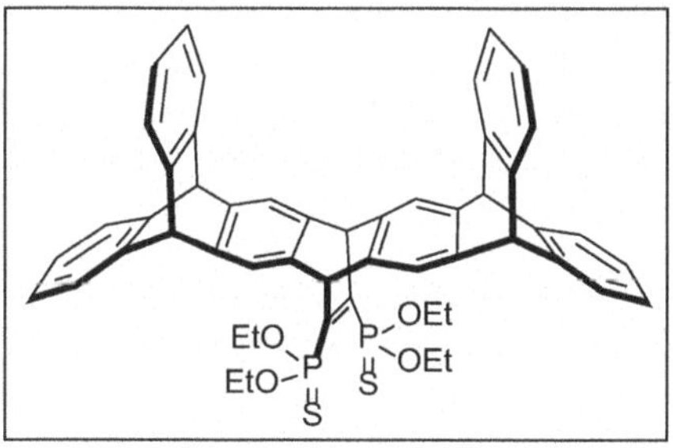

The Ipt-phosphonothioate (**19**) was attempted by following a procedure by Kiyokawa et al. [23] Lawesson's reagent (97 mg, 0.24 mmol) was added to 1 equiv. (0.12 mmol) of IptP(O)(OEt)$_2$ in toluene (5 mL) and left to reflux under argon for 20 h [23]. Multiple TLC compound spots were visible through various solvent eluent systems such as combinations of EtOAc/Hex/MeOH/$CHCl_3$/CH_2Cl_2. The reaction was concentrated under vacuum for analysis of the crude mixture of only 15 mg and attempted on a substrate for ART. No further analysis followed.

7.9 Field-Effect Transistor Fabrication and Testing

7.9.1 LED Photolithography Patterning

To an ART silicon substrate was applied OFPR-800 LB photoresist (Tokyo Ohka Kogyo Co., Ltd.) at 5000 rpm/60 s on an Opticoat Mikasa Spin Coater MSB100, and immediately, the sample was baked at 90 °C for 3 min. Sample was kept away from any white light by wrapping substrate holder in aluminum. The samples patterned best when dried in a vacuum chamber for minimum 3 h beforehand. Patterning occurred on an Olympus LEXT OLS5000 connected with ARMs-UTA-IIIA UV LED patterning system. The range of values for the exposure are were as followed: 90 pixel left side position, 140 pixel upper position, 1120 pixel width and 710 pixel height. The exposures for electrode patterning were done first at 100 x to obtain a 1.5 μm gap for the nanotubes to cross at 0.8 s exposure, then 20x squares were patterned at 4–8 s exposure with minimum size of 50 μm such that the electrodes could be deposited.

The patterning was first confirmed with exposure to safe red light, then white light. Patterning was developed by immersing the sample in NMD-3 (tetramethyl ammonium hydroxide, 2.85%, Wako chemicals) solution for 25 s and then immediately in MilliQ water for 60 s, then dried with $N_{2(g)}$. Samples were wrapped in aluminum foil to keep them in the dark and stored under vacuum and submitted to the Namiki Foundry Clean Room (National Institute for Materials Science, Tsukuba) for metal deposition with an e-beam heating vacuum evaporator by ULVAC, Inc. within 24 h. Samples were subjected to electron beam deposition of 5 nm Ti and 50 nm of Au or 1 nm Cr/ 10 nm Pd/ 50 nm Au to form the electrodes. Lift-off was performed in acetone at 45 °C for 1–10 h with a ratio of 20 mL acetone/cm^2. Sample was cooled in the acetone beaker and then, as quickly as possible, rinsed with IPA for 30 s and dried with $N_{2(g)}$. The process is depicted below in Fig. 7.3.

7.9.2 Electron Beam Lithography Patterning

The EBL patterning and lift-off was performed by the Namiki Foundry Clean Room staff (National Institute for Materials Science, Tsukuba). The E-beam writer was an Elionix ELS-7500 DEX using an electron acceleration voltage of 50 kV and a beam current of 600 pA. For the large electrode pads, a 575 μC/cm^2 electron irradiation dose was used, whereas for the smaller portions of the electrodes in contact with the CNTs, a 600 μC/cm^2 electron irradiation dose was used. The positional resolution

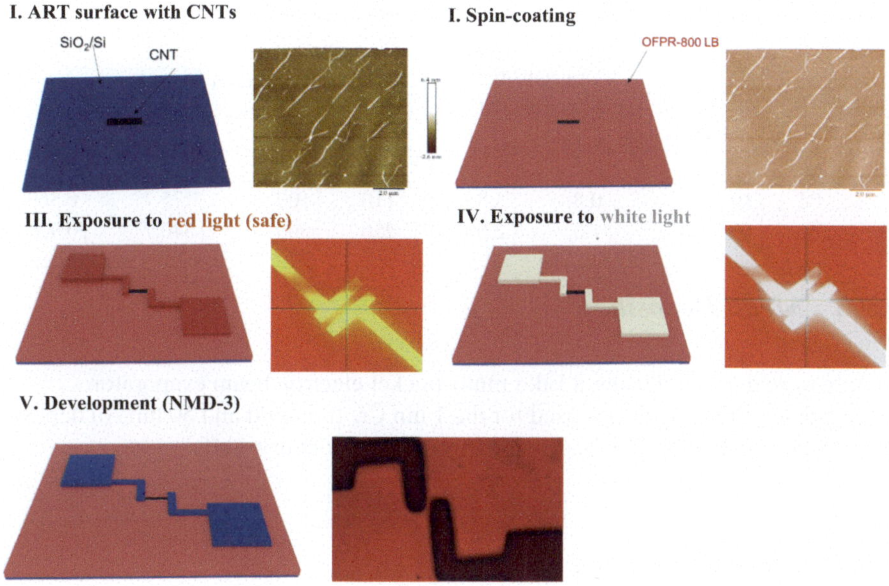

Fig. 7.3 LED Photolithography patterning

was 10 nm × 10 nm (x,y), corresponding to single pixel size. The irradiation time per pixel was 0.95 μs for the 575 μC/cm^2 electron irradiation doses and 1 μs for the 600 μC/cm^2 electron irradiation dose to limit charge on the electron beam resist.

A 950 PMMA-A4 (MicroChem Corp.) patterning resist and Espacer-300Z (Showa Denko K.K). top electric conductive layer were used. The 950 PMMA resist was spin coated at 2000 RPM and dried at 180 °C for 3 min. The Espacer 300Z was placed on the surface after the PMMA back by spin coating at 2000 RPM and no additional baking. The Espacer 300Z was removed first after the patterning of the substrate by dipping in MilliQ H_2O for 10–15 s. The PMMA resist was removed in a 3:1 solution of methyl isobutyl ketone (MIBK): IPA for 45 s then rinsed immediately with IPA for 30 s and dried under nitrogen. The surface layer was connected to the sample holder (ground layer) using standard conductive tape. The metals of 1 nm Cr/ 10 nm Pd/ 50 nm Au were deposited using an e-beam heating vacuum evaporator by ULVAC, Inc. The metal lift-off was carried out in ZDMAC liquid and then rinsed with IPA to create EBL CNTFET devices.

For both the top gate (TG) LED and EBL fabricated devices, a 20 nm Al_2O_3 dielectric layer was applied through ALD (SUNALER-R-150 ALS, Picosun). The deposition occurred using 170 cycles of switching TMA and water at 200 °C. The film thickness was confirmed using spectroscopic ellipsometry (DekTak 6 M, Bruker) to be 20.9 nm on a separate Si (100) sample that was loaded with the device samples.

Table 7.1 Sample conditions for the metal electron beam deposition

Layer No.	Metal	Thickness[nm]	Targeted rate [Å/s]	Pressure before dep. [Pa]	Current [mA]	Deposition amount [Å]	Deposition time [s]	Average rate [Å/s]
1	Cr	1	0.2	1.5×10^{-5}	10	10	50	0.200
2	Pd	10	0.8	5.8×10^{-6}	110	100	175.7	0.569
3	Au	50	1.0	2.0×10^{-5}	250	500	540.7	0.925

7.9.3 Metal Deposition

Metals were deposited using a Eiko multi-pocket electron beam evaporator.

Example of the conditions used for the 1 nm Cr, 10 nm Pd and 50 nm Au deposition are shown in Table 7.1 as provided by Namiki Foundry staff.

7.9.4 Electrical Testing

The SWNT devices were characterized electrically on a Keithley 2636B system source meter with Agilent 4284A for ambient testing (Tanaguchi Lab, NIMS) 20 Hz-1 MHz precision lCR meter and Appoloware M20 micro positioners. Measurements were conducted in ambient conditions at room temperature (20–24 °C). Devices with CNTs connected between the source and drain were first identified using HR-SEM. Testing in vacuum occurred with a Semi-auto Agilent 4156C (Tsukagoshi Lab, NIMS) measurement system at room temperature. Vacuum pressure was at ~1×10^{-5} Pa. The bottom gate was constructed by either scratching the surface or using conductive silver paste on the sides of the wafer adhering to a copper platform.

7.9.5 Probe-Test

Using a 4200A SCS parameter analyzer, a sample of bare silica was tested with two-probes touching the surface 1 mm apart to measure the resistance. Then, a sample with aligned SWNTs from ART was measured with the probes 1 mm apart at different angles to observe the difference in conductivity. Based on the different angles, it was estimated this form of testing could identify the overall alignment of the SWNTs. The results are shown in the Appendix for Chap. 5, however, only one sample was tested. From the preliminary data, it appears that varying the angle on a wafer of aligned SWNTs with ART does show a deference depending on if the nanotubes are aligned or not, but further verification is required by corroborating with AFM/SEM results and testing different samples with the probes.

References

1. Selmani, S., Schipper, D.: Orientation control of molecularly functionalized surfaces applied to the simultaneous alignment and sorting of carbon nanotubes. Angew. Chem. Int. Ed. **130**(9), 2423–2427 (2018)
2. Osorio, A.G., Silveira, I.C.L., Bueno, V.L., Bergmann, C.P.: H2SO4/HNO3/HCl—Functionalization and its effect on dispersion of carbon nanotubes in aqueous media. Appl. Surf. Sci. **255**(5, Part 1), 2485–2489 (2008)
3. Chalifoux, W.A.: The synthesis of non-planar, helically coiled graphene nanoribbons. Angew. Chem. Int. Ed. **56**(28), 8048–8050 (2017)
4. Yang, W., Lucotti, A., Tommasini, M., Chalifoux, W.A.: Bottom-up synthesis of soluble and narrow graphene nanoribbons using alkyne Benzannulations. J. Am. Chem. Soc. **138**(29), 9137–9144 (2016)
5. Jun, Y.-S., Park, M.G., Um, J.G., Habibpour, S., Sy, S., Park, C.B., Yu, A.: The conductivity of polydimethylsiloxane/graphene nano-ribbon foam composite with elongation. Carbon. **162**, 328–338 (2020)
6. Kosynkin, D.V., Higginbotham, A.L., Sinitskii, A., Lomeda, J.R., Dimiev, A., Price, B.K., Tour, J.M.: Longitudinal unzipping of carbon nanotubes to form graphene nanoribbons. Nature. **458**(7240), 872–876 (2009)
7. Jorio, A., Saito, R., Hafner, J., Lieber, C., Hunter, D., McClure, T., Dresselhaus, G., Dresselhaus, M.: Structural (n, m) determination of isolated single-wall carbon nanotubes by resonant Raman scattering. Phys. Rev. Lett. ***86***(6), 1118 (2001)
8. Kataura, H., Kumazawa, Y., Maniwa, Y., Ohtsuka, Y., Sen, R., Suzuki, S., Achiba, Y.: Diameter control of single-walled carbon nanotubes. Carbon. **38**(11–12), 1691–1697 (2000)
9. Weisman, R.B., Bachilo, S.M.: Dependence of optical transition energies on structure for single-walled carbon nanotubes in aqueous suspension: an empirical Kataura plot. Nano Lett. **3**(9), 1235–1238 (2003)
10. Sato, K., Saito, R., Jiang, J., Dresselhaus, G., Dresselhaus, M.: Discontinuity in the family pattern of single-wall carbon nanotubes. Phys. Rev. B. **76**(19), 195446 (2007)
11. Thomsen, C., Reich, S.: Raman Scattering in Carbon Nanotubes, pp. 115–234. Springer, Berlin (2007)
12. Brandsma, L.: Preparative Acetylenic Chemistry, 2nd edn. Elsevier, Amsterdam (1988)
13. Hanik, N., Kilbinger, A.F.M.: Narrowly distributed homotelechelic polymers in 30 minutes: using fast in situ pre-functionalized ROMP initiators. J. Polym. Sci. A Polym. Chem. **51**(19), 4183–4190 (2013)
14. Hart, H., Bashir-Hashemi, A., Luo, J., Meador, M.A.: Iptycenes: extended triptycenes. Tetrahedron. **42**(6), 1641–1654 (1986)
15. Acheson, R.M., Ansell, P.J.: The synthesis of diethyl p-tolylsulphonylethynylphosphonate and related acetylenes, and their reactions with nucleophiles, pyridinium-1-dicyanomethylides, and dienes. J. Chem. Soc. Perkin Trans. **1**(0), 1275–1281 (1987)
16. Saltiel, J., Redwood, C.E., Kumar, V.K.R.: The photoisomerization of cis,trans-1,2-dideuterio-1,4-diphenyl-1,3-butadiene in solution. No bicycle-pedal. Photochem. Photobiol. Sci. **18**(9), 2174–2179 (2019)
17. Bouffard, J., Eaton, R.F., Müller, P., Swager, T.M.: Iptycene-derived Pyridazines and Phthalazines. J. Org. Chem. **72**(26), 10166–10180 (2007)
18. Snowdon, M.R., Wang, S., Mashmoushi, N., Hopkins, S.W., Schipper, D.J.: Carboxylic acids as anchoring components on aluminum oxide for the alignment relay technique of single-walled carbon nanotubes. New J. Chem. **45**(12), 5340–5349 (2021)
19. McNamara, Y., Bright, S., Byrne, A., Cloonan, S., McCabe, T., Williams, D., Meegan, M.: Synthesis and antiproliferative action of a novel series of maprotiline analogues. Eur. J. Med. Chem. **71**, 333–353 (2014)
20. Yang, Y., Cheng, K., Lu, Y., Ma, D., Shi, D., Sun, Y., Yang, M., Li, J., Wei, J.: A Polyaromatic Nano-nest for hosting fullerenes C60 and C70. Org. Lett. **20**(8), 2138–2142 (2018)

21. Hutchins, D.O., Weidner, T., Baio, J., Polishak, B., Acton, O., Cernetic, N., Ma, H., Jen, A.K.Y.: Effects of self-assembled monolayer structural order, surface homogeneity and surface energy on pentacene morphology and thin film transistor device performance. J. Mater. Chem. C. **1**(1), 101–113 (2013)
22. Volle, J.-N., Filippini, D., Midrier, C., Sobecki, M., Drag, M., Virieux, D., Pirat, J.-L.: Revisited synthesis of aryl-H-phosphinates. Synthesis. **2011**(15), 2490–2494 (2011)
23. Kiyokawa, K., Suzuki, I., Yasuda, M., Baba, A.: Synthesis of cyclopropane-containing phosphorus compounds by radical coupling of Butenylindium with Iodo phosphorus compounds. Eur. J. Org. Chem. **2011**(11), 2163–2171 (2011)

Appendices

Abstract Additional experimental information and characterization results that were used in this work are shown for each individual chapter. This information includes Raman spectroscopy graphs in both red and green wavelengths, the analysis for nanotube chiralities and microscopy images (AFM and SEM), including topography and phase analysis images. Calculations on the alignment are also included, as well as characteriazation details for the CNTFETs and their fabrication.

Keywords Characterization; Fabrication; Nanotubes; Carbon; Lithography; Synthesis; Alignment; Transistor; Raman spectroscopy; Microscopy

M. Snowdon, *Methods of the Alignment-Relay Technique for Nanosystems*,
Springer Theses, https://doi.org/10.1007/978-3-030-84413-4

Appendix Chapter 2

Sonication Experiments [1]

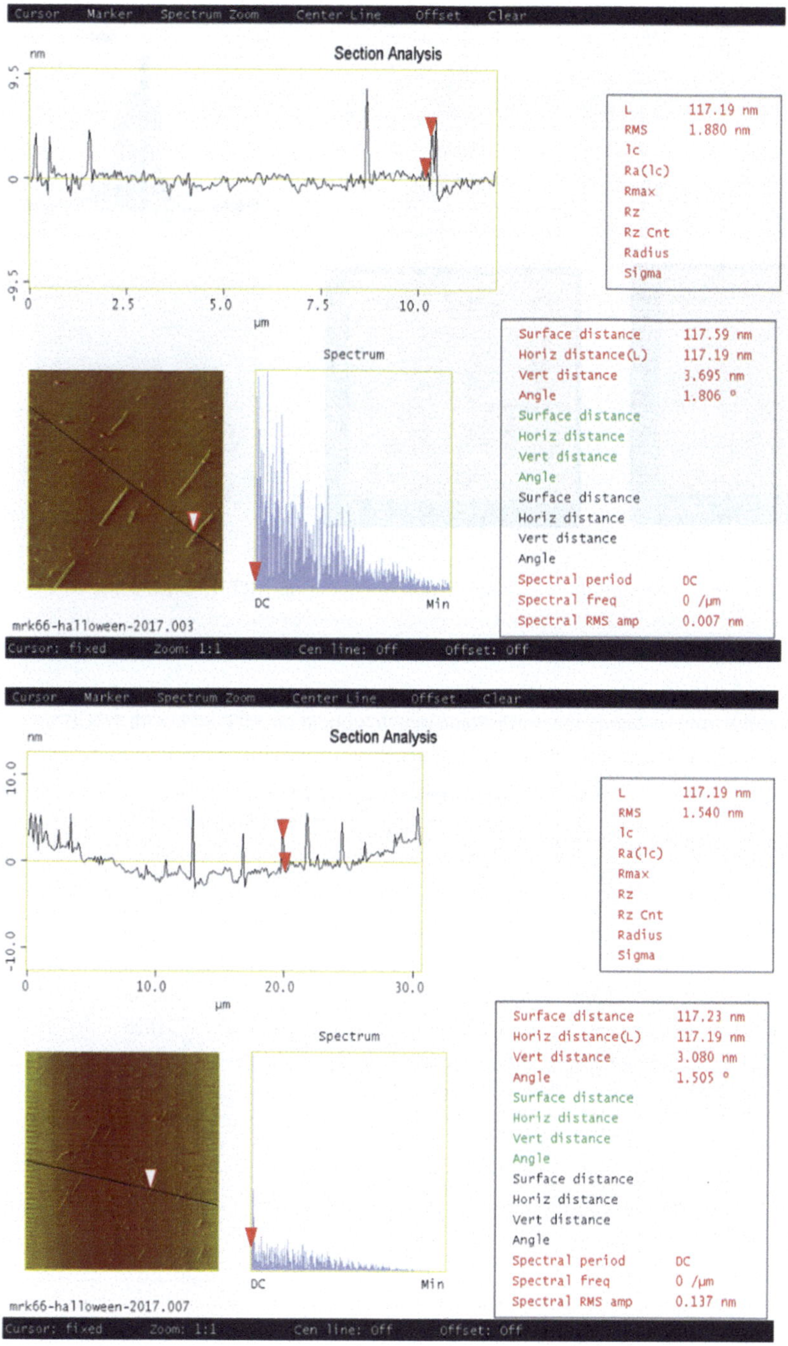

Fig. A.1 AFM height data depicted in the right-hand side under vertical distance for the 10-min sonicated sample. In (**a**), the height is double of what expected in Raman, likely due to increased surfactant wrapping on the tubes due to the longer sonication time and exposure to surfactant molecules. In (**b**) the sample is shown on a 30 × 30 µm scale demonstrating the alignment of the tubes but also that the other tubes are about 3 nm in height off the surface and that this observation is consistent across the sample surface. The AFM tip also became blunt

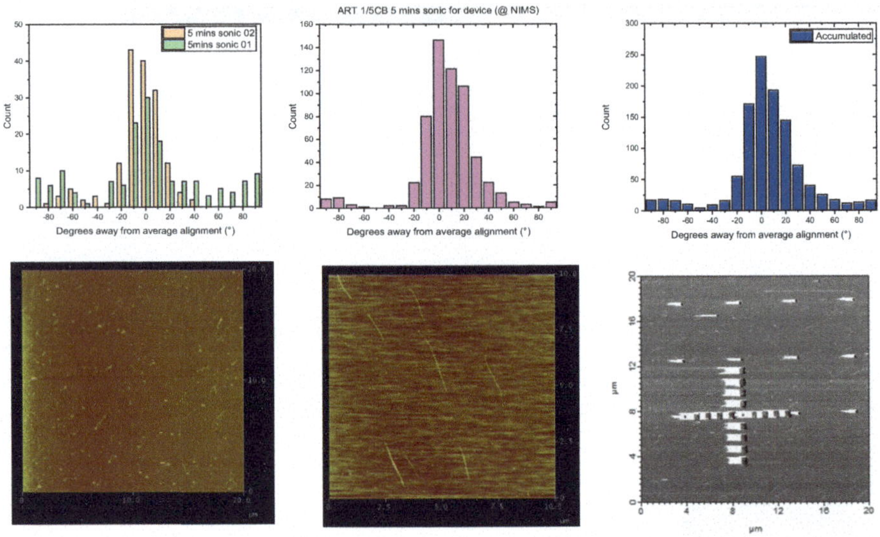

Fig. A.2 Results from repeating the 5-min sonication treatment on ART. (5 mins sonic 01 from Sect. 2.1.1.2)

Table A.1 Results from repeating the 5-min sonication treatment on ART-SiO_2 with 1/5CB

5-mins sonicated sample	SD (°)	$\frac{\text{Aligned}\, SWNTs\left(0 \pm 10^{\circ}\right)}{\text{Total}\, SWNTs} \times 100(\%)$	Density ($SWNTs/\mu m^2$)
−01	20.7	76.7	0.13
−02 (section	38.3	43.8	0.28
No LBL (Sect. 2.2.1)[a]	30.0	55.7	0.25
Device −01 (Chap 5)[a]	25.9	61.9	0.35
Average	29.6	58.7	0.25

Full Raman spectra (Including Si-calibration peak and G-band)

[a]Conducted at National Institute for Materials Science, Tsukuba

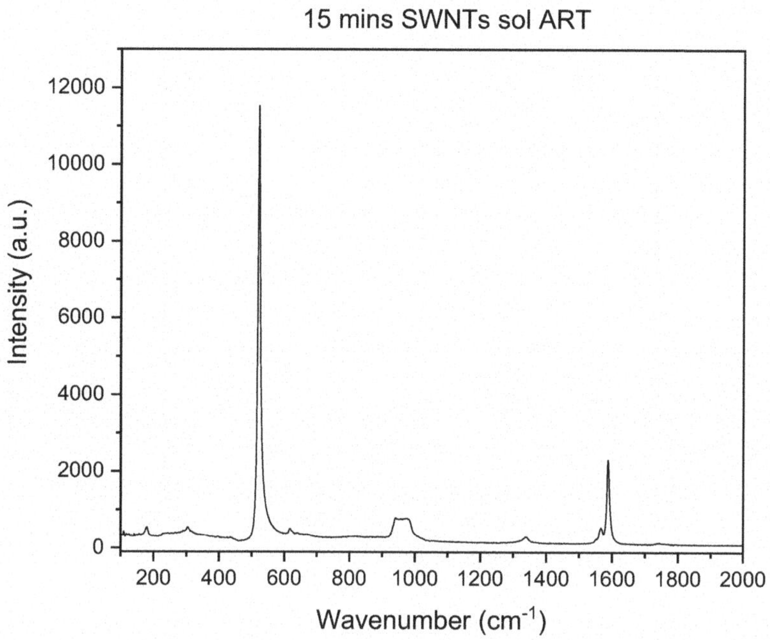

Fig. A.3 Raman spectrum at 632 nm of the 3x15 min sonicated sample in SWNTs-solution

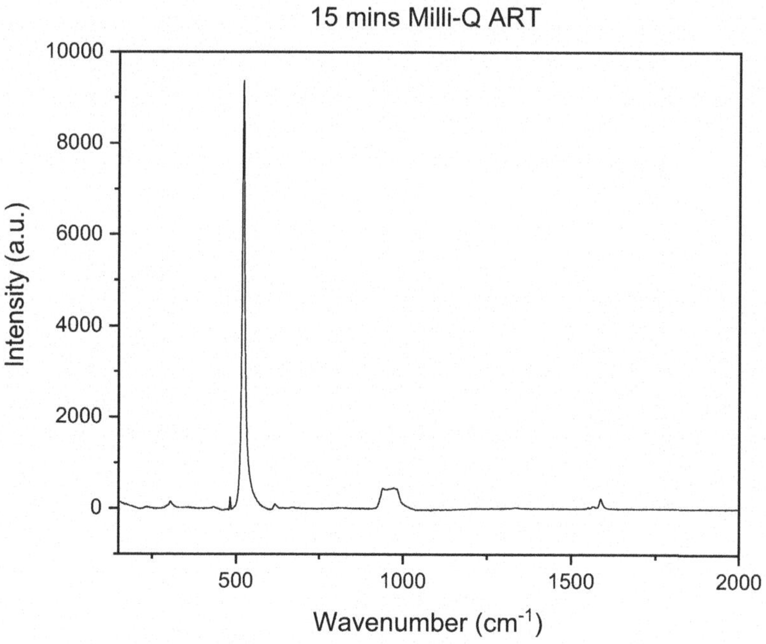

Fig. A.4 Raman spectrum at 632 nm of the 3x15 min sonicated sample in Milli-Q water

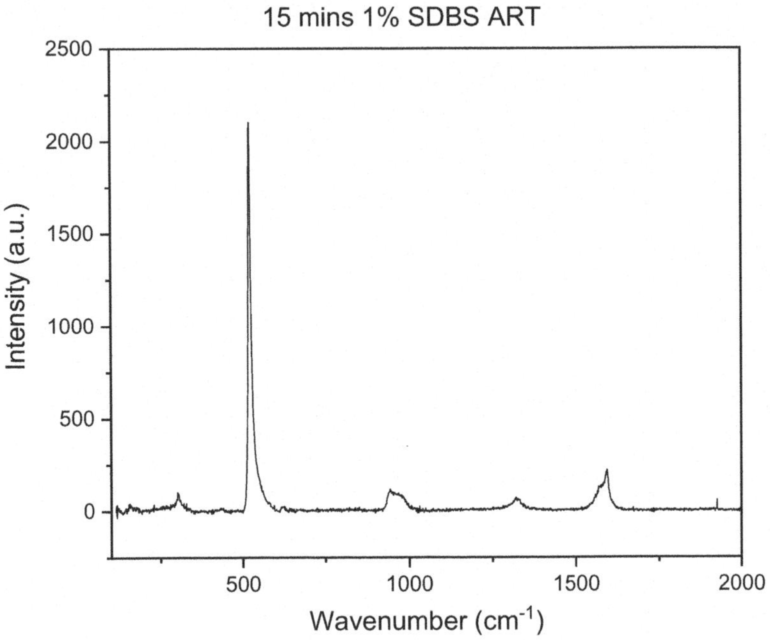

Fig. A.5 Raman spectrum at 632 nm of the 3x15 min sonicated sample in 1% SDBS aqueous solution

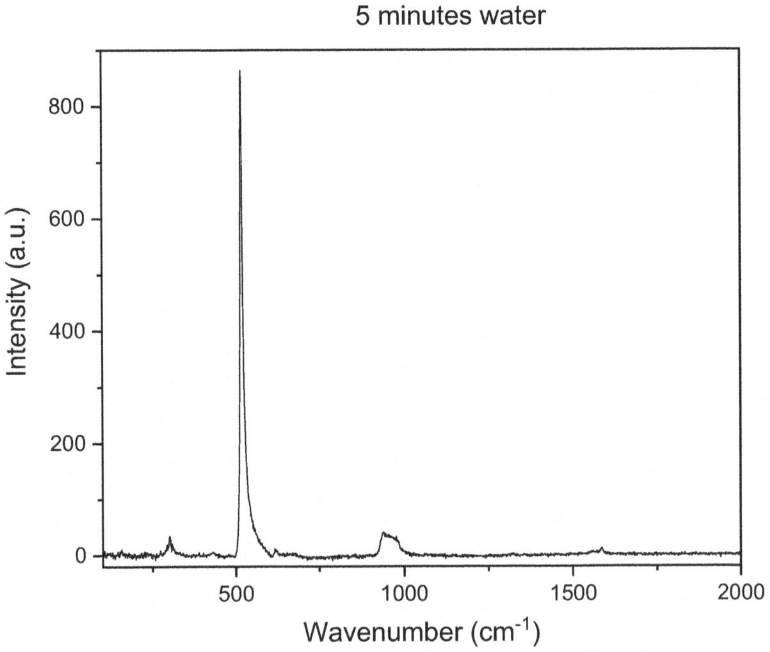

Fig. A.6 Raman spectrum at 632 nm of the 5 min sonicated sample in Milli-Q water. The very low intensity was obtained for the G-band, which substantially decreased for the samples that were sonicated in water longer

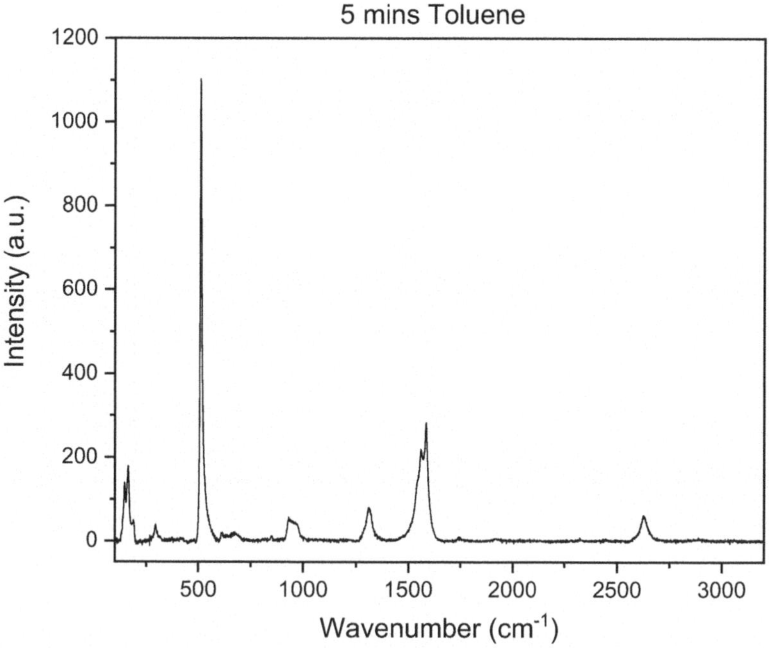

Fig. A.7 Raman spectrum at 632 nm of the 5 min sonicated sample in toluene

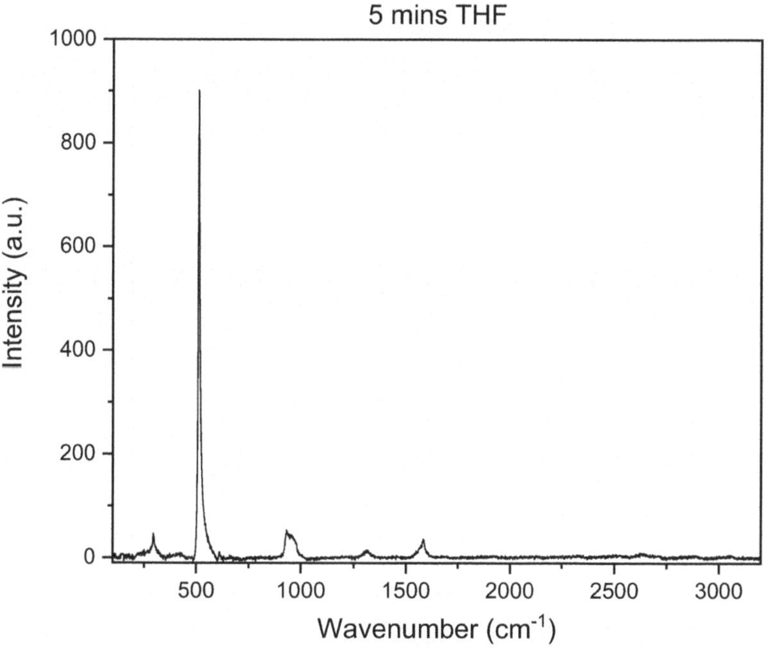

Fig. A.8 Raman spectrum at 632 nm of the 5 min sonicated sample in THF

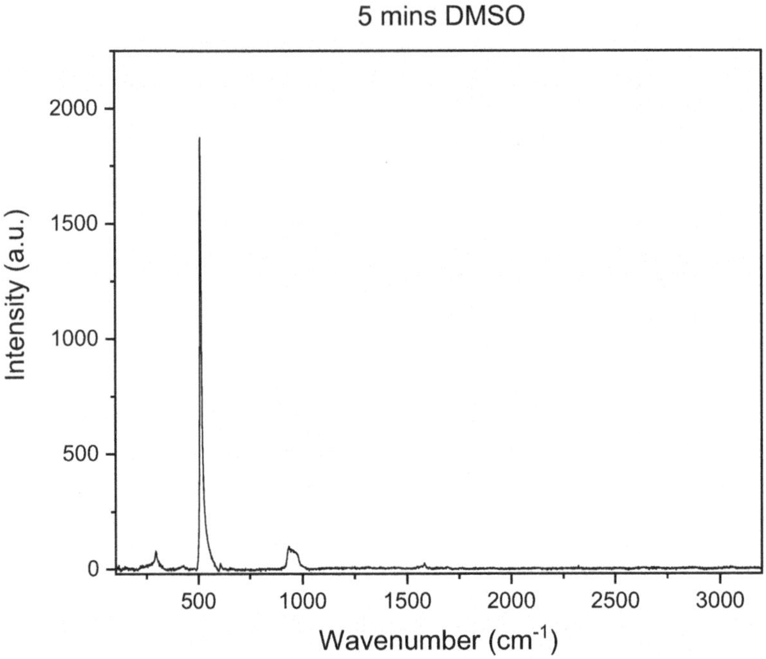

Fig. A.9 Raman spectrum at 632 nm of the 5 min sonicated sample in DMSO

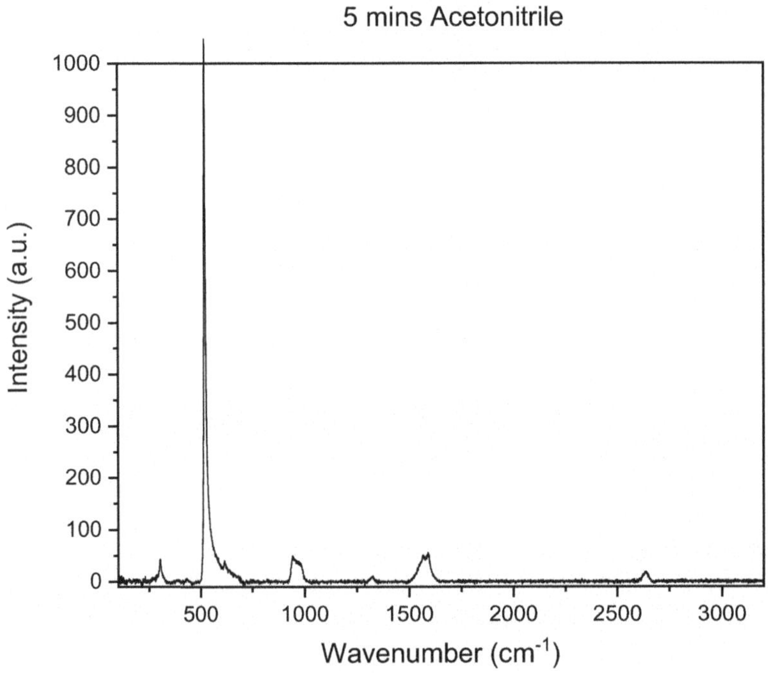

Fig. A.10 Raman spectrum at 632 nm of the 5 minsonicated sample in acetonitrile

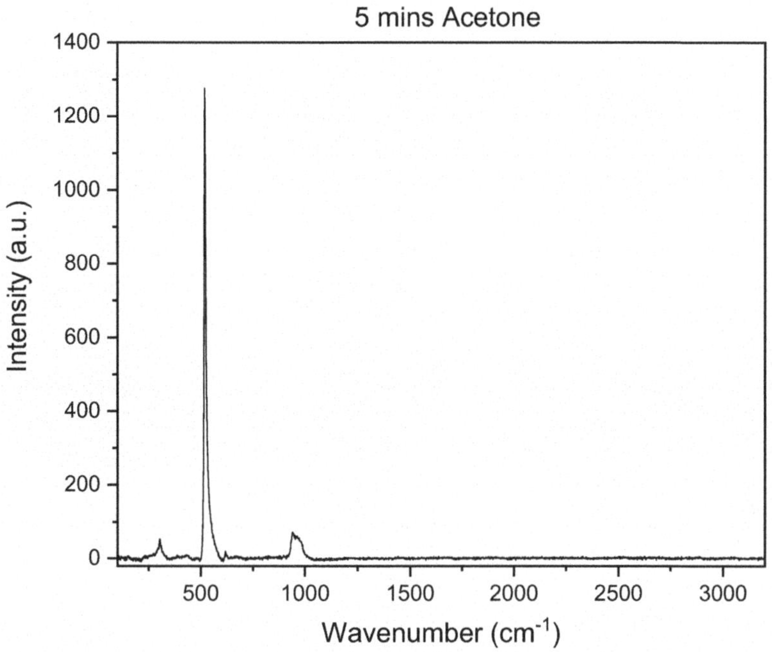

Fig. A.11 Raman spectrum at 632 nm of the 5 min sonicated sample in acetone

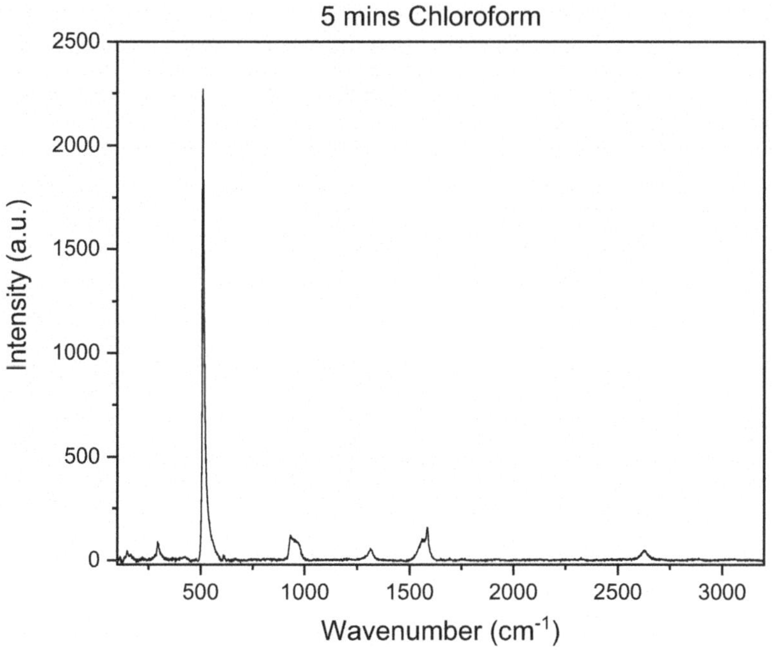

Fig. A.12 Raman spectrum at 632 nm of the 5 min sonicated sample in chloroform

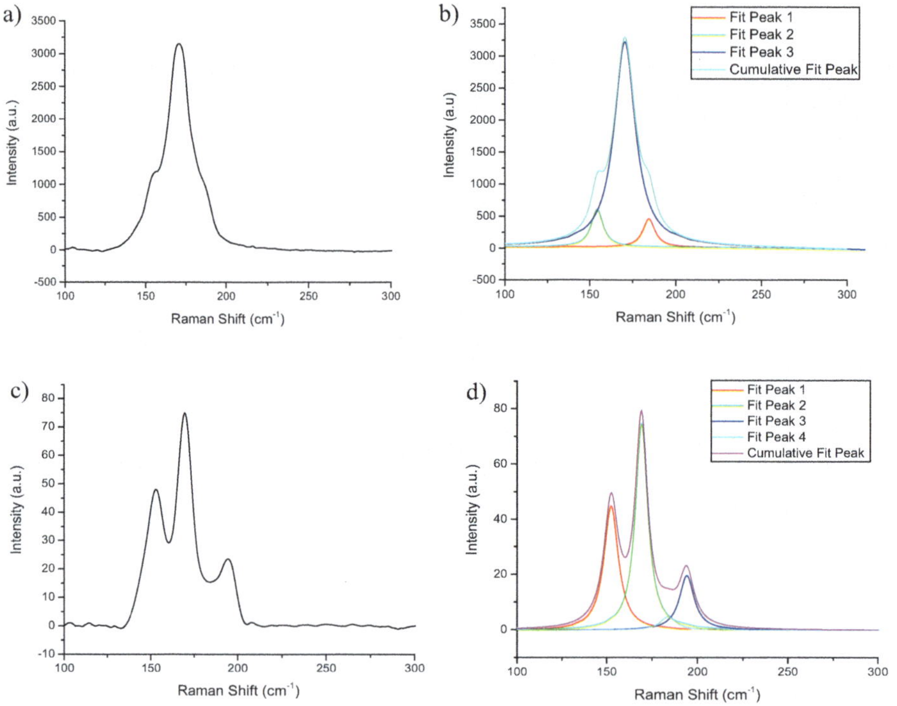

Fig. A.13 Radial breathing mode region analysis for the dropcast 90% pure sc-SWNTs solution in the red (633 nm) and green (533 nm) frequencies. The original raw RBM data is plotted alongside the fitted data so that it is easier to observe. (**a**) Green 533 nm normalized RBM spectrum for dropcast SWNTs on silicon, (**b**) Green 533 nm Lorentzian fitted peaks and generated cumulative fitted peak for dropcast SWNTs, (**c**) Red 633 nm normalized RBM spectrum for dropcast SWNTs on silicon, d) Red 633 nm Lorentzian fitted peaks and generated cumulative fitted peak for dropcast SWNTs

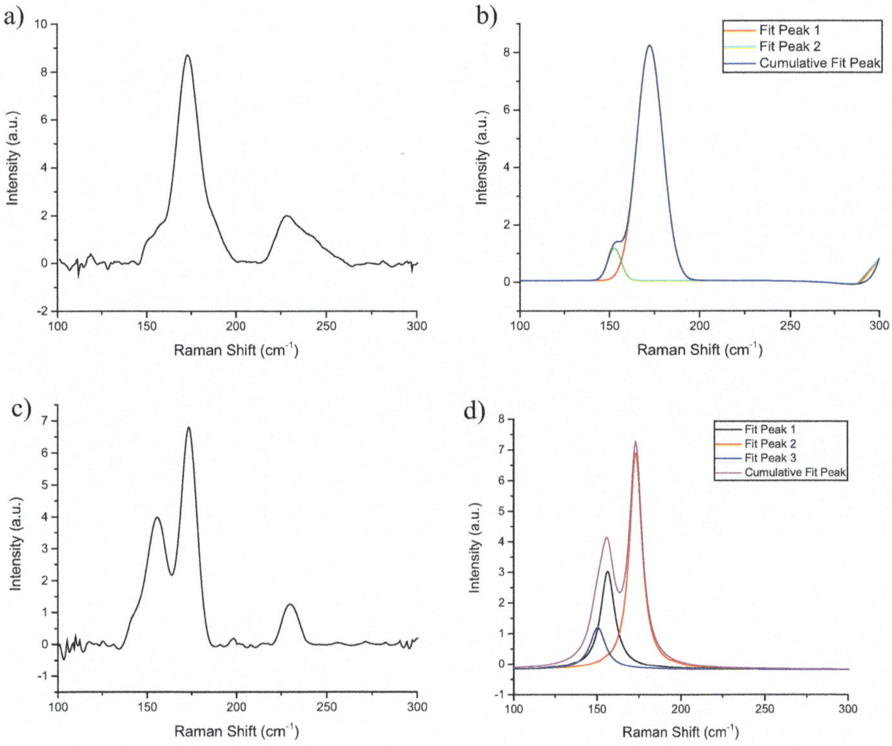

Fig. A.14 Radial breathing mode region analysis for ART on silicon control sample that was not subjected to any sonication red (633 nm) and green (533 nm) frequencies. The original raw RBM data is plotted alongside the fitted data so that it is easier to observe. (**a**) Green 533 nm normalized RBM spectrum for non-sonicated SWNTs on silicon with ART, (**b**) Green 533 nm Lorentzian fitted peaks and generated cumulative fitted peak for non-sonicated SWNTs on silicon with ART, (**c**) Red 633 nm normalized RBM spectrum for non-sonicated SWNTs on silicon with ART, (**d**) Red 633 nm Lorentzian fitted peaks and generated cumulative fitted peak for non-sonicated SWNTs on silicon with ART

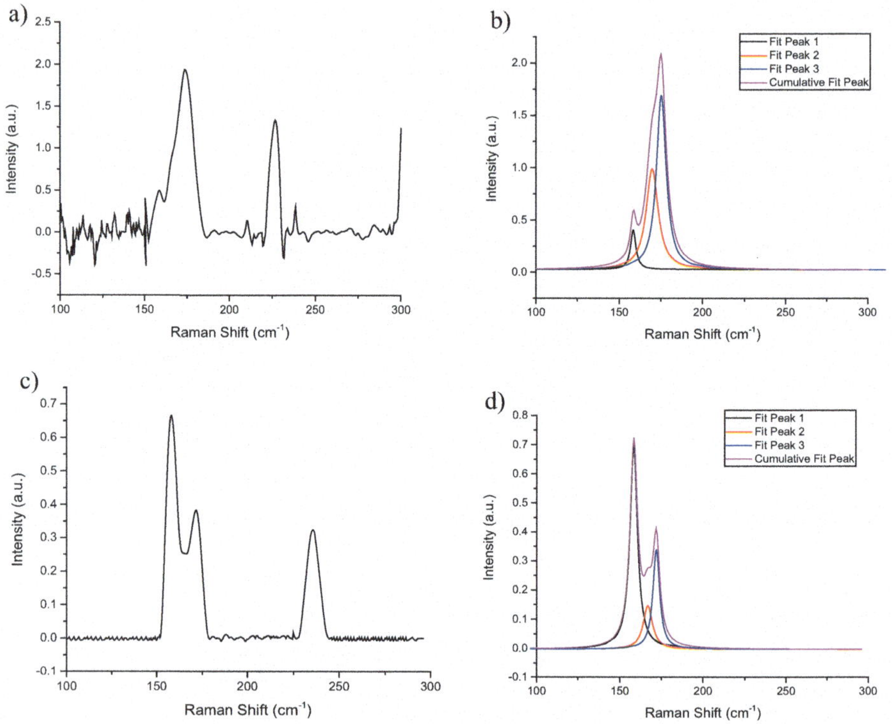

Fig. A.15 Radial breathing mode region analysis for ART silicon sample that was subjected to 5 min of sonication in red (633 nm) and green (533 nm) frequencies. (**a**) Green 533 nm normalized RBM spectrum for 5 min sonicated SWNTs on silicon with ART, (**b**) Green 533 nm Lorentzian fitted peaks and generated cumulative fitted peak for 5 min sonicated SWNTs on silicon with ART, (**c**) Red 633 nm normalized RBM spectrum for 5 min sonicated SWNTs on silicon with ART, (**d**) Red 633 nm Lorentzian fitted peaks and generated cumulative fitted peak for 5 min sonicated SWNTs on silicon with ART

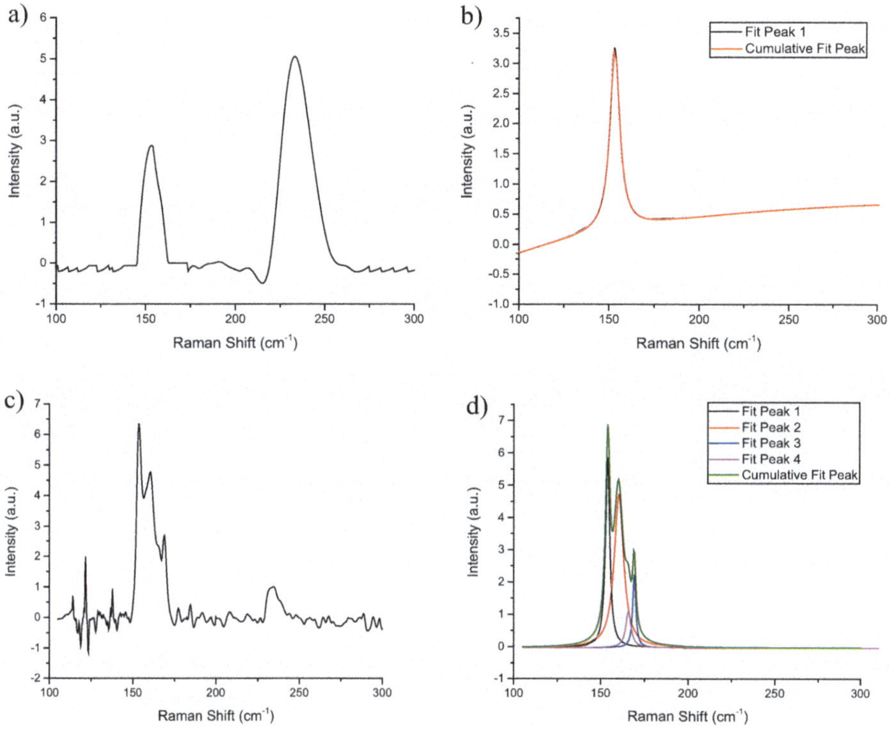

Fig. A.16 Radial breathing mode region analysis for ART silicon sample that was subjected to 10 min of sonication in red (633 nm) and green (533 nm) frequencies. (**a**) Green 533 nm normalized RBM spectrum for 10 min sonicated SWNTs on silicon with ART, (**b**) Green 533 nm Lorentzian fitted peaks and generated cumulative fitted peak for 10 min sonicated SWNTs on silicon with ART, (**c**) Red 633 nm normalized RBM spectrum for 10 min sonicated SWNTs on silicon with ART, (**d**) Red 633 nm Lorentzian fitted peaks and generated cumulative fitted peak for 10 min sonicated SWNTs on silicon with ART

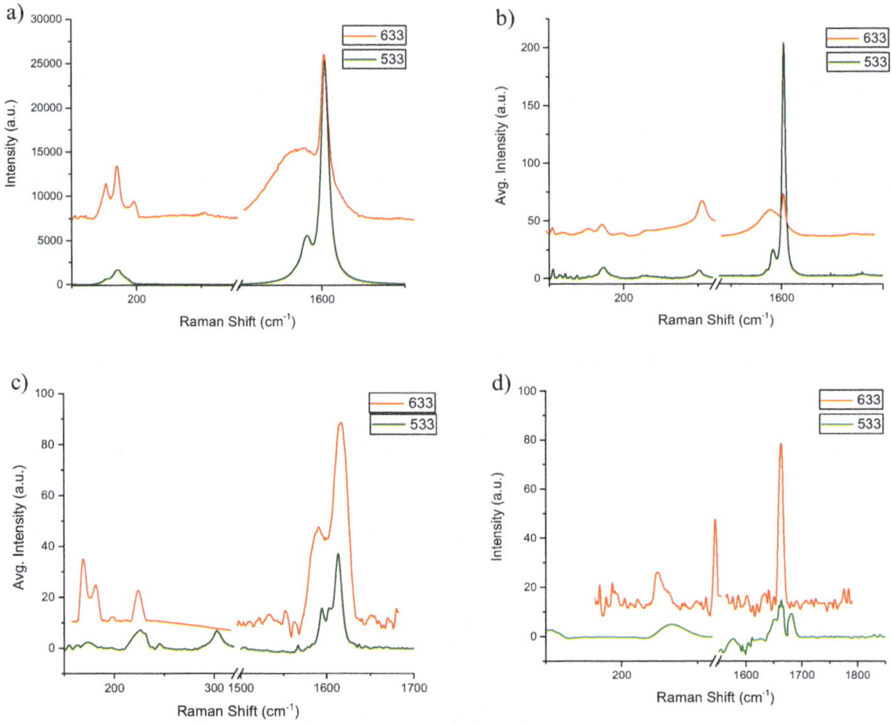

Fig. A.17 Raman spectrum containing the RBM and corresponding G-Band to indicate the presence of SWNTs on the silicon surface at both excitation wavelengths. (**a**) SWNTs only, (**b**) no sonication, (**c**) 5 mins sonication, (**d**) 10 mins sonication

LBL and LB Experiments

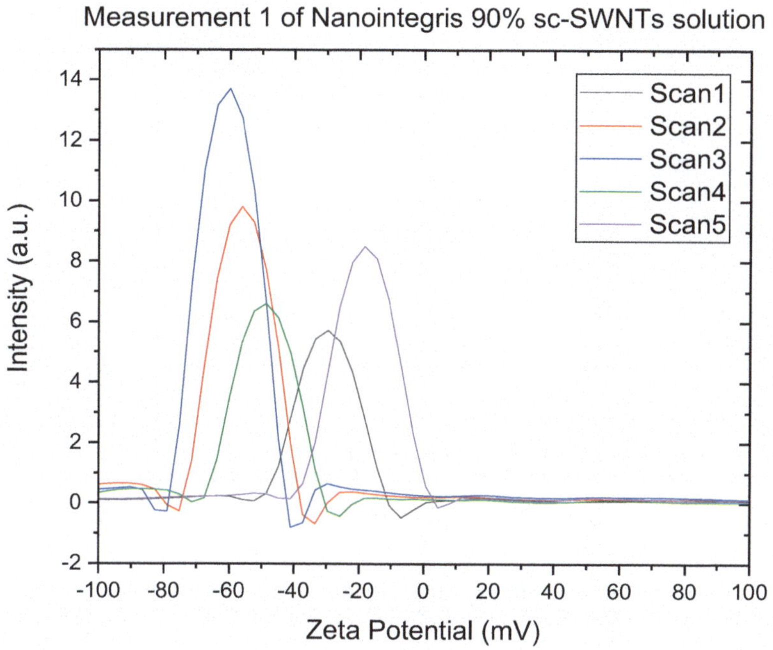

Fig. A.18 The five scans of the first Zeta potential-Intensity plot for the nanotubes in the surfactant solution, Average Zeta Potential(mV): –29.6, Average Mobility(cm^2/Vs) -2.454E-4

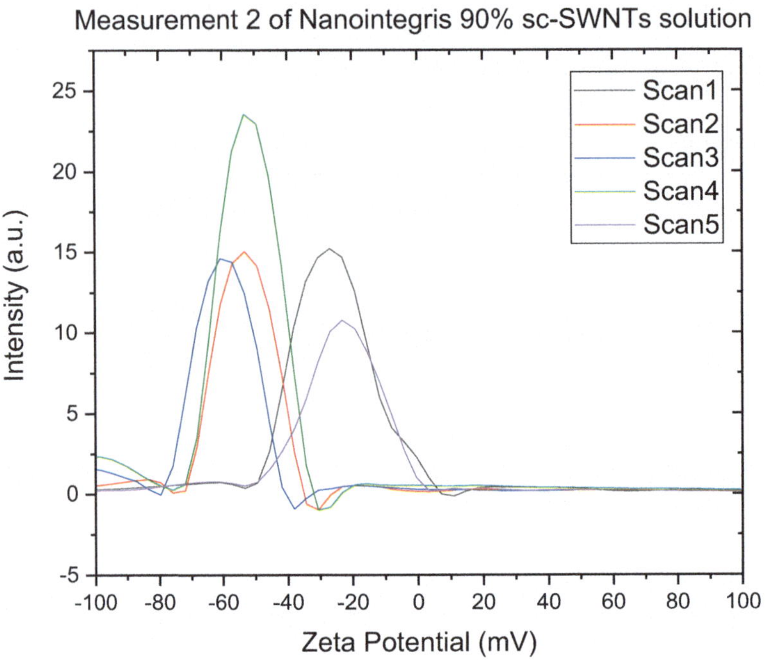

Fig. A.19 The five scans of the second Zeta potential-Intensity plot for the nanotubes in the surfactant solution, Average Zeta Potential(mV): −30.28, Average Mobility(cm^2/Vs) -2.519E-4

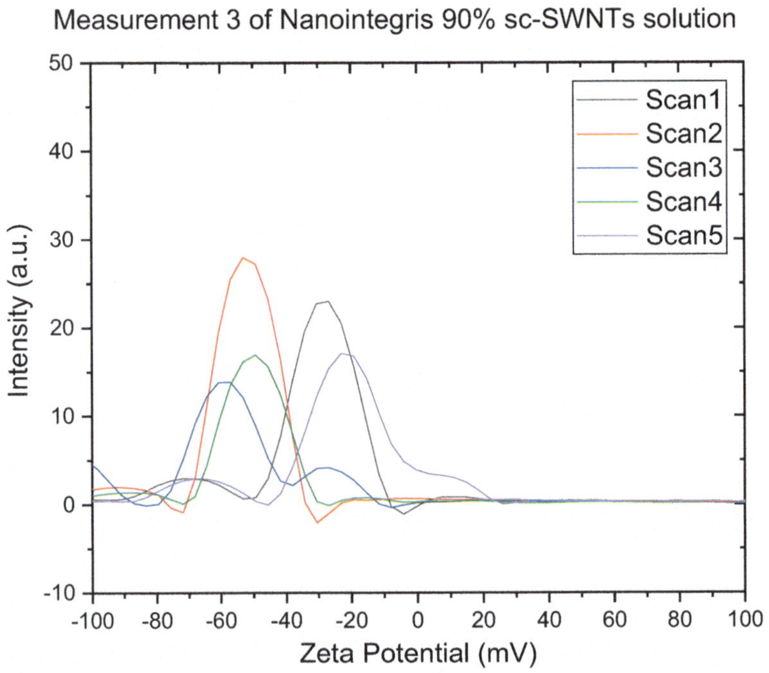

Fig. A.20 The five scans of the third Zeta potential-Intensity plot for the nanotubes in the surfactant solution, Average Zeta Potential(mV): −30.03, Average Mobility(cm^2/Vs) −2.499E-4

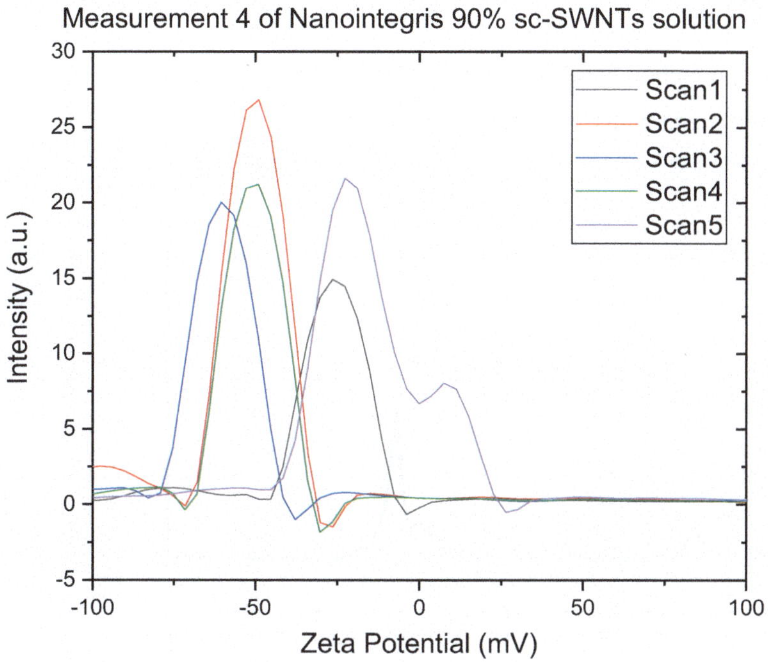

Fig. A.21 The five scans of the fourth Zeta potential-Intensity plot for the nanotubes in the surfactant solution, Average Zeta Potential(mV): −28.3, Average Mobility(cm^2/Vs) −2.355E-4

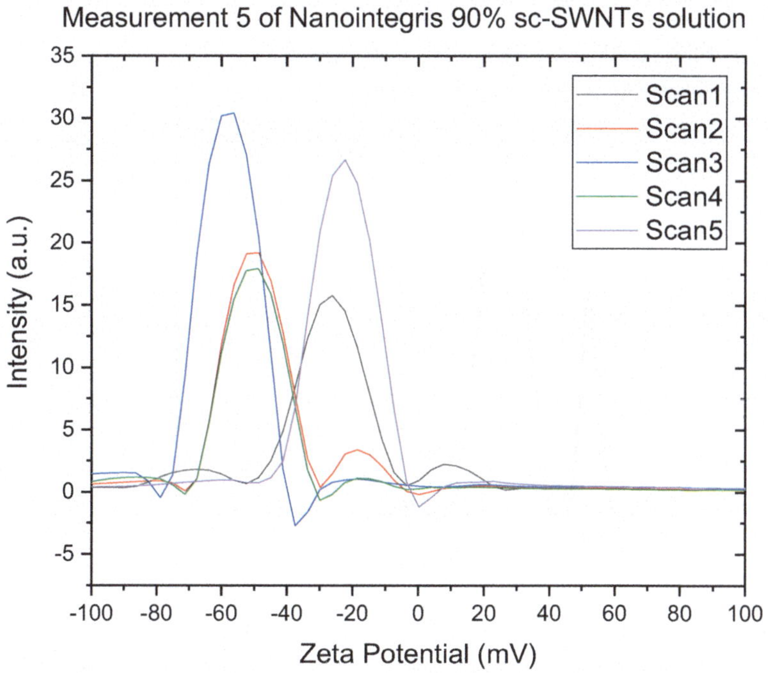

Fig. A.22 The five scans of the fifth Zeta potential-Intensity plot for the nanotubes in the surfactant solution, Average Zeta Potential(mV): −29.67, Average Mobility(cm^2/Vs) −2.469E-4

Filler Experiments

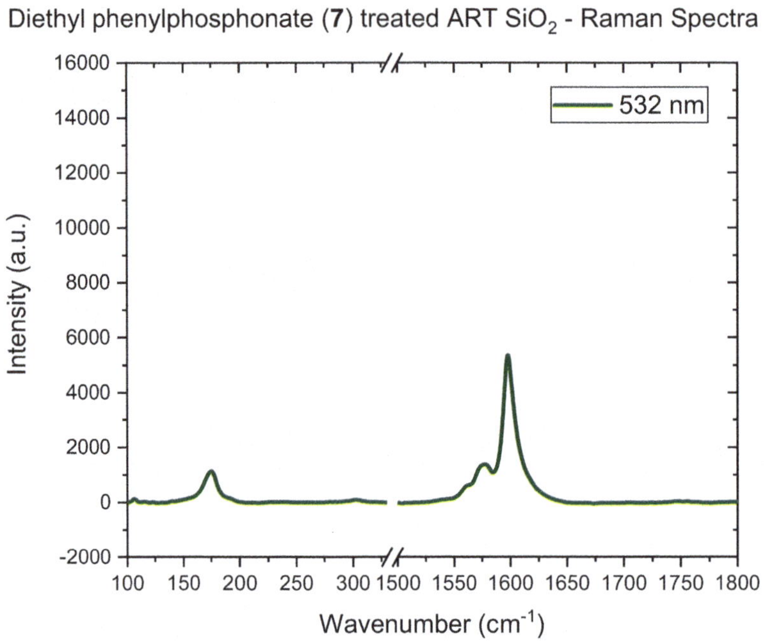

Fig. A.23 Raman spectra in red and green wavelengths showing the RBM and G-band of the diethyl phenylphosphonate (**7**) treated sample

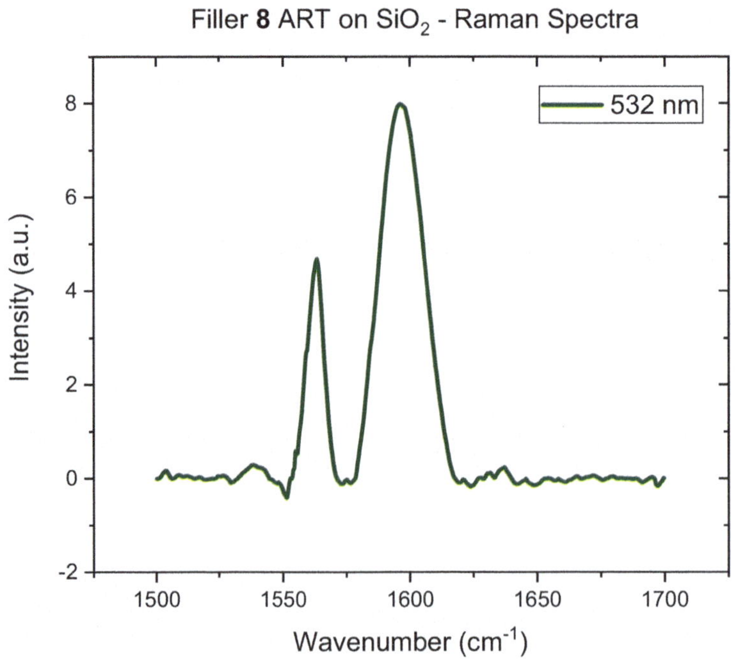

Fig. A.24 Raman spectra in red and green wavelengths showing the RBM and G-band of compound **8** treated sample

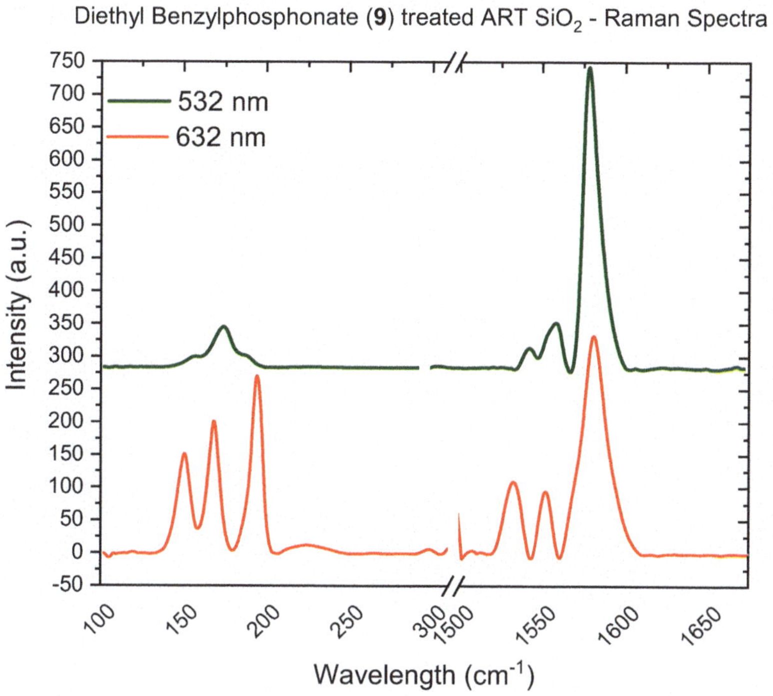

Fig. A.25 Raman spectra in red and green wavelengths showing the RBM and G-band of diethyl benzylphosphonate (**9**) treated sample

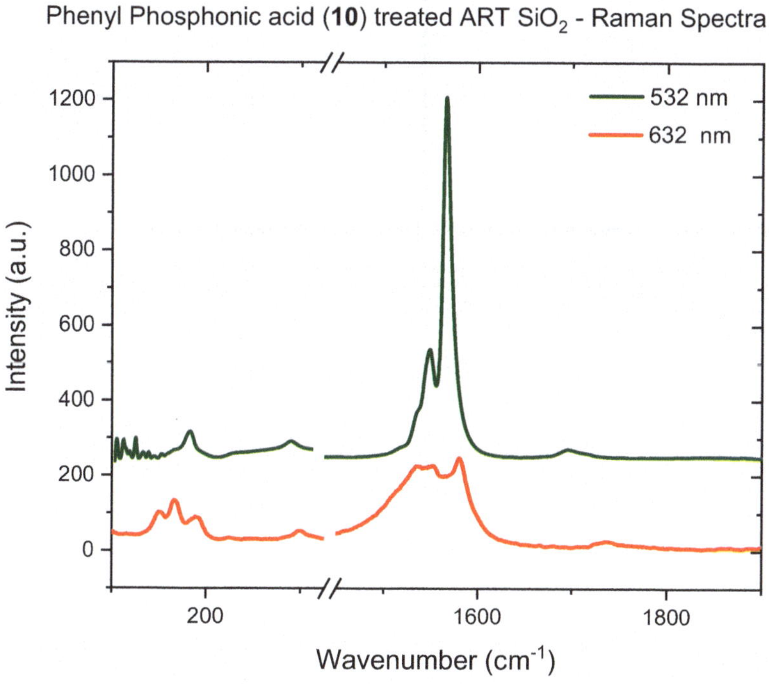

Fig. A.26 Raman spectra in red and green wavelengths showing the RBM and G-band of phenyl phosphonic acid (**10**) treated sample

Appendix Chapter 3

a)

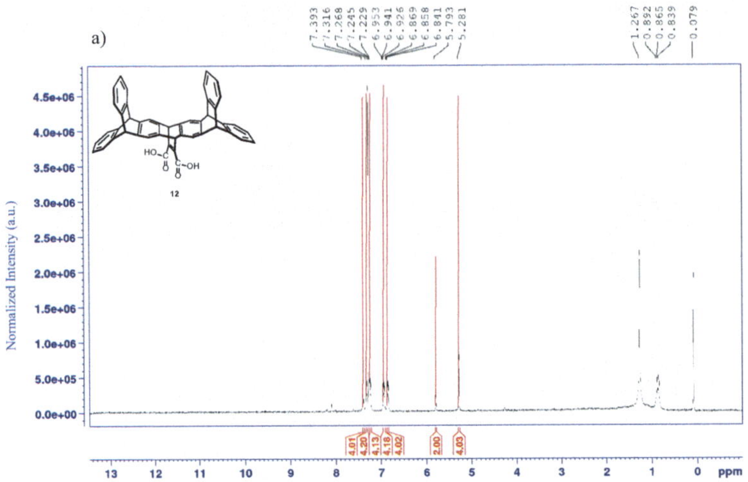

b)

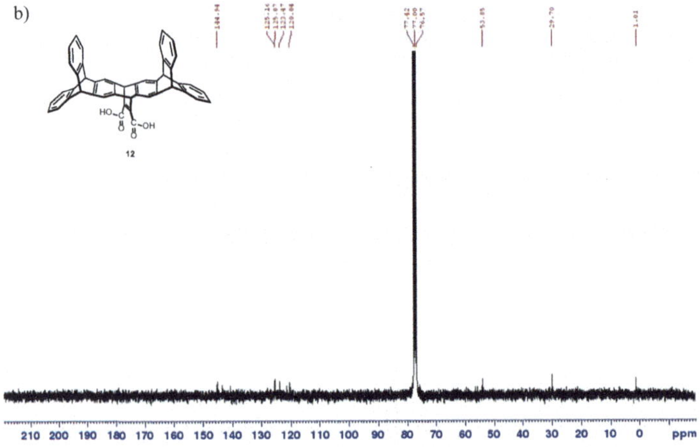

c)

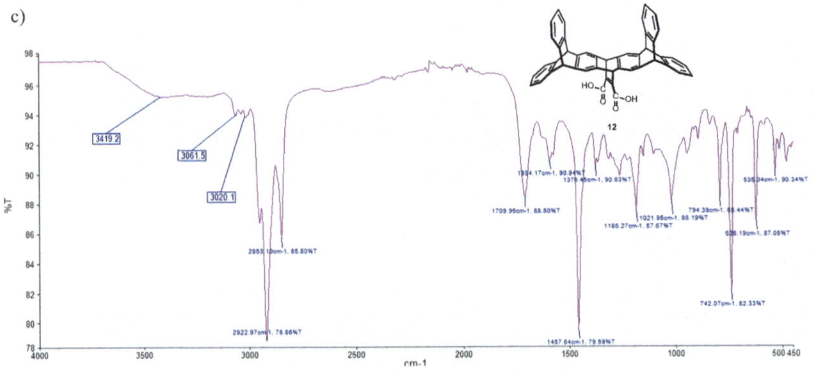

d)

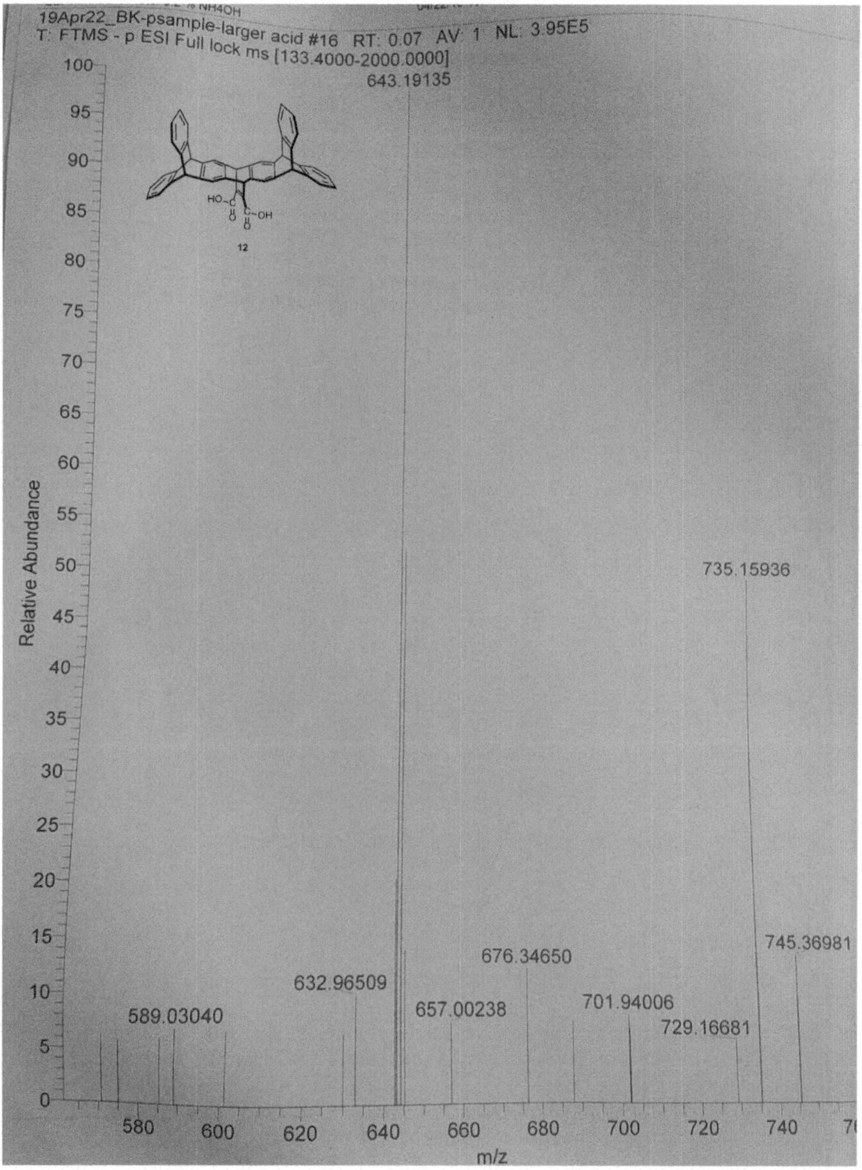
19Apr22_BK-psample-larger acid #16 RT: 0.07 AV: 1 NL: 3.95E5
T: FTMS - p ESI Full lock ms [133.4000-2000.0000]
643.19135
735.15936
745.36981
676.34650
632.96509
657.00238
701.94006
729.16681
589.03040
Relative Abundance
m/z
12

e)

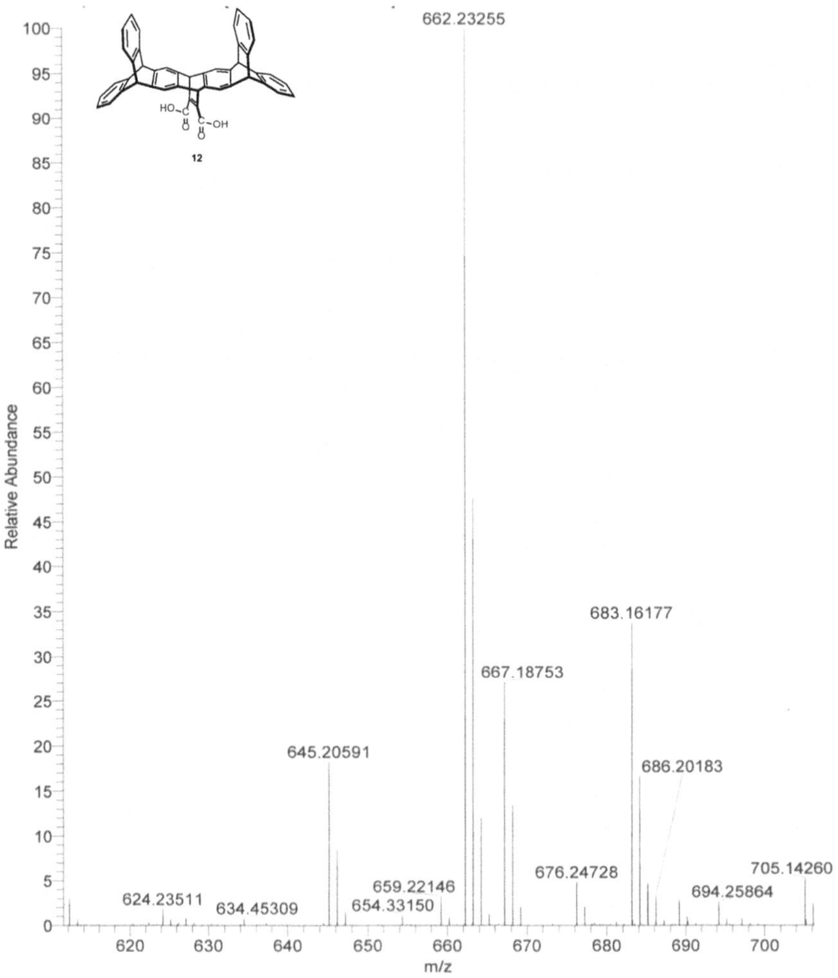

Fig. A.27 Characterization for IptCOOH [2] – compound **12**. (**a**) ^{1}H NMR (300 MHz, $CDCl_3$,7.26 ppm): δ 7.42 (br s, 4 H), 7.39–7.35 (m, 4 H), 7.32–7.29 (m, 4 H), 6.95–6.92 (d, 4 H, $J = 4.7$ Hz), 6.86–6.84 (d, 4 H, $J = 4.7$ Hz), 5.79 (s, 2 H), 5.28 (s, 4 H). 1.27 (water), 0.86 (hexanes), 0.07 (TMS) (**b**) ^{13}C NMR (75.5 MHz, $CDCl_3$, 77.0 ppm): 167.8, 144.9, 143.3, 140.6, 125.2, 125.1, 123.5, 123.0, 53.9, 29.7. (low sample concentration) (**c**) FTIR neat (cm^{-1}): 3419 (-OH), 3061–3020 (C-H, sp^2), 2922–2853 (C-H, sp^3), 1709 (C=O). (**d**) HR ESI MS of compound 12 in anionic mode. Performed in 1:1 MeOH: H2O + 0.1% NH_4OH Calculated for $C_{46}H_{29}O_4$ [M-H] - 643.19039, found 643.19135; (**e**) HR ESI MS of compound 12 in cationic mode. Performed in 1:1 MeOH: H_2O + 0.1% Formic Acid. Calculated for $C_{46}H_{29}O_4$ [M + H] + 645.20604, found 645.20591; calculated for C46H29O4 [M + NH_4] + 662.24504, found 662.23255

a)

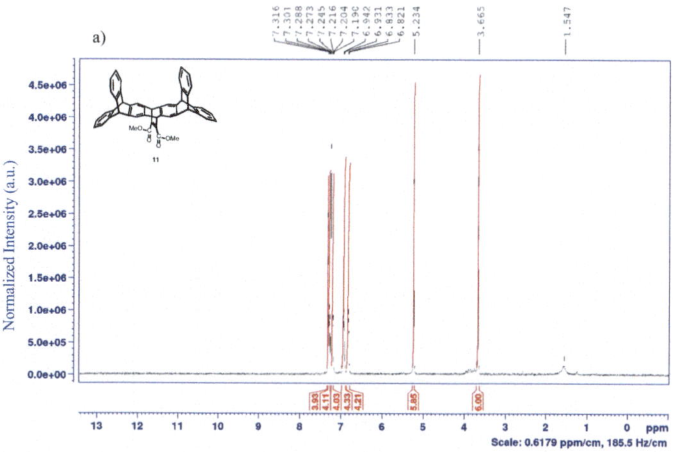

b)

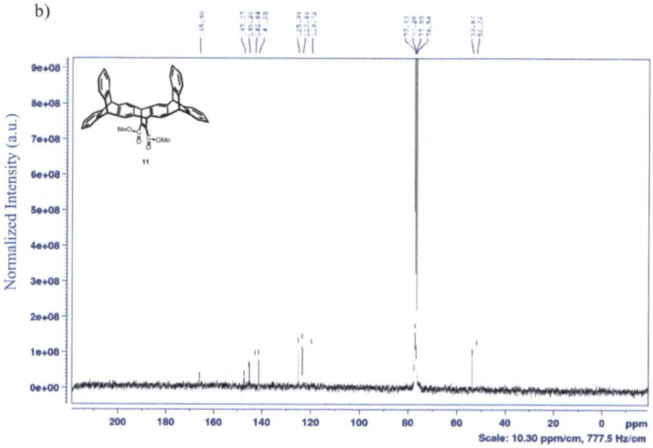

c)

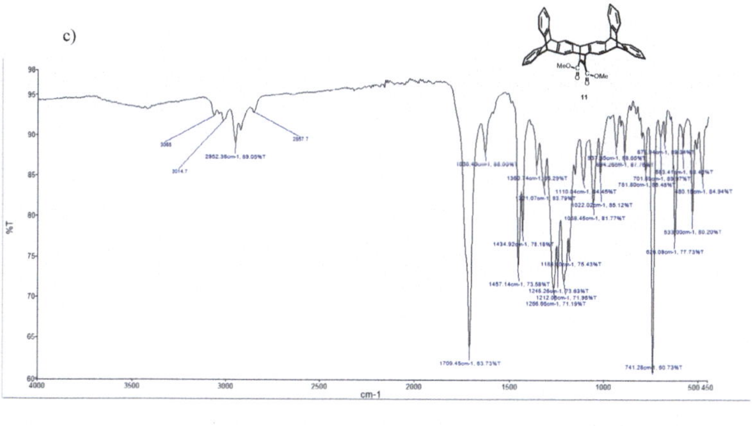

Fig. A.28 Characterization for compound **11**, precursor to IptCOOH. (**a**) ^{1}H NMR (300 MHz, CDCl3, 7.26 ppm): δ = 7.32–7.30 (br s, 4 H), 7.28 (t, 4 H, $J = 3.7$ Hz), 7.20 (t, 4 H, $J = 3.7$ Hz), 6.94 (t, 4 H, $J = 3.5$ Hz), 6.82 (t, 4 H, $J = 3.5$ Hz), 5.24–5.22 (br s, 6 H), 3.66 (s, 6 H), 1.54 (water). (**b**) ^{13}C NMR (75.5 MHz, CDCl3, 77.0 ppm): 165.9, 147.3, 142.8, 141.3, 125.1, 123.4, 123.4, 119.7, 76.6, 53.9, 52.2. (**c**) FTIR neat (cm^{-1}): 3065–3014 (C-H, sp^2), 2952–2857 (C-H, sp^3), 1709 (C=O)

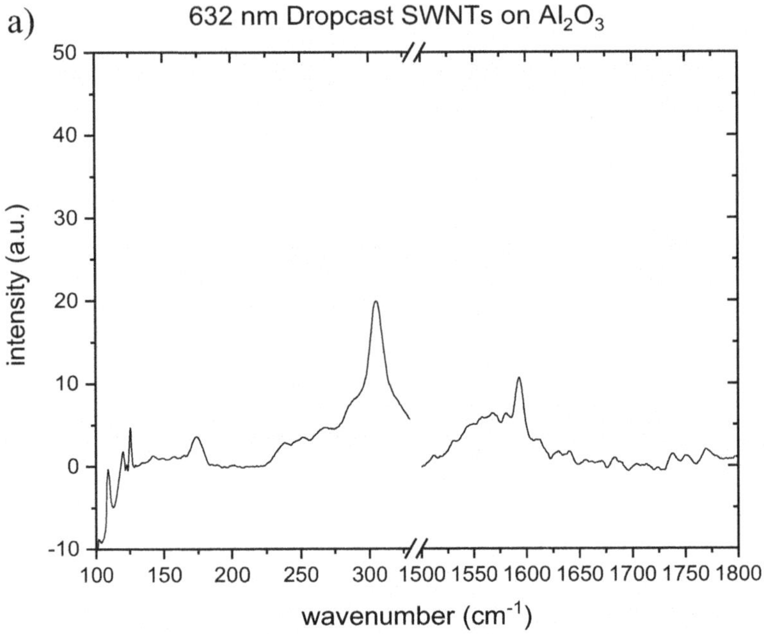

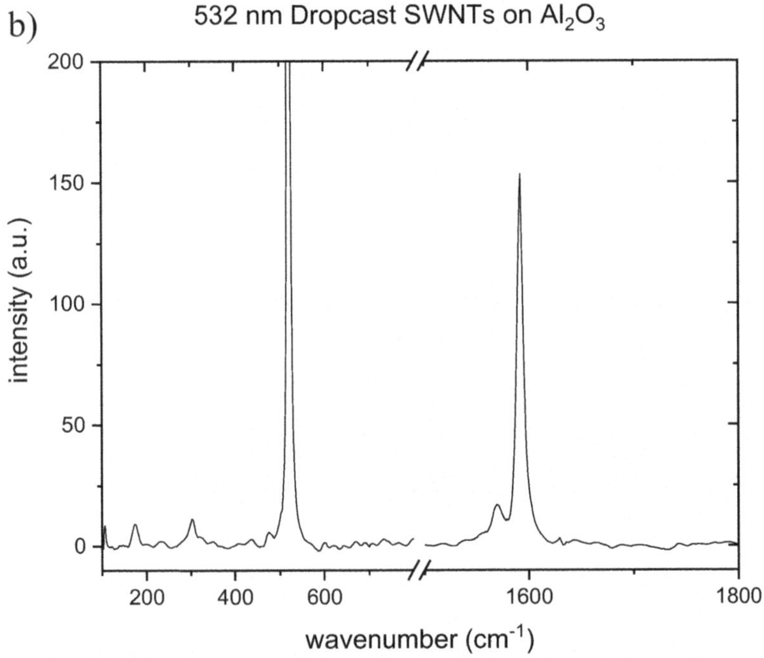

Fig. A.29 (**a**) Red ($\lambda = 632$ nm) laser wavelength spectrum of the NanoIntegris Inc. 90% pure single-walled carbon nanotubes dropcast on the TMA-ALD Al_2O_3 treated silicon wafer, (**b**) Green ($\lambda = 532$ nm) laser wavelength spectrum of the NanoIntegris Inc. 90% pure single-walled carbon nanotubes dropcast on the TMA-ALD Al_2O_3 treated silicon wafer. The silicon peaks dominate the spectrum and are used for calibration purposes

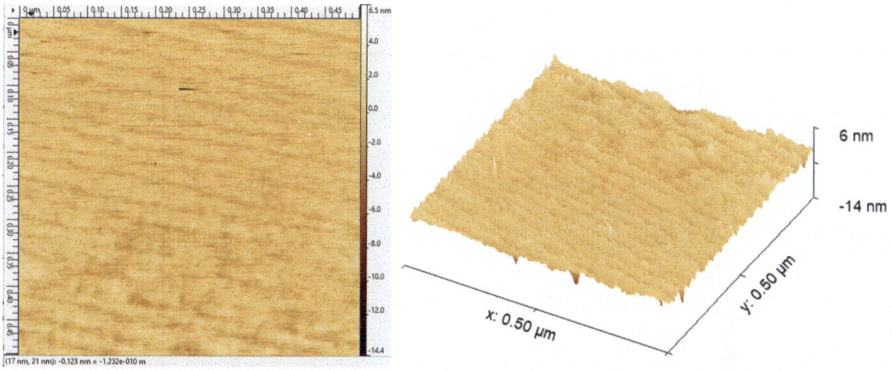

Fig. A.30 α-Al_2O_3 500 × 500 nm before UV-Ozone

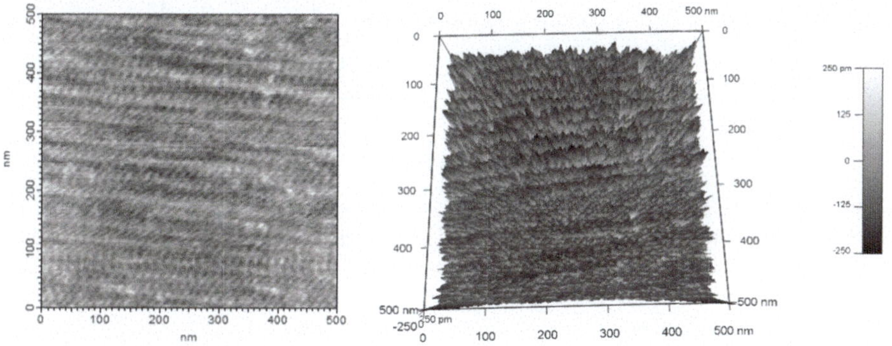

Fig. A.31 α-Al_2O_3 500 × 500 nm after UV-Ozone

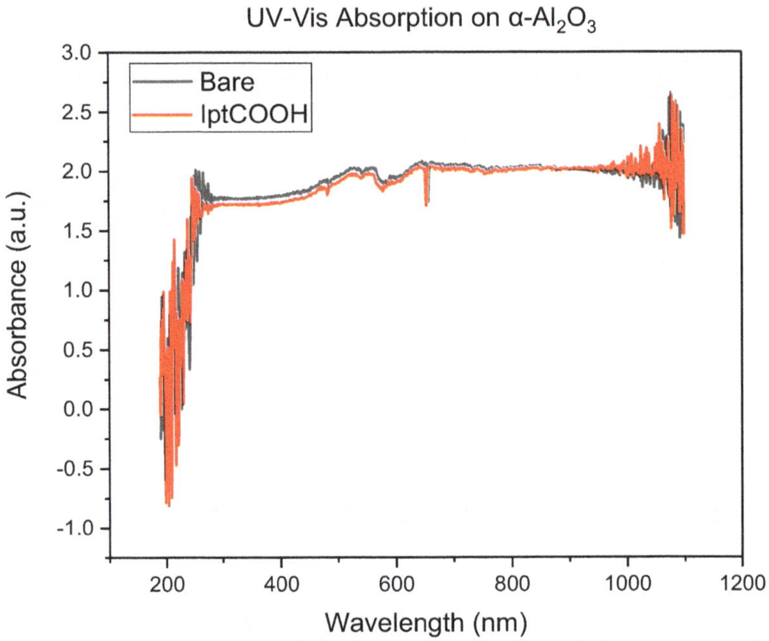

Fig. A.32 UV-vis on blank sapphire and IptCOOH treated sapphire indicating too much scattering

a)

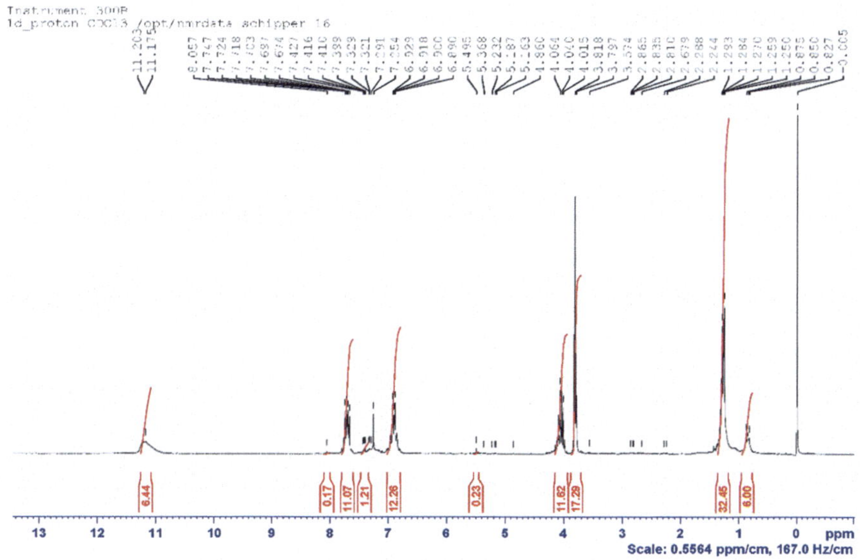

b)

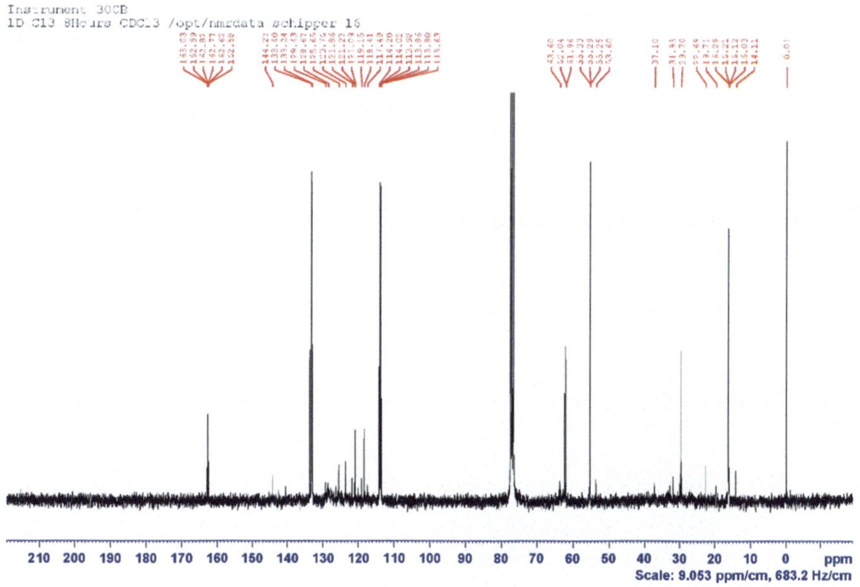

Fig. A.33 Resulting crude NMR spectra for the attempted synthesis of compound **19** (a) ^{1}H proton, (**b**) ^{13}C NMR. Both starting materials of **1** and Lawesson's reagent have some peaks present, alongside other impurities and possibly **19**. Sample should be revisited and purified further

Appendix Chapter 4

SWNTs-COOH

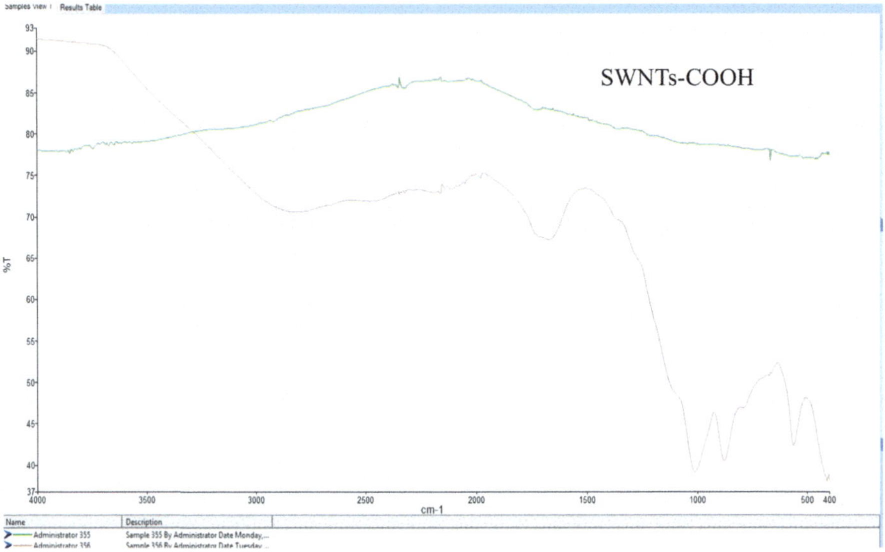

Fig. A.34 Additional IR of the treated vs. untreated SWNTs testing if the carboxylic acid carbonyl stretch is there

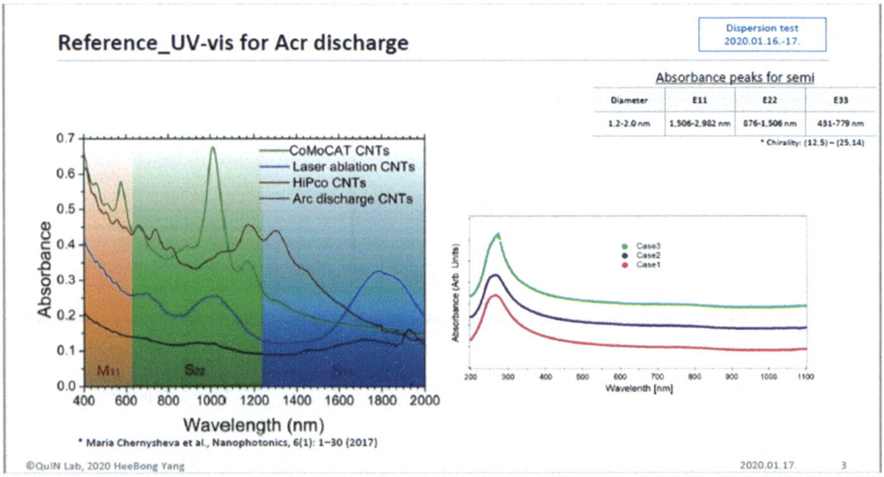

Fig. A.35 The Tuball brand SWNTs UV-Vis analysis spectrum. The UV-visible spectrum was also collected for the raw untreated SWNTs (Supplemental Data) by Heebong Yang of the Quantum Innovation (QuIN) laboratory (Prof Kim Lab) at the University of Waterloo to compare the findings

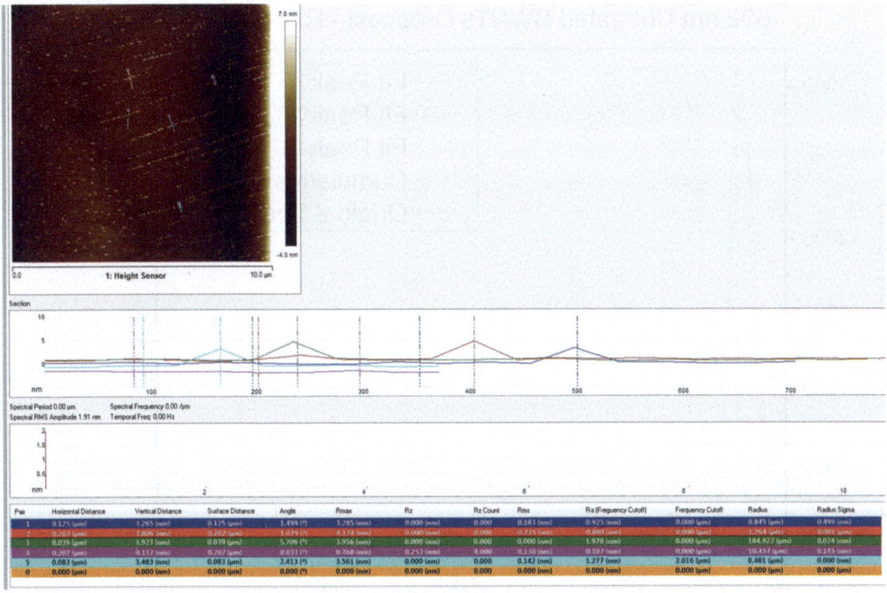

Fig. A.36 AFM topography height data for SWNTs-COOH treated with ART in water

Untreated SWNTs Dropcast Raman Data

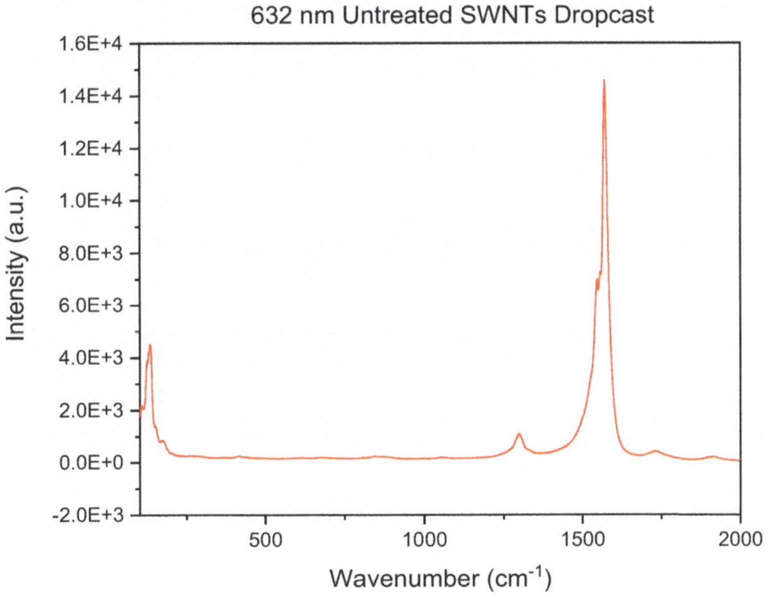

Fig. A.37 Full spectrum of Tuball SWNTs dropcasted on Si using a red laser

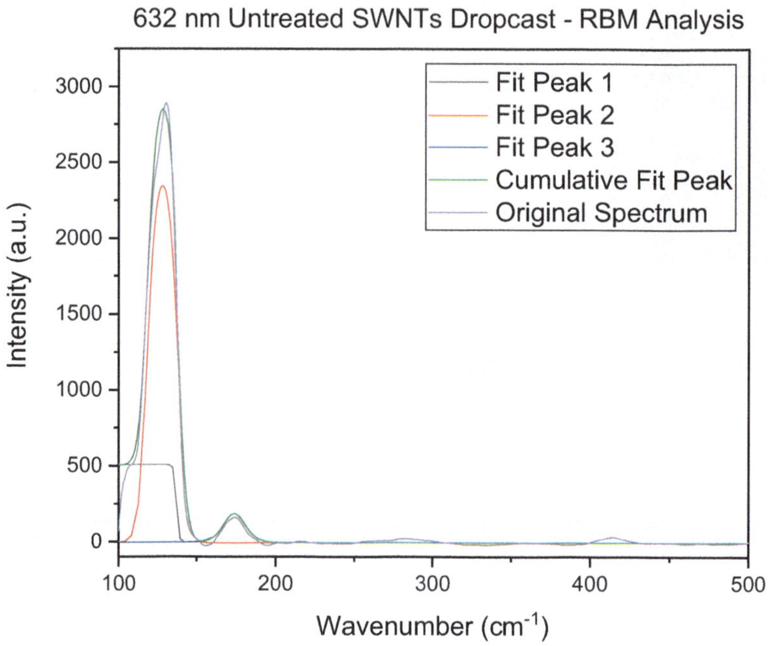

Fig. A.38 Cumulative fit peak data of RBM region in the red laser wavelengths

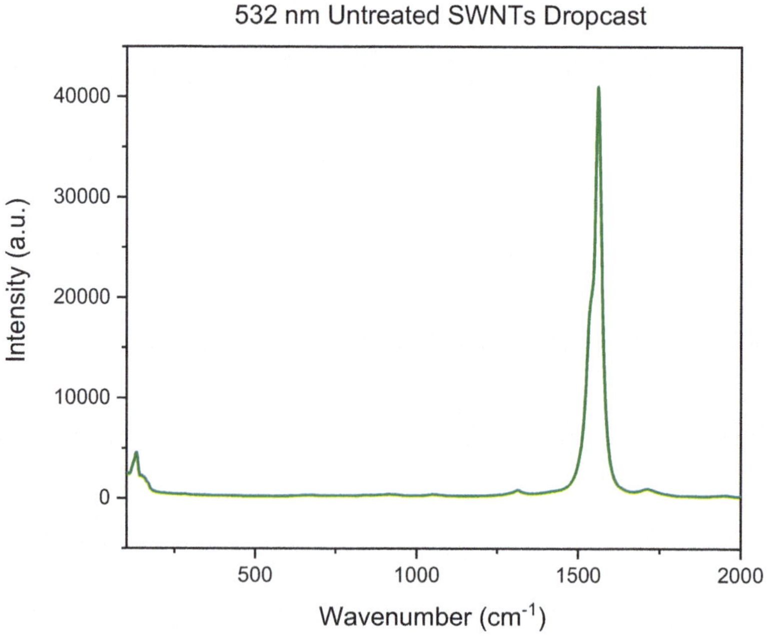

Fig. A.39 Full spectrum of Tuball SWNTs dropcasted on Si using a green laser

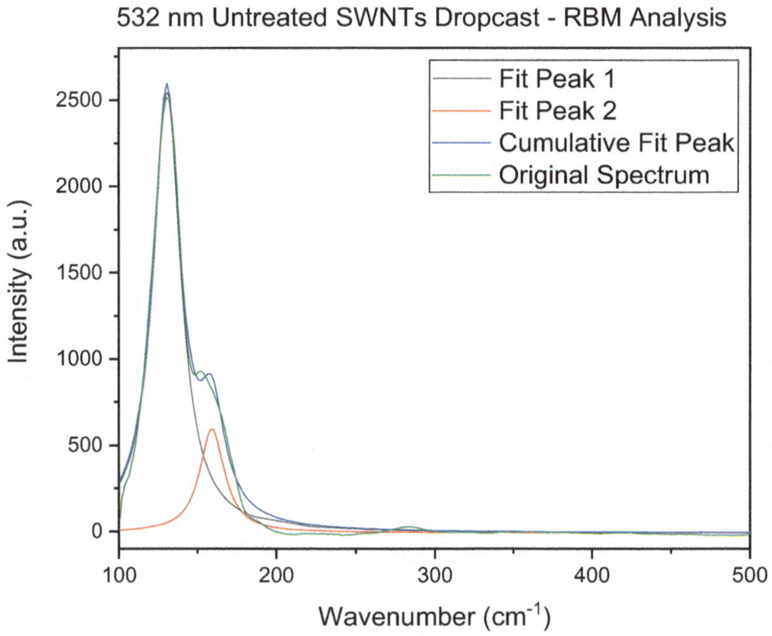

Fig. A.40 Cumulative fit peak data of RBM region in the red laser wavelengths

Untreated SWNTs with ART

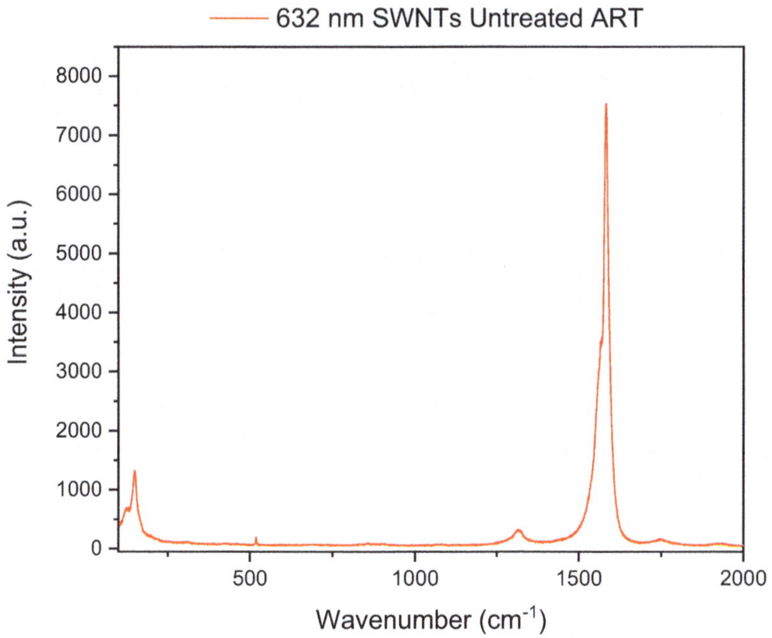

Fig. A.41 Full spectrum of ART + Tuball brand SWNTs on Si using a red laser

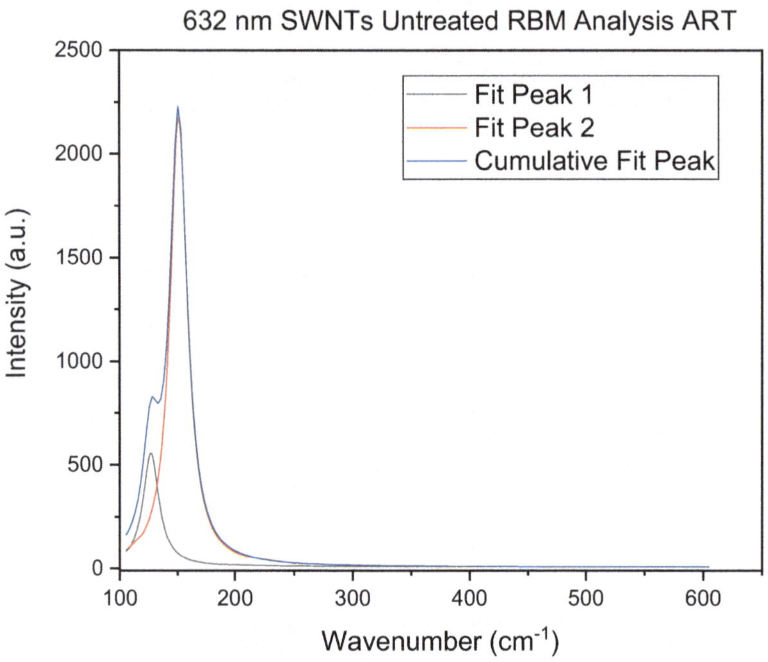

Fig. A.42 Cumulative fit peak data of RBM region in the red laser wavelengths with ART

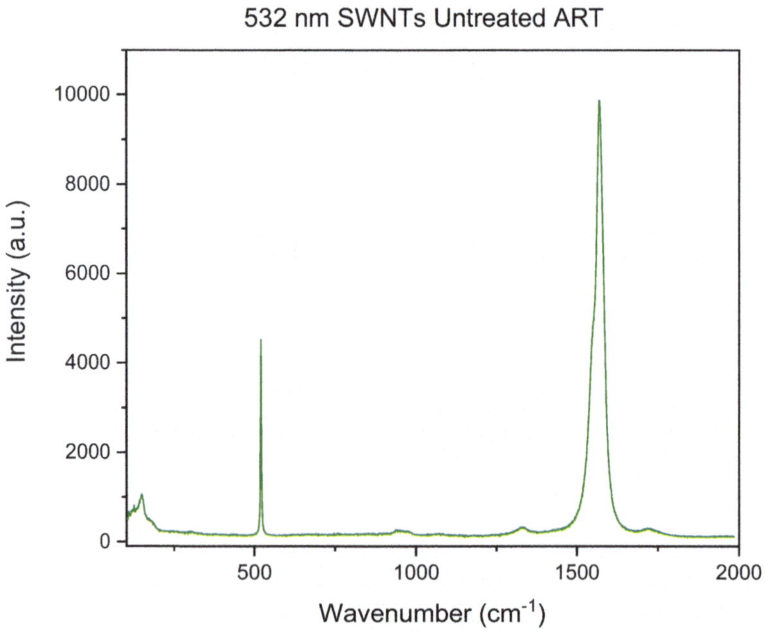

Fig. A.43 Full spectrum of ART + Tuball brand SWNTs on Si using a green laser

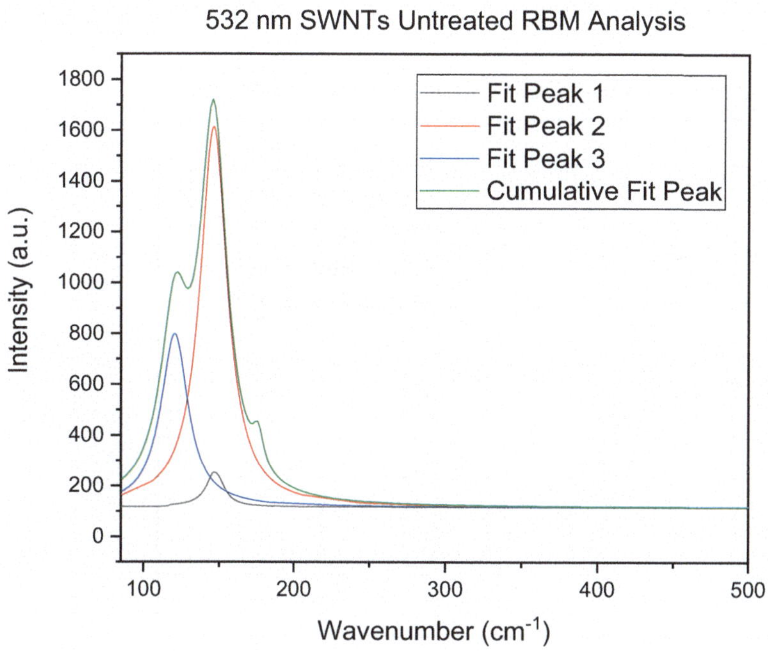

Fig. A.44 Cumulative fit peak data of RBM region in the green laser wavelengths with ART

The Treated SWNTs (SWNTs-COOH) Dropcast Raman Data

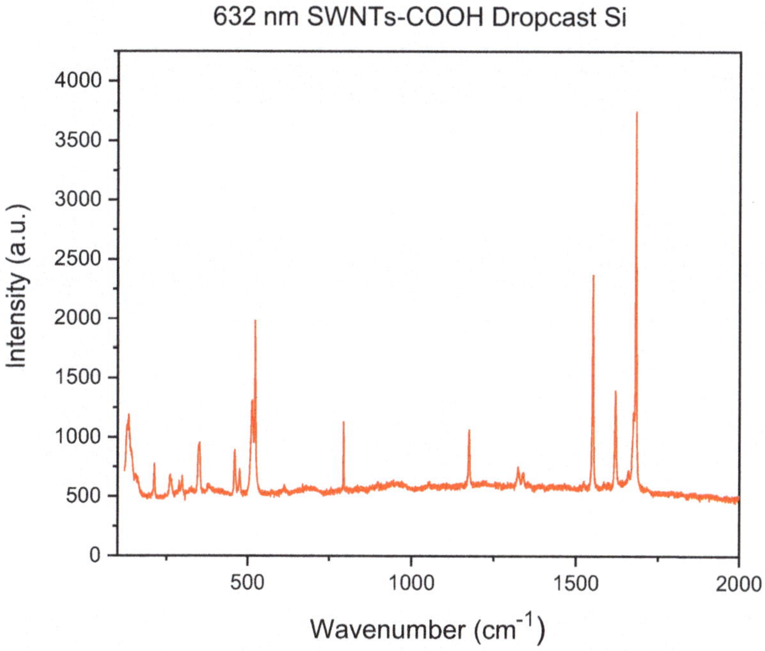

Fig. A.45 Full spectrum of SWNTs-COOH dropcast on Si using a red laser

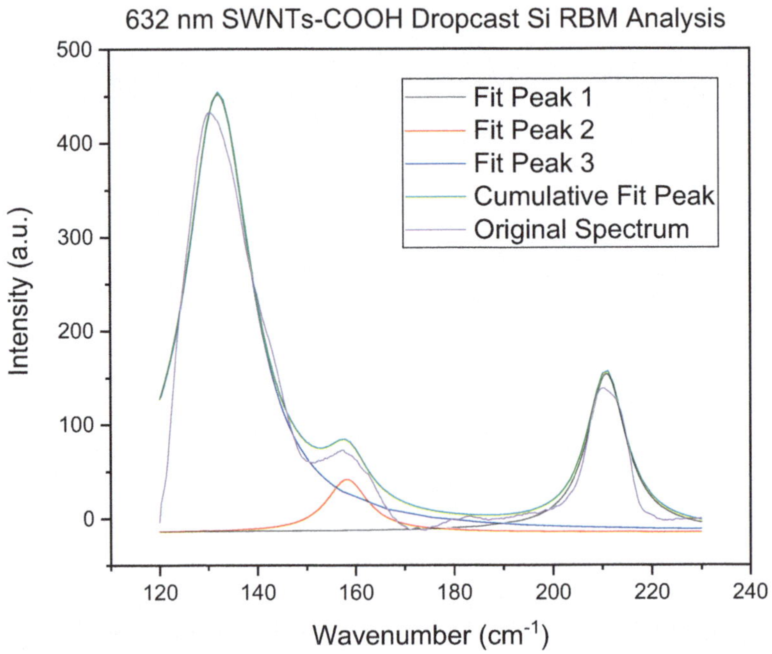

Fig. A.46 Cumulative fit peak data of RBM region in the red laser wavelengths for SWNTs-COOH dropcast on Si

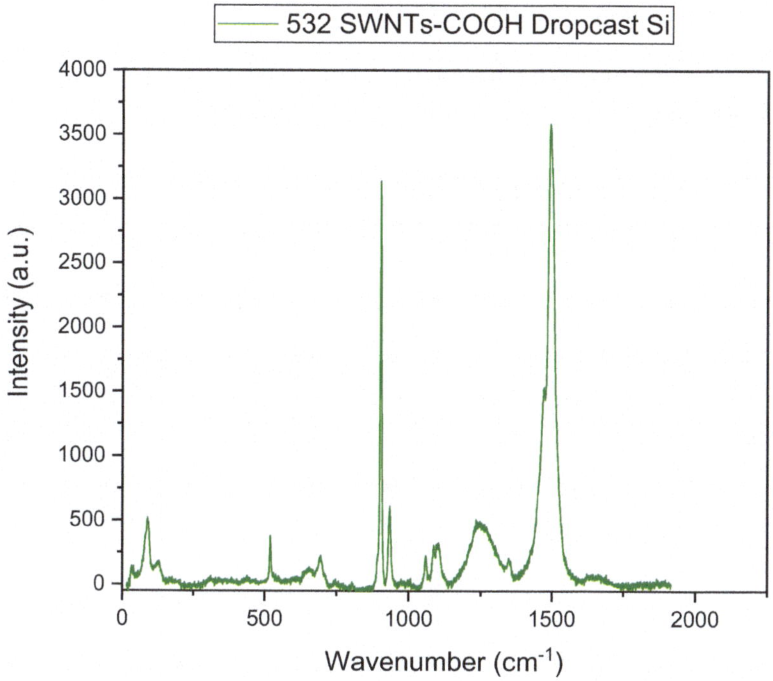

Fig. A.47 Full spectrum of SWNTs-COOH dropcast on Si using a green laser

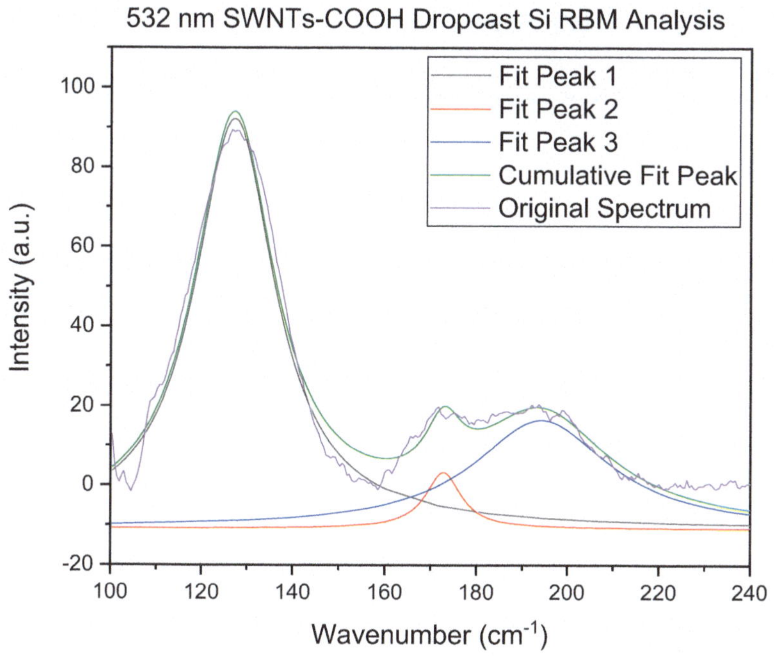

Fig. A.48 Cumulative fit peak data of RBM region in the green laser wavelengths for SWNTs-COOH dropcast on Si

Alignment Relay Technique + SWNTs-COOH

1. Raman Data in Red and Green Laser Wavelengths for the Surfactant Dispersed SWNTs-COOH/ ART

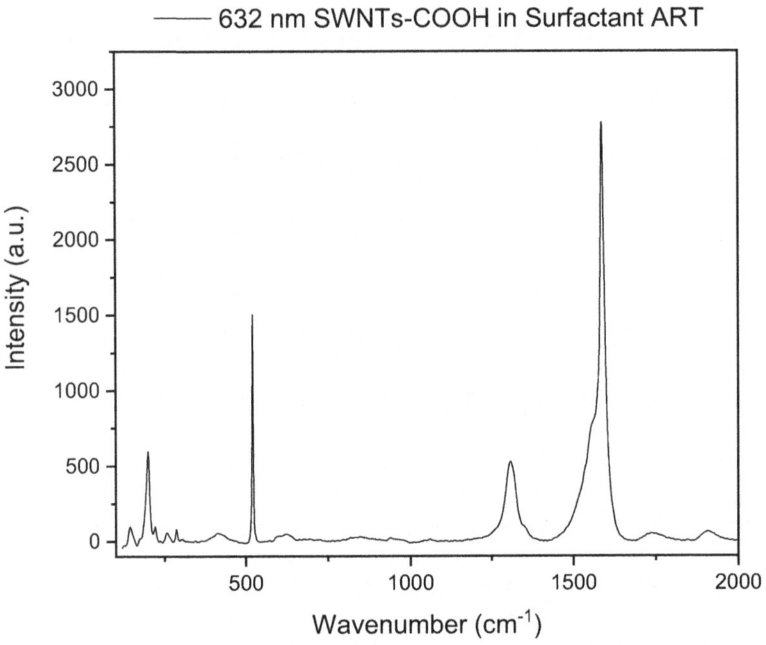

Fig. A.49 Full spectrum of SWNTs-COOH on ART surface on Si using a red laser in surfactant

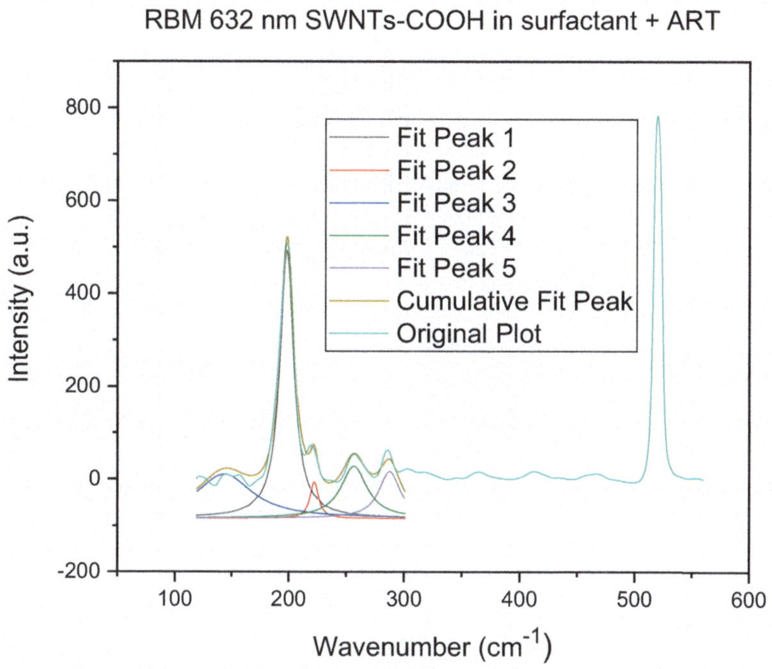

Fig. A.50 Cumulative fit peak data of RBM of ART treated SWNTs-COOH region in the red laser wavelengths in surfactant

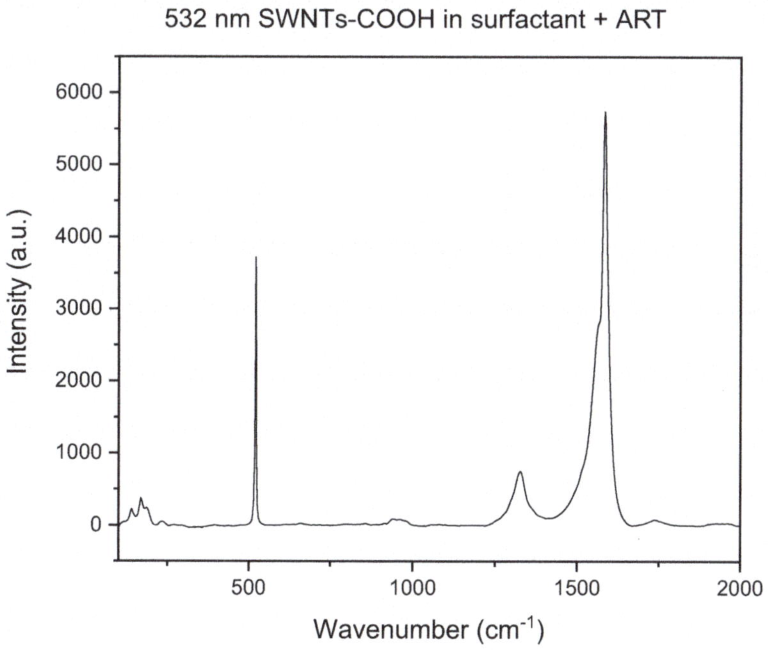

Fig. A.51 Full spectrum of ART treated SWNTs-COOH on Si using a green laser in surfactant

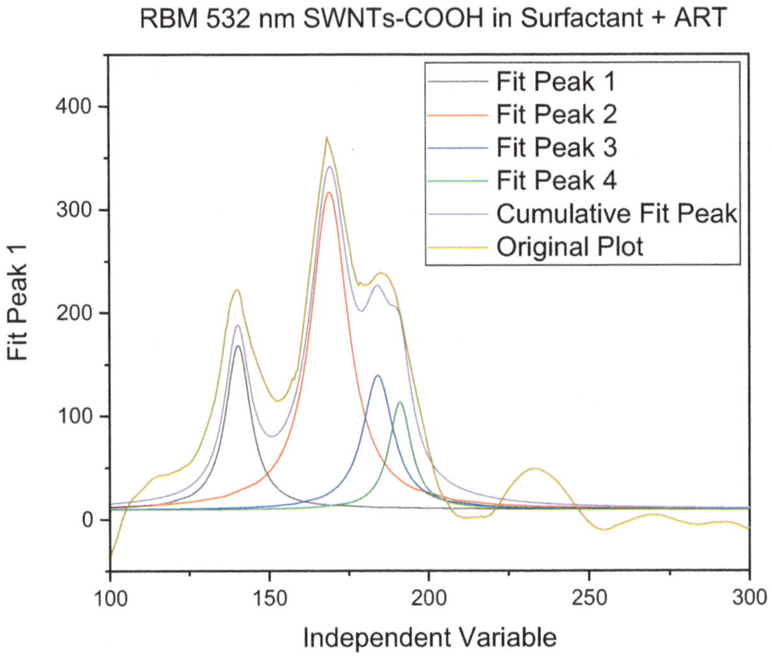

Fig. A.52 Cumulative fit peak data of RBM region in the Green laser wavelengths for ART treated SWNTs-COOH in surfactant

Raman Data in Red and Green Laser Wavelengths for the Water Dispersed SWNTs-COOH/ ART

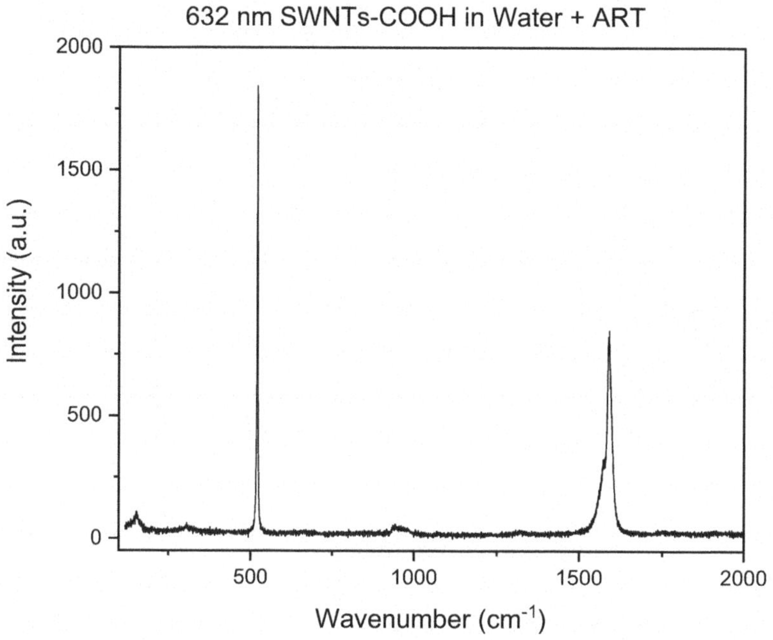

Fig. A.53 Full spectrum of SWNTs-COOH on ART surface on Si using a red laser in water

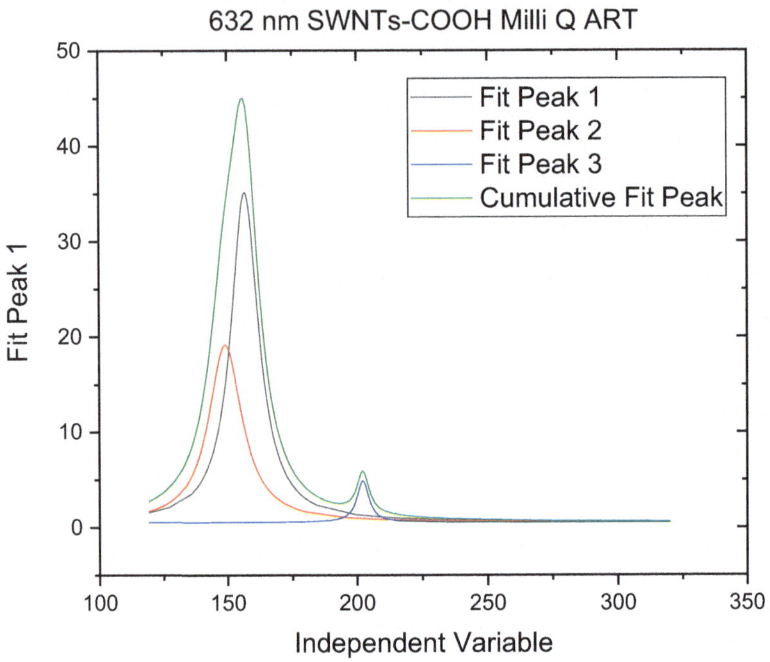

Fig. A.54 Cumulative fit peak data of RBM of ART treated SWNTs-COOH region in the red laser wavelengths in water

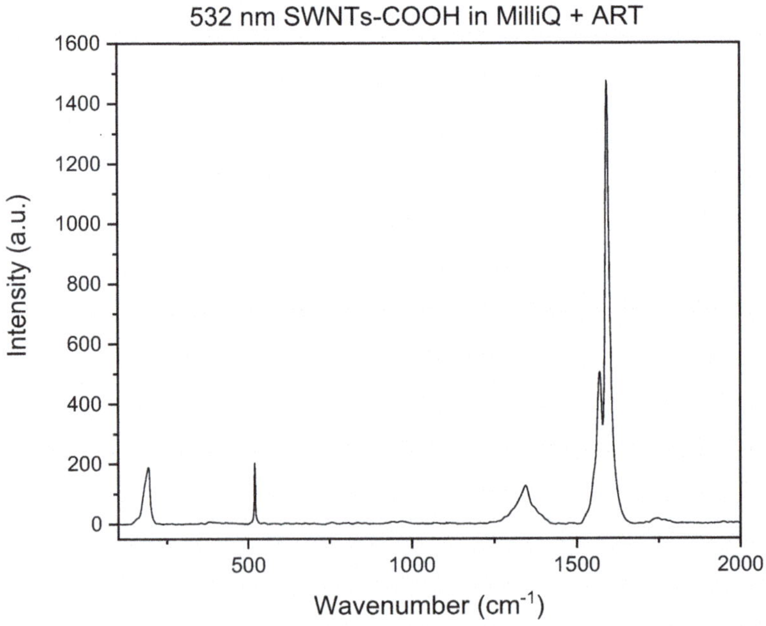

Fig. A.55 Full spectrum of SWNTs-COOH on ART surface on Si using a green laser in water

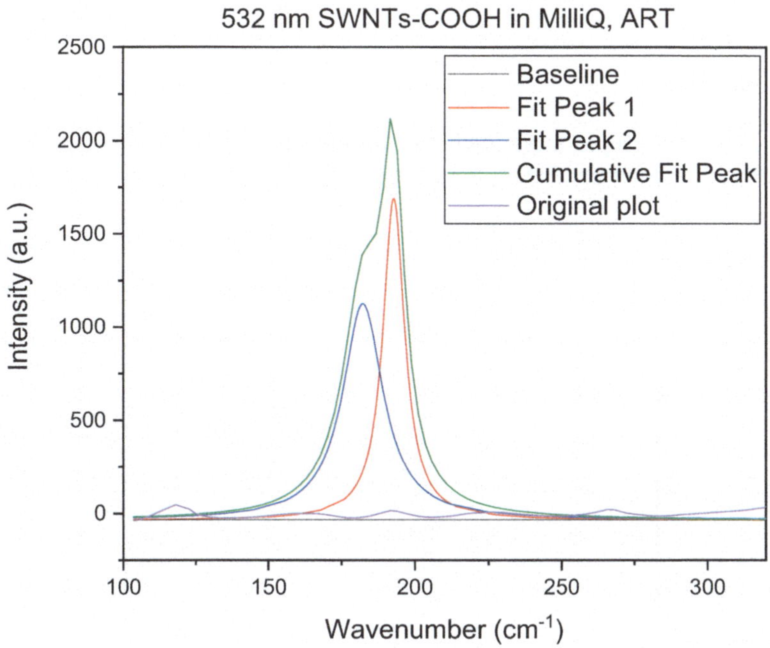

Fig. A.56 Cumulative fit peak data of RBM of ART treated SWNTs-COOH region in the green laser wavelengths in water

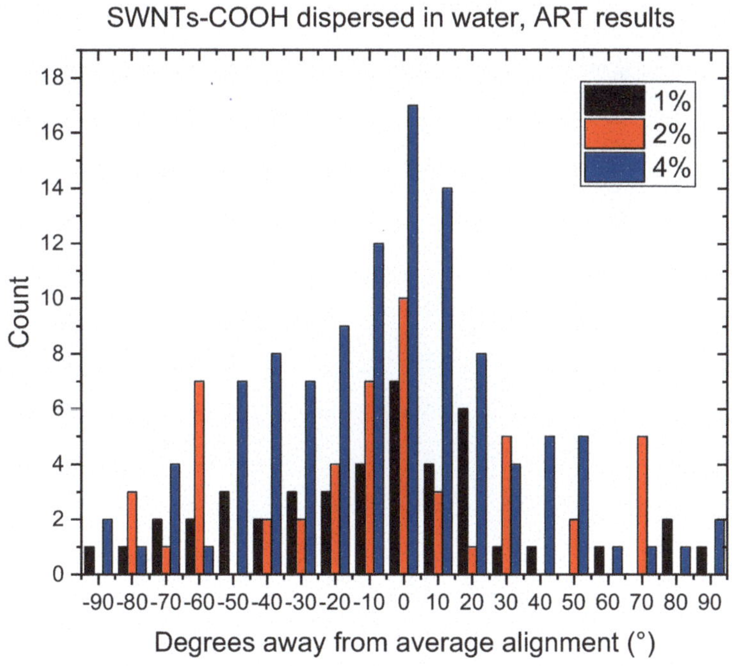

Fig. A.57 Comparison of degrees away from average alignment for the data at different concentrations (wt/v) of the SWNTs-COOH in water. The 4% had the higher number of nanotubes from the samples tested so it was chosen for further analysis

Longer Length SWNTs

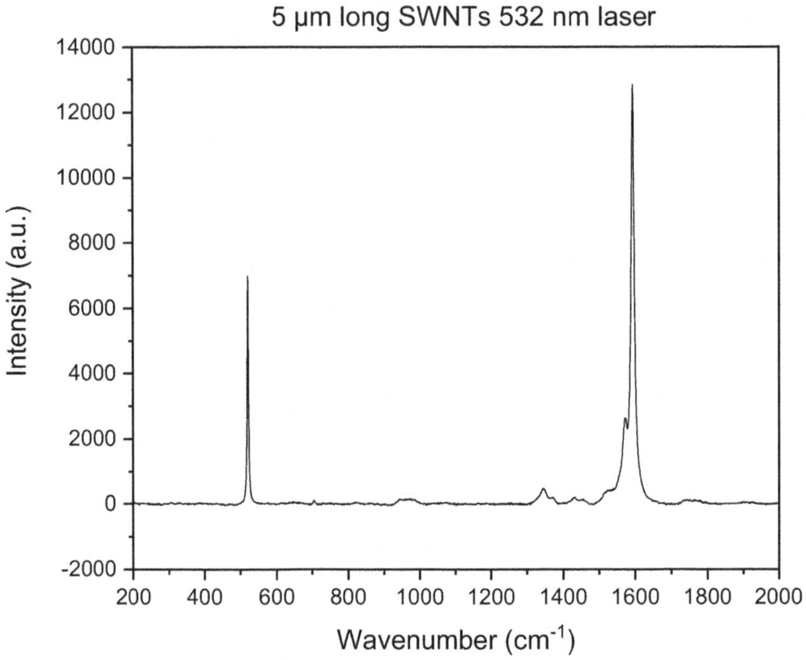

Fig. A.58 Full spectrum at 532 nm wavelength for the sample of longer SWNTs dropcat on Si

Smaller Diameter SWNTs

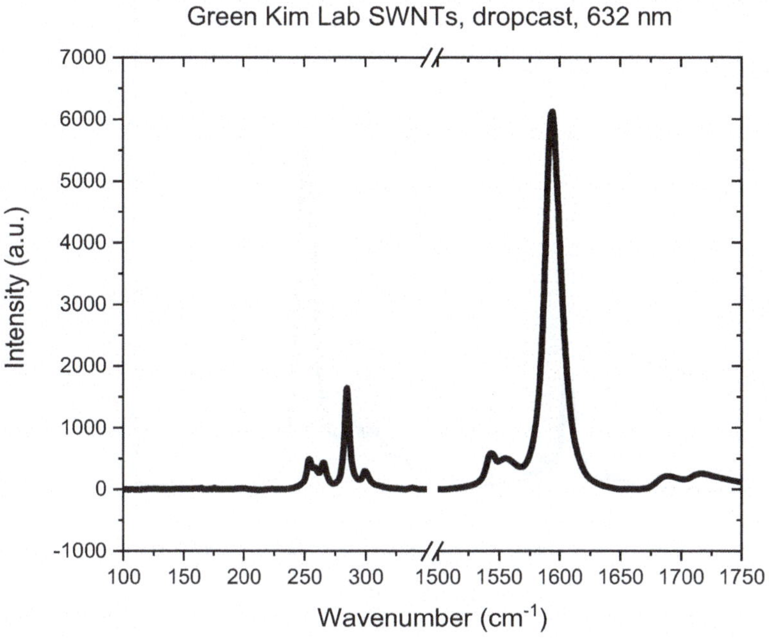

Fig. A.59 Spectrum of smaller diameter green SWNTs dropcast on Si using a red laser

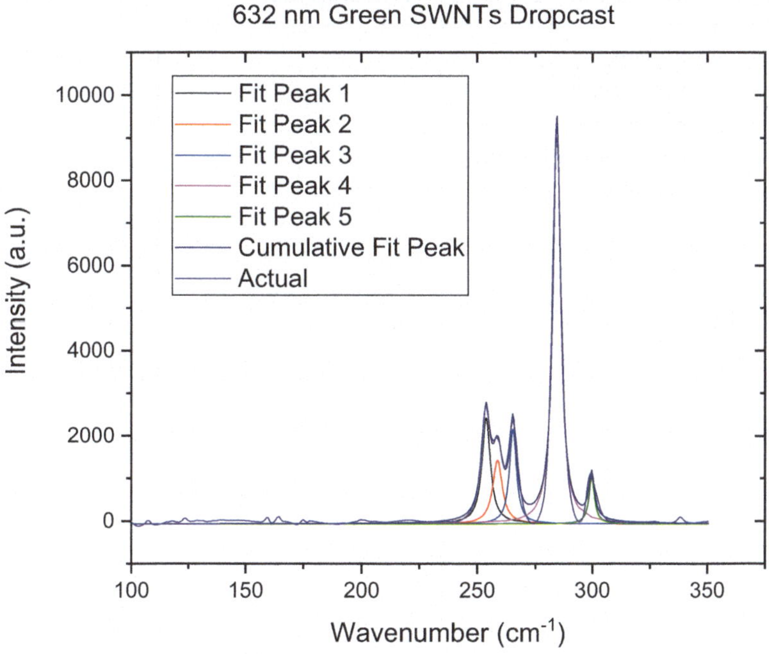

Fig. A.60 RBM deconvoluted peaks of the smaller diameter green SWNTs dropcast on Si using a red laser

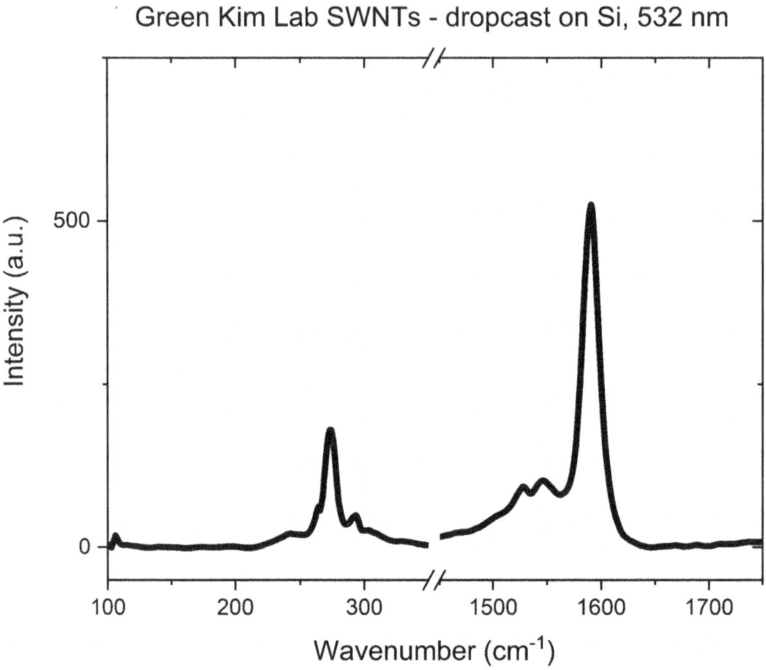

Fig. A.61 Spectrum of smaller diameter green SWNTs dropcast on Si using a green laser

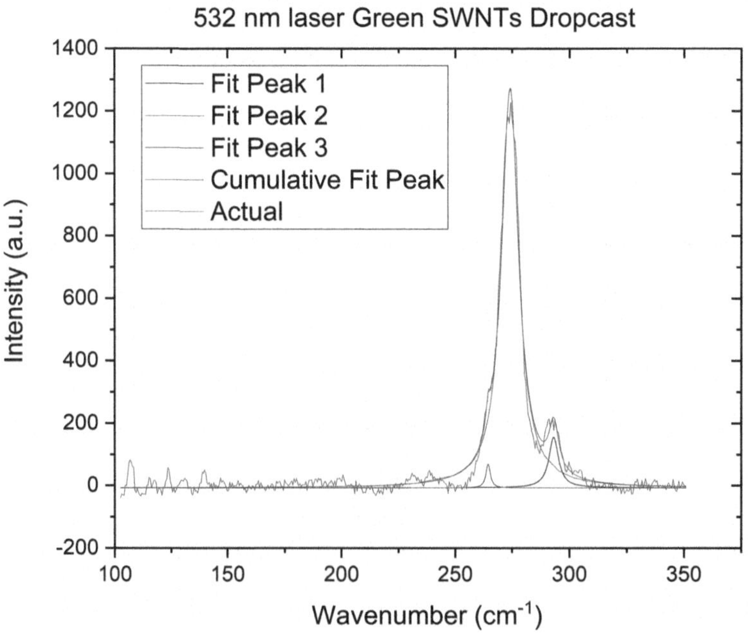

Fig. A.62 RBM deconvoluted peaks of the smaller diameter green SWNTs dropcast on Si using a green laser

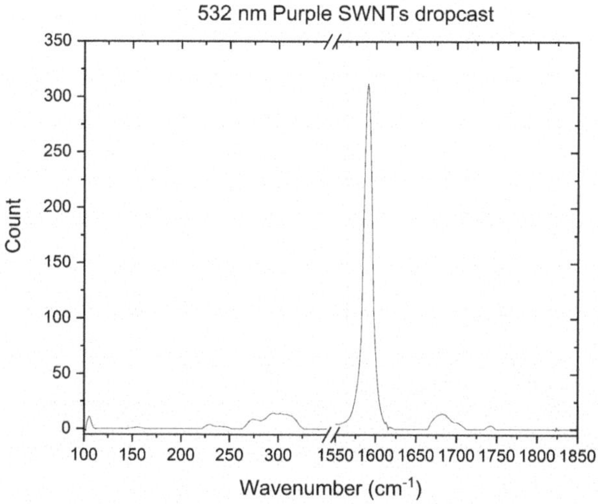

Fig. A.63 Spectrum of smaller diameter purple SWNTs dropcast on Si using a red laser

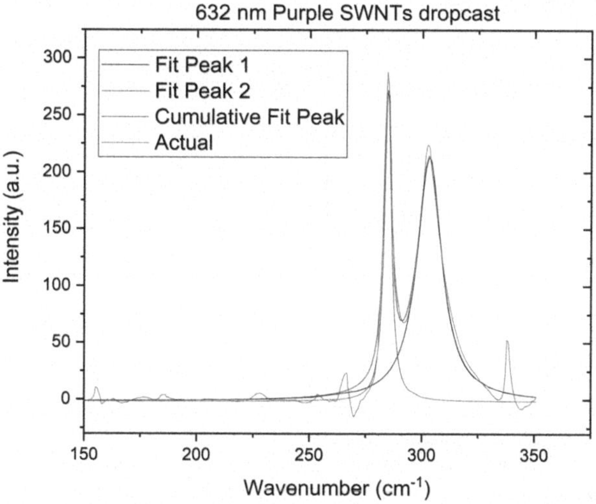

Fig. A.64 RBM deconvoluted peaks of the smaller diameter purple SWNTs dropcast on Si using a red laser

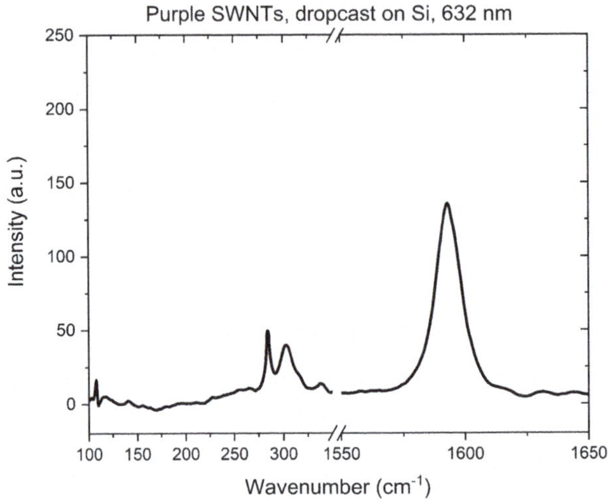

Fig. A.65 Spectrum of smaller diameter purple SWNTs dropcast on Si using a green laser

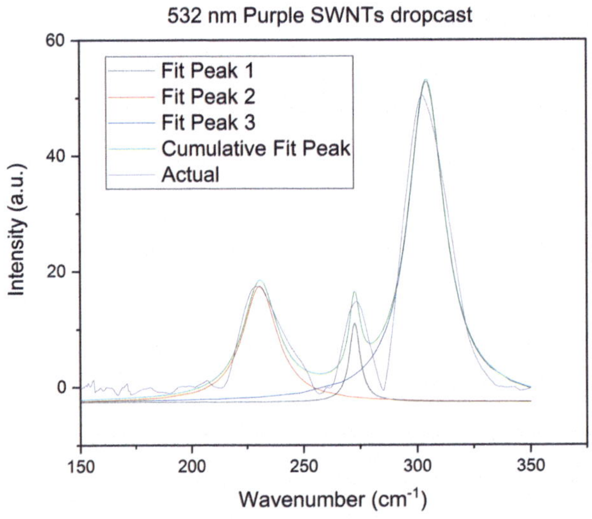

Fig. A.66 RBM deconvoluted peaks of the smaller diameter purple SWNTs dropcast on Si using a green laser

Bottom-Up Synthesized GNRs Raman

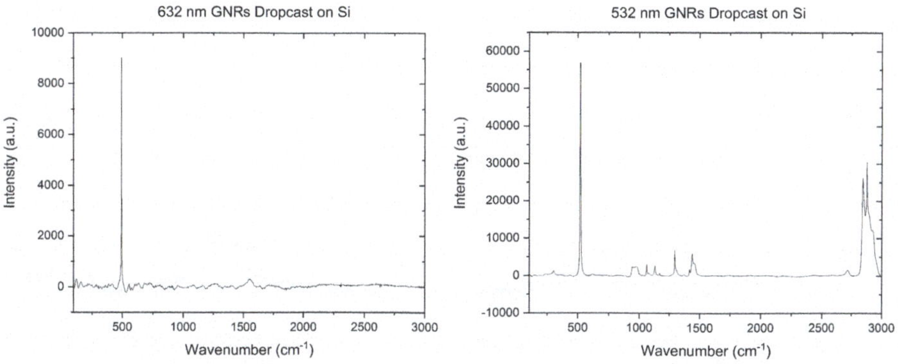

Fig. A.67 Raman in (**a**) red and (**b**) green excitation wavelengths of the synthesized GNRs

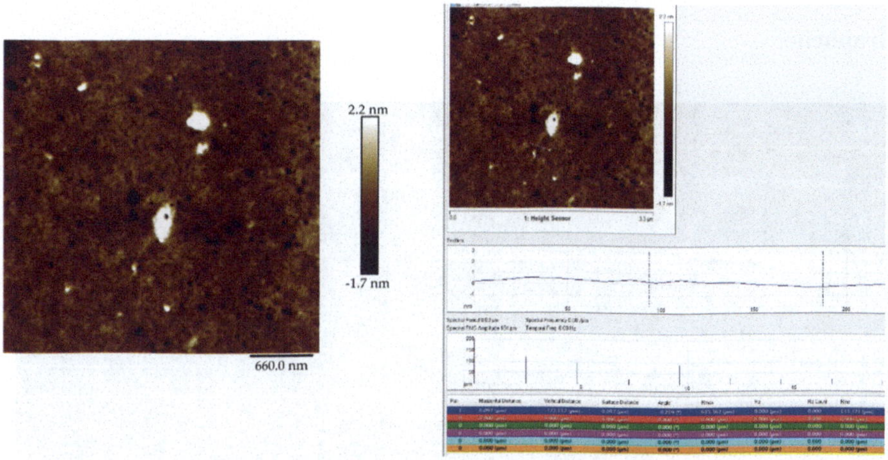

Fig. A.68 The results are shown what could possibly be a very small rGNR however, the sample should have much larger ones of the surface

MWNTs

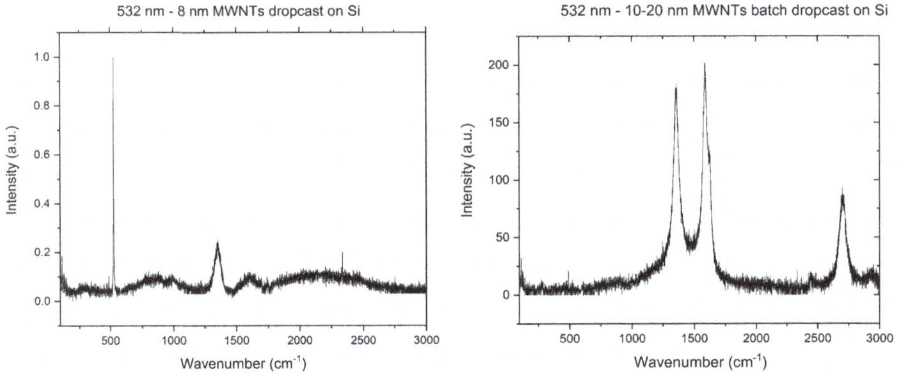

Fig. A.69 Dropcast of MWNTs on the surface of Si, both diameters samples (**a**) M1 (8 nm) and (**b**) M2 (10–20 nm)

Graphene

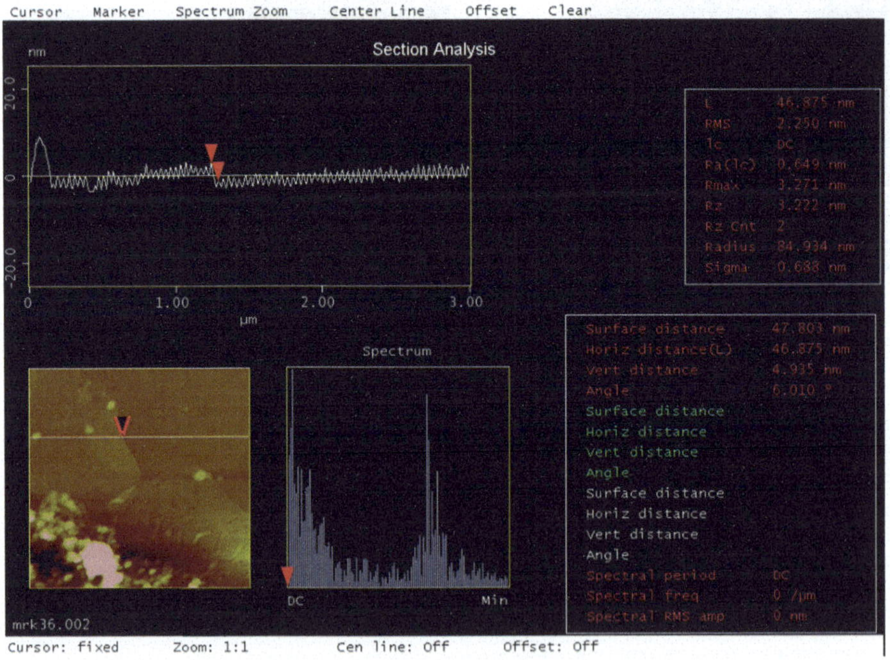

Fig. A.70 Possible graphene flakes on ART surface, height measured to be 5 nm

Appendix Chapter 5

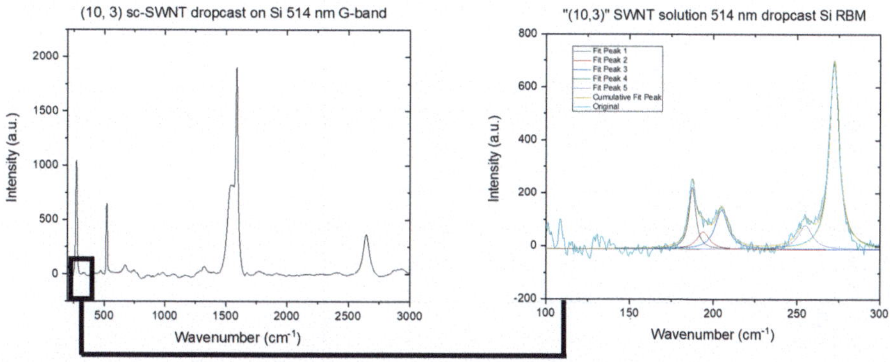

Fig. A.71 Raman of (10,3) sc-SWNTs, G-band and RBM region

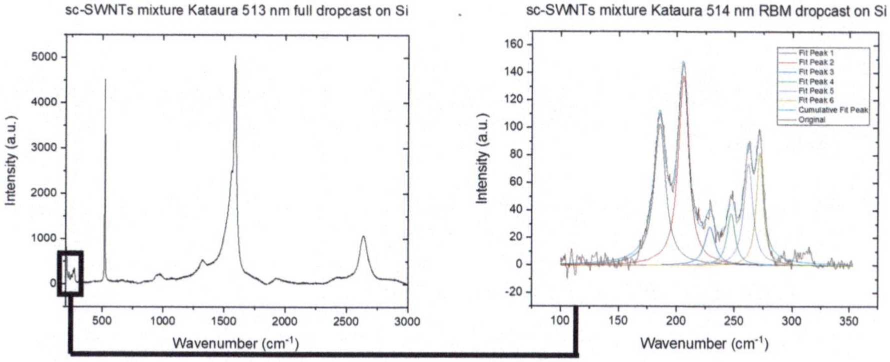

Fig. A.72 Raman of the mixture of sc-SWNTs, G-band and RBM region

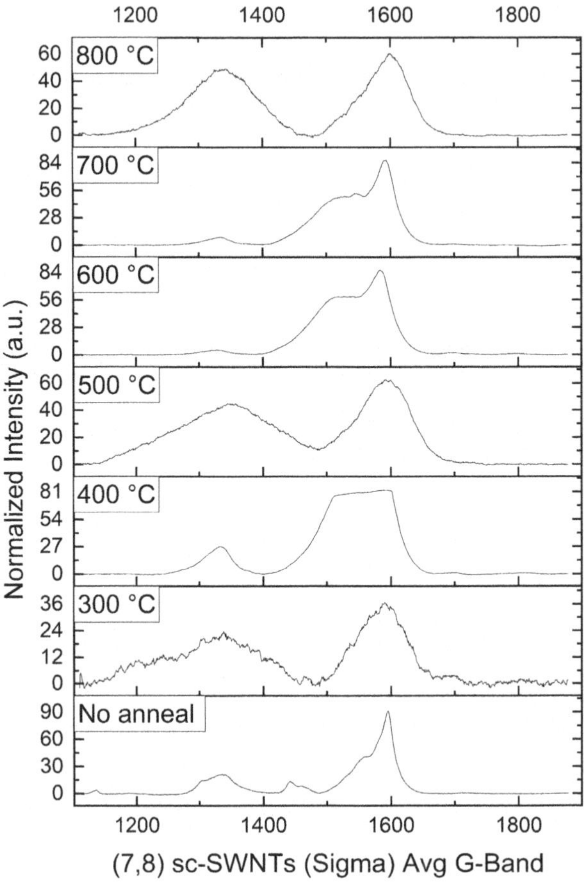

Fig. A.73 Sigma Aldrich high purity SWNTs also used for the annealing studies but not further pursued for the ART treatment

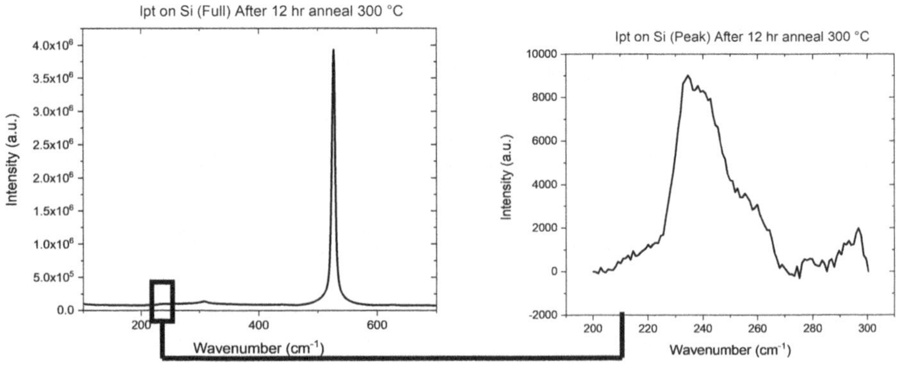

Fig. A.74 Raman at 532 nm of the iptycene monolayer only after annealing for 300 °C for 12 h

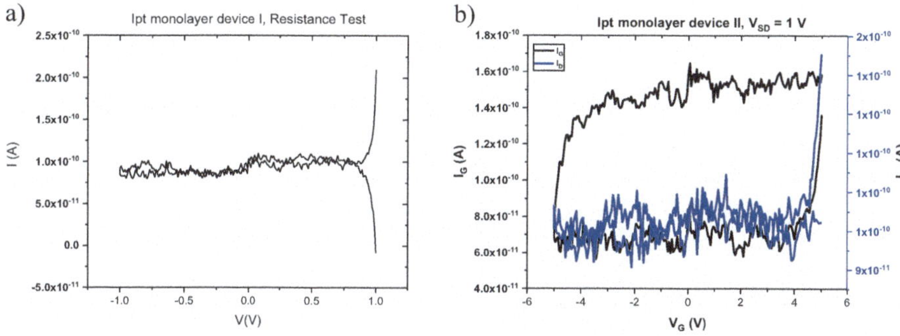

Fig. A.75 (**a**) Resistor test and (**b**) transport data of the iptycene monolayer as a device

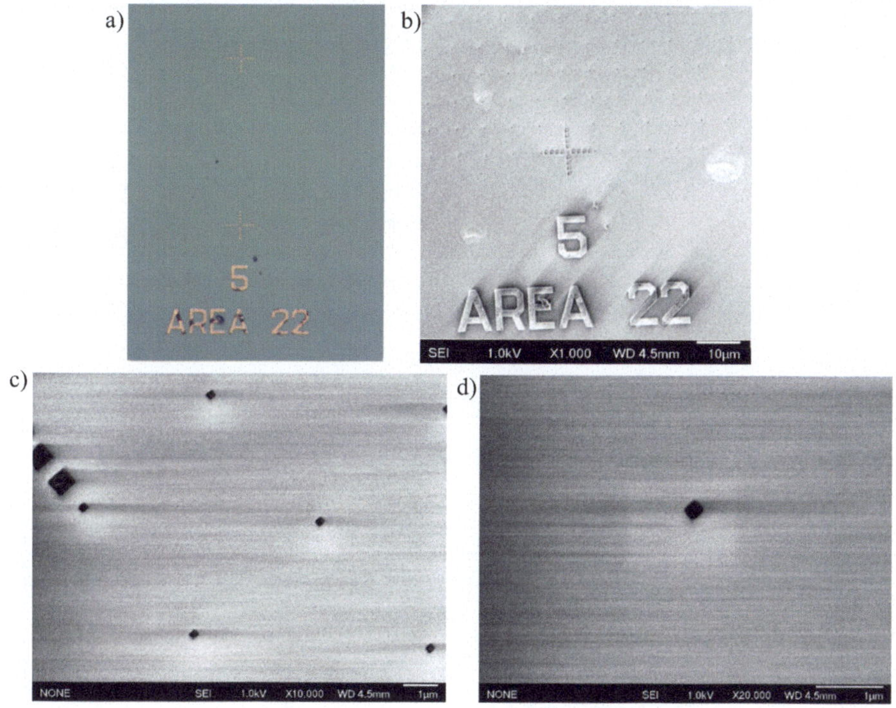

Fig. A.76 Marked substrate of Ti/Au after undergoing 600 °C annealing, (**a**) optical microscope, (**b–d**) HR-SEM at different resolutions

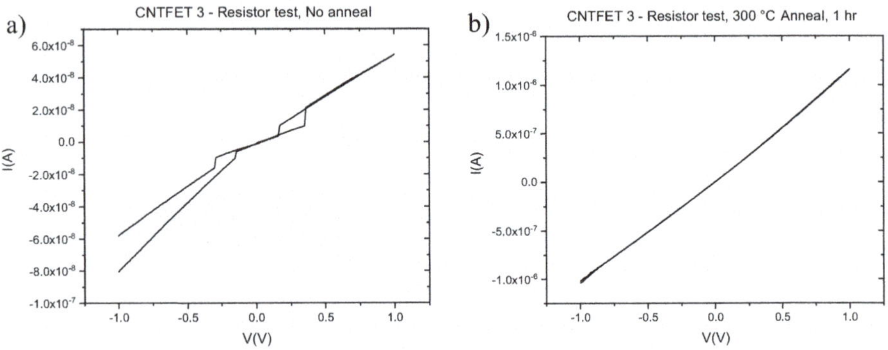

Fig. A.77 Resistor test (**a**) before annealing and (**b**) after annealing for CNTFET device #3

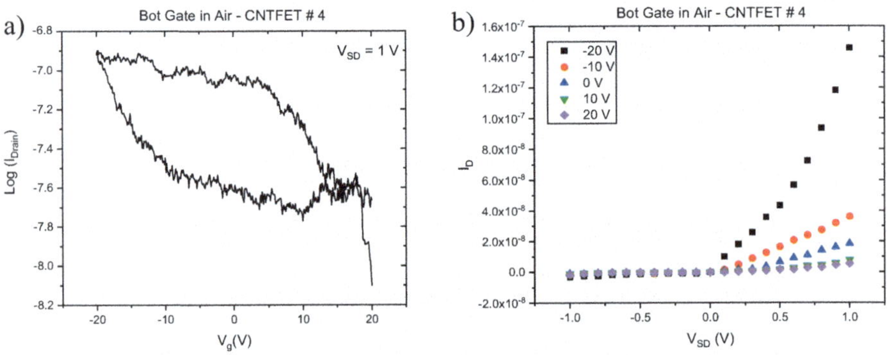

Fig. A.78 BG in air (**a**) transport data, $V_{SD} = 1$ V and (**b**) output data for CNTFET device #4

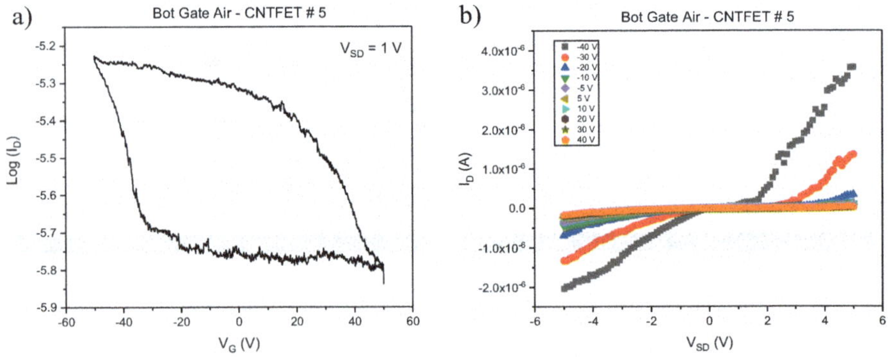

Fig. A.79 BG in air (**a**) transport data, $V_{SD} = 1$ V and (**b**) output data for CNTFET device #5

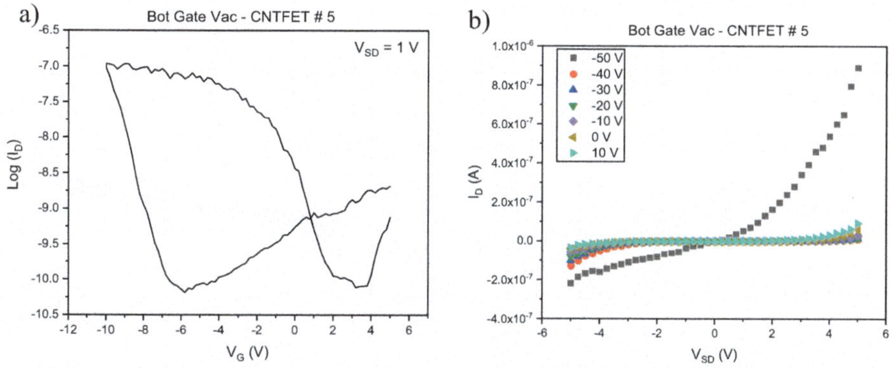

Fig. A.80 BG in vacuum (**a**) transport data, V_{SD} = 1 V and (**b**) output data for CNTFET device #5

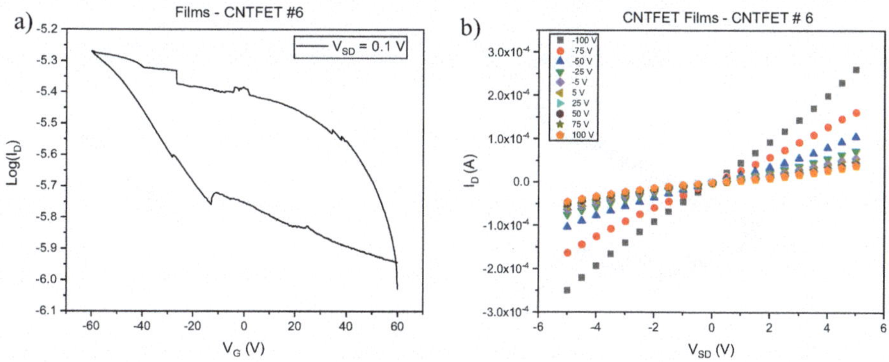

Fig. A.81 BG in vacuum for a device of thin films (**a**) transport data, V_{SD} = 0.1 V and (**b**) output data for CNTFET device #6

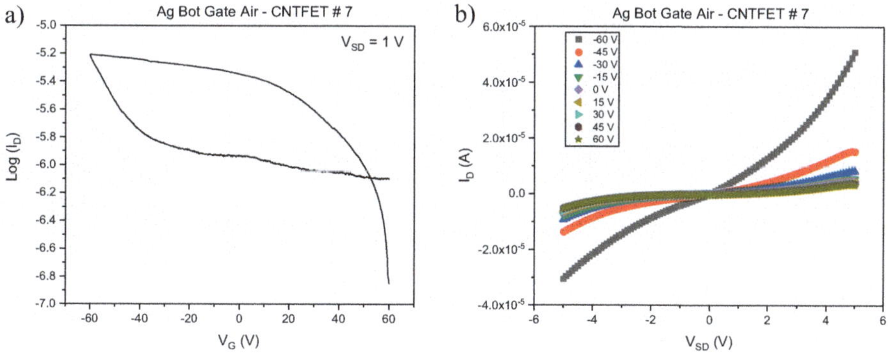

Fig. A.82 BG in air using silver conductive past on the surface of a scratched area at the Si/SiO_2 interface (**a**) transport data, V_{SD} = 1 V and (**b**) output data for CNTFET device #7

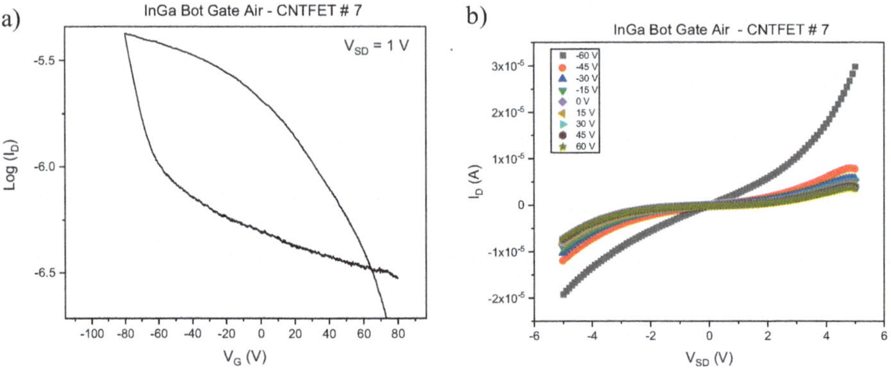

Fig. A.83 BG in air using InGa eutectic on the surface of a scratched area at the Si/SiO_2 interface (**a**) transport data, V_{SD} = 1 V and (**b**) output data for CNTFET device #7

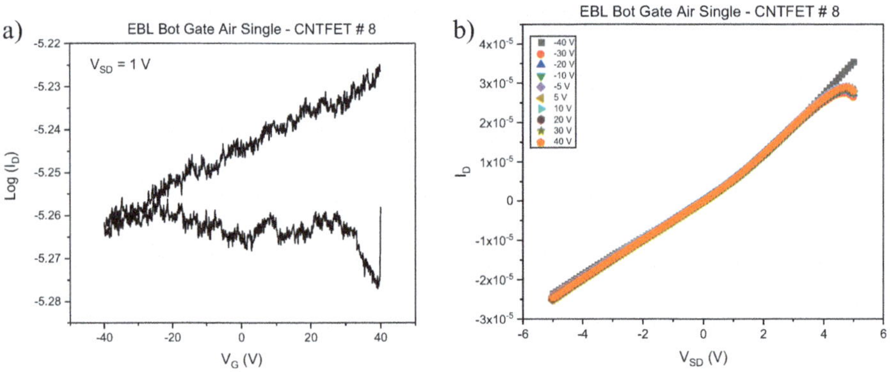

Fig. A.84 BG in air for (**a**) transport data, V_{SD} = 1 V and (**b**) output data for CNTFET device #8

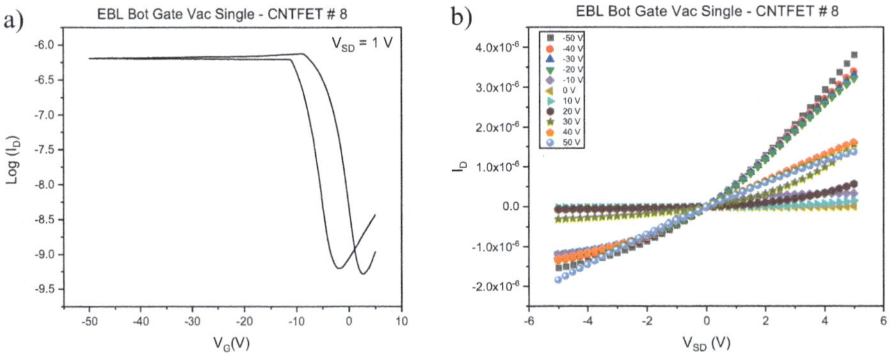

Fig. A.85 BG in vacuum for (**a**) transport data, V_{SD} = 1 V and (**b**) output data for CNTFET device #8

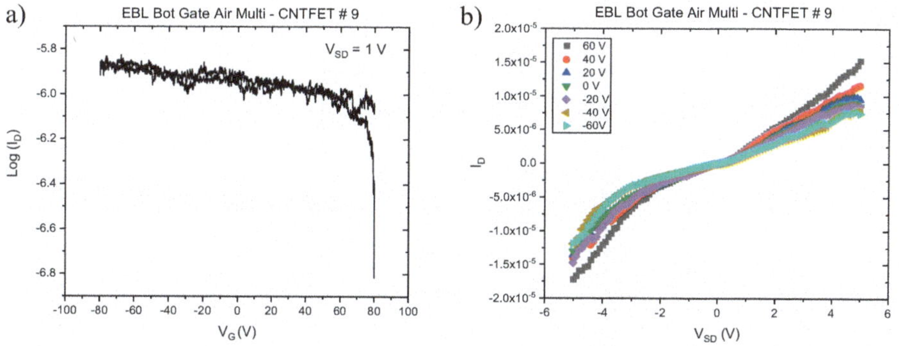

Fig. A.86 BG in air for (**a**) transport data, $V_{SD} = 1$ V and (**b**) output data for CNTFET device #9

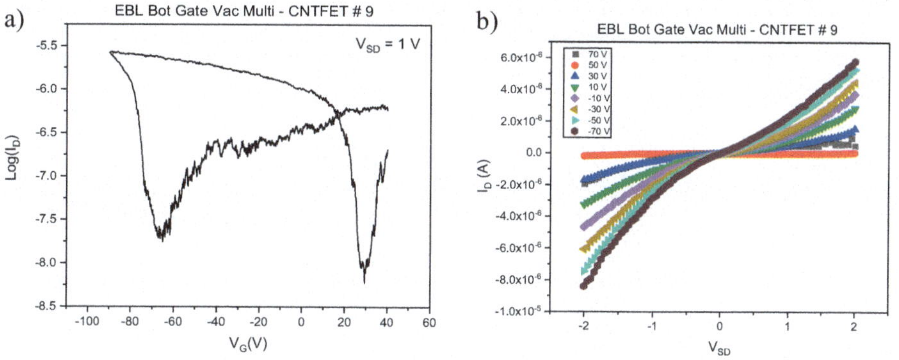

Fig. A.87 BG in vacuum for (**a**) transport data, $V_{SD} = 1$ V and (**b**) output data for CNTFET device #9

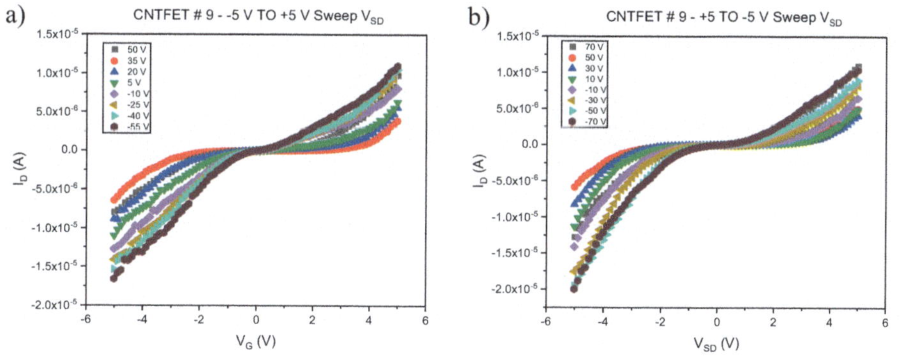

Fig. A.88 Change in sweeping direction of output data for CNTFET #9 (**a**) sweep -5 V to +5 V, (**b**) sweep +5 V to −5 V

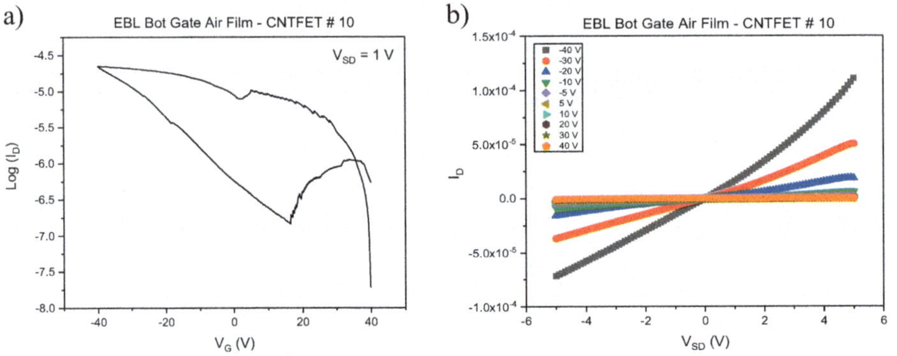

Fig. A.89 BG in air for (**a**) transport data, V_{SD} = 1 V and (**b**) output data for CNTFET device #10

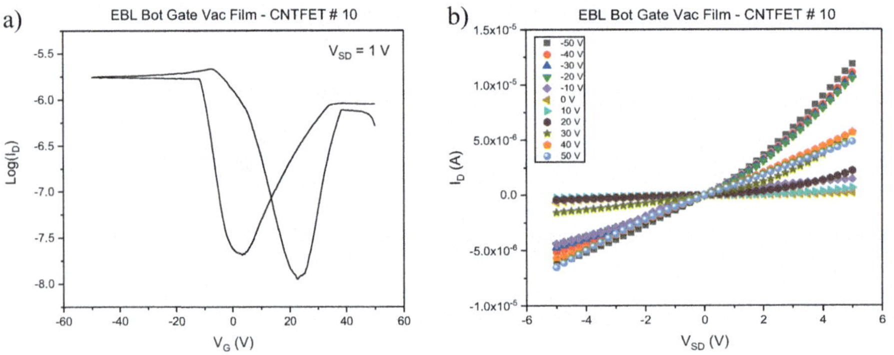

Fig. A.90 BG in vacuum for (**a**) transport data, V_{SD} = 1 V and (**b**) output data for CNTFET device #10

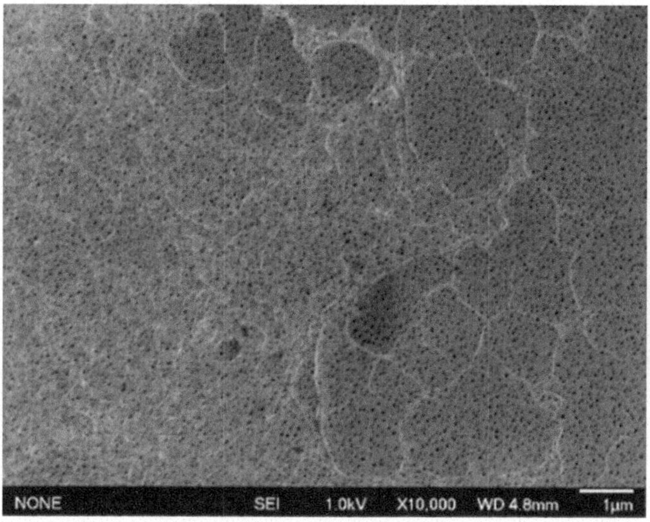

Fig. A.91 Pores in a TEM grid window with CNTs overlaid on the surface

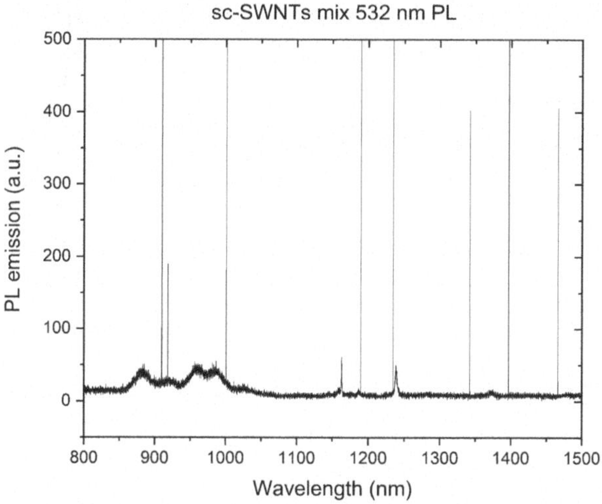

Fig. A.92 Micro PL at 532 nm wavelength for the mixture of sc-SWNTs dropcast on a silicon wafer. The instrument was a Horiba LabRamHR-PL NF from the Namiki Foundry Common facilities tested at room temperature

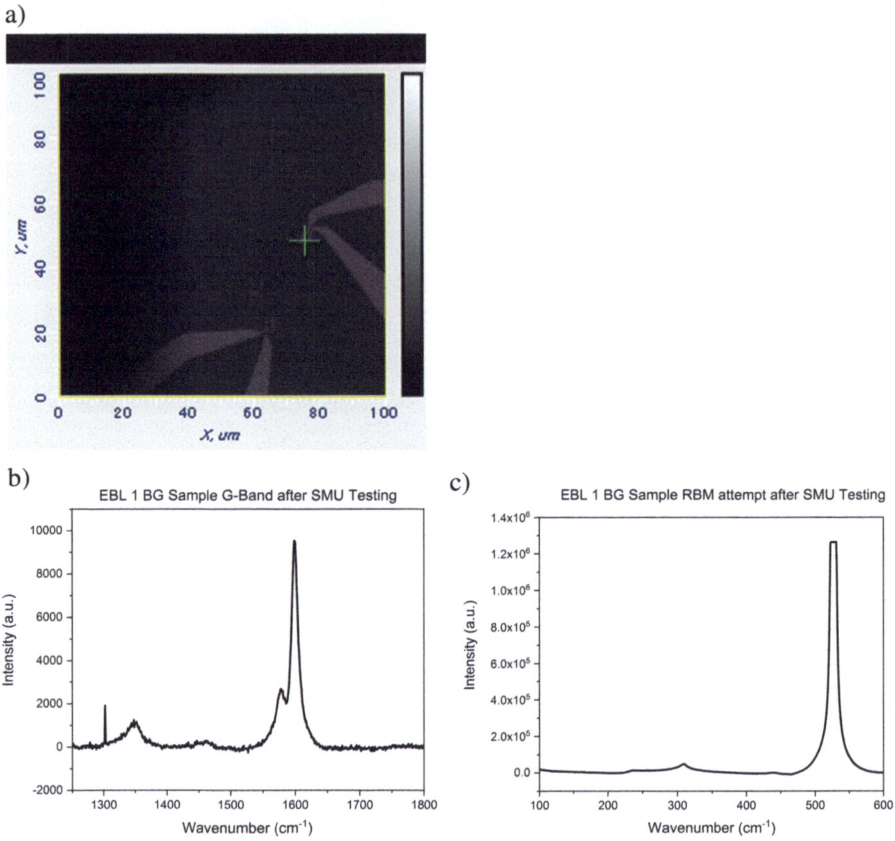

Fig. A.93 (**a**) Example of area selection for Raman spectrum, (**b**) BG G-band of EBL fabricated device after all the sample testing and annealing, (**c**) corresponding RBM region with only a small signal for the iptycene and no CNTs noticeable. Unfortunately, even in areas of a film there was no RBM signal only G-band signals were responsive at 532 nm wavelength (Tsukagoshi Lab Raman)

a)

b)

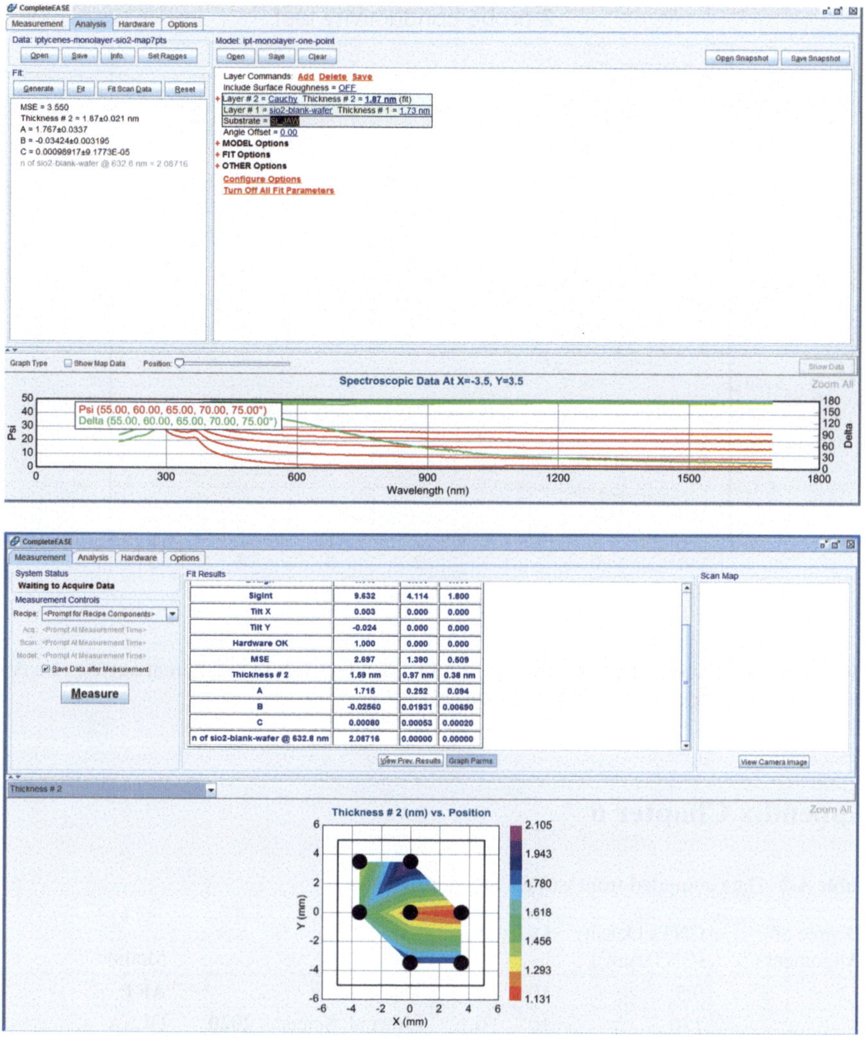

Fig. A.94 Ellipsometry raw data screenshot, (**a**) single scan, (**b**) map scan of 7 points

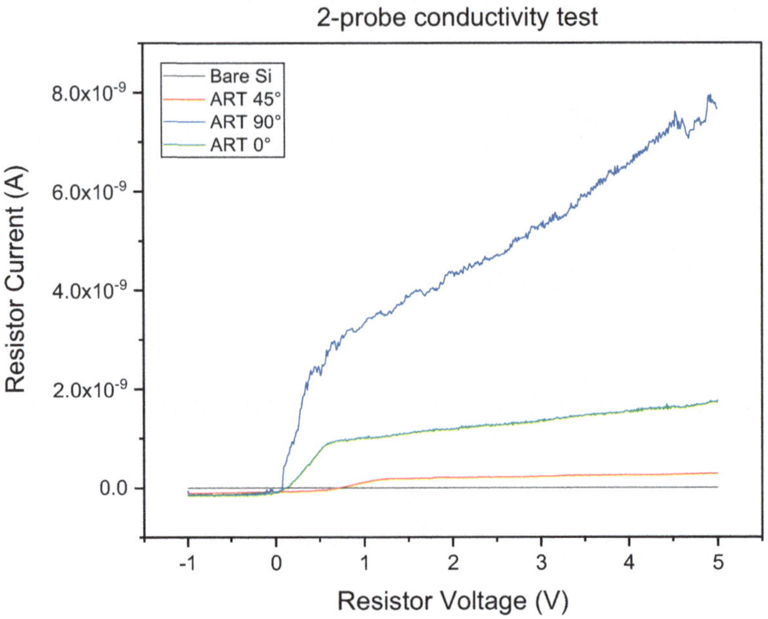

Fig. A.95 Two-probe testing on a surface of ART oriented SWNTs at different angles, observing that at different angles the conductivity changed

Appendix Chapter 6

Table A.2 Data compiled from various literature sources

Degree of Alignment (°)	CNTs Density (CNTs/μm²)	I_{ON}/ I_{OFF}	V_{SD} (V)	Ref	Method
12	0.5	10^6	1	This work	ART
8	120	10^5	0.1	Liu et al. Science, **2020**. 368: 850–856	DLSA
10	15	10	2	Si et al. ACS Nano., **2018**. 12: 627–634.	Directional shrinking
16	37	10^4	1	Zhu et al. Adv. Mater., **2018**. 30: 1707068	Shrinking alignment
25	20	N/A	–	Ma et al., Appl. Phys. Lett., **2016** 108: 063114	Evaporation induced self-assembly
15	50	10^6	1	Joo et al. Langmuir, **2014**. 30: 3460–3466	FESA

(continued)

Degree of Alignment (°)	CNTs Density (CNTs/μm^2)	I_{ON}/ I_{OFF}	V_{SD} (V)	Ref	Method
17	500	10^2	0.5	Cao et al., Nat. Nanotechnol., **2013** 8: 180–186.	Langmuir-Schaefer
10	30	10^4	0.5	Shekhar et al. ACS Nano., **2011**. 5: 1739–1746	Ultra high-density assembly

The ION/IOFF ratio is taken from the highest recorded in the source paper, regardless of LCH

The SWNTs' density in Liu et al. was reported as CNTs/μm

The degree of alignment was from the source papers or as calculated by Liu et al. in the supplemental information

References

1. Snowdon, M.R., Selmani, S., Schipper, D.J.: Sonication-enhanced alignment relay technique for the orientation of single-walled carbon nanotubes. ACS Appl. Nano Mater. **2**(10), 6637–6645 (2019)
2. Snowdon, M.R., Wang, S., Mashmoushi, N., Hopkins, S.W., Schipper, D.J.: Carboxylic acids as anchoring components on aluminum oxide for the alignment relay technique of single-walled carbon nanotubes. New J. Chem. **45**(12), 5340–5349 (2021)

M. Snowdon, *Methods of the Alignment-Relay Technique for Nanosystems*,
Springer Theses, https://doi.org/10.1007/978-3-030-84413-4

The manufacturer's authorised representative in the EU is Springer Nature Customer Service Centre GmbH, Europaplatz 3, 69115 Heidelberg, Germany. If you have any concerns regarding our products, please contact ProductSafety@springernature.com

Printed and bound by CPI Group (UK) Ltd, Croydon, CR0 4YY
30/09/2026
02225506-0002